Steels

Wei Sha

Steels

From Materials Science to Structural Engineering

Springer

Wei Sha
Belfast
UK

ISBN 978-1-4471-4871-5 ISBN 978-1-4471-4872-2 (eBook)
DOI 10.1007/978-1-4471-4872-2
Springer London Heidelberg New York Dordrecht

Library of Congress Control Number: 2012955041

Printed on acid-free paper

Springer is part of Springer Science+Business Media (www.springer.com)

Preface

Alloy steels are an extremely active field of research. Each year, included in the Science Citation Index, more than 3,000 research papers are published on the alloy steels topic, written by more than 9,000 authors, from more than 2,000 organisations in more than 80 countries.

A quick search in Google Books will show no shortage of books on steels. There are already a number of books discussing steels and steel-based microstructures, for example:

(1) Steels: Microstructure and Properties, Harry Bhadeshia and Robert Honeycombe
(2) Steels: Metallurgy and Applications, D. Llewellyn and R. C. Hudd
(3) The Book of Steel, G. Beranger, G. Henry, and G. Sanz (Editors)
(4) Advanced Steels: The Recent Scenario in Steel Science and Technology, Yuqing Weng, Han Dong, and Yong Gan (Editors)
(5) ASM Specialty Handbook: Carbon and Alloy Steels, J. R. Davis (Editor).

So what is unique about this new book?

The author has been a Professor of Materials Science in the School of Planning, Architecture and Civil Engineering, since 2004. Previous to that, he was a Lecturer and then Reader in Construction Materials, since 1995.

Since 2000, indexed in the Science Citation Index and Conference Proceedings Citation Index-Science, he has published more than 15 papers in each of the following subject areas: materials science; metallurgy and metallurgical engineering; physics; science and technology; chemistry; engineering. A list of some of these papers relevant to this book is given separately at the end of this preface. It shows evidence of the significant experience in a wide range of steel-based topics.

His unique, wide research experience brings a unique feature of the book *Steels: From Materials Science to Structural Engineering*. None of the existing steels books covers this spectrum of topics.

This book is intended to be a research monograph, cumulating the experience of 25 years' research by the author. It includes the more recent results, since 2000, but also covers the relevant, recent work by other researchers around the world. The content of the book has a strong focus on fire-engineering, heat-resistant and fire-resistant steels. The book also includes manufacturing and microstructure engineering-related subjects. These become more and more important for extending the performance in critical applications.

The research papers are the backbone of this book, but the underlining structure of the book is based on the types of alloy steels and the areas of steel applications in structural engineering. The idea is to bring together years of personal experience so the book will provide an interesting compendium of such knowledge. In addition, other researchers' work is reviewed and the major results presented and discussed.

The author is a world authority in the steel research field:

- Among the some 1,200 research papers on maraging steels published since 1970 in the Science Citation Index, two of the top ten most cited papers were written by Sha. The journal impact factors are assembled based on Science Citation Index data.
- The two refereed journals that alloy steels papers are most published in are *Materials Science and Engineering A* and *Surface and Coatings Technology*. Sha has published 15 papers in the former and 9 in the latter. Since 2007, Sha has refereed, on invitation, 28 manuscripts for *Materials Science and Engineering A*, and 54 manuscripts for *Surface and Coatings Technology*, being a member of both Editorial Boards.

Computer-based modelling is a fast growing field in materials science as well as structural engineering, demonstrated by the large recent literature. Combining modelling and experimental studies into one book is one idea behind this book, which, I think, is its strength and selling point.

The author of this book is in a unique position, in his combined expertise in steels research and model development. Though modelling is only a part of it, this book will fill a gap of the book literature in the application of modelling techniques in steels, both hot topics in contemporary materials studies, whilst at the same time a document on the latest research in this area. A large chunk of this latest research will be from the author himself, based on his world leading position in this area, but, as stated before, the book will also cover important relevant research by others. A lot of the structural engineering modelling is about the fire resistance of structural steelwork, one of the two most important issues for constructional steels, the other being corrosion.

The book is primarily intended for researchers studying steels. These include postgraduate students and researchers and lecturers and steel experts. The materials expert, however, will be able to learn modelling and apply this increasingly important technique in their steel materials research and development. As the statistics given earlier has shown, the research base in steel and its application areas is huge, so we hope that this book will have a big impact.

Funding from the following is acknowledged:

- Engineering and Physical Sciences Research Council, UK, under Grant No. GR/N08971, for the project of 'Computer Modelling of the Evolution of Microstructure During Processing of Maraging Steels'
- The Royal Academy of Engineering's Global Research Award Scheme

- Institution of Structural Engineers Research Award 2011, for the project of 'Full-Scale Fire Tests on a Cold-Formed Steel Portal Frame Building'
- Nullifire Ltd, Coventry, UK., and British Steel (now Tata Steel)
- Steel Construction Institute, Ascot, UK
- National Natural Science Foundation of China (No. 51001102)
- National Basic Research Program of China (Nos. '973' 2010CB630800, 2008CB 717802)
- National '863' Sophisticated Technology Foundation of China (High-tech Project No. 2006AA03Z530)
- National High Tech Program, China (No. 2009GB109002)
- Knowledge Innovative Program of The Chinese Academy of Sciences (No. KJCX2-YW-N35)
- Liaoning Province, China (No. 20081011), for a Doctoral Research Fund
- Division of Materials Sciences, US Department of Energy, under contract DE–AC05–96OR22464 with Lockheed Martin Energy Research Corporation and through the SHaRE Program under contract DE-AC05-76OR00033 with Oak Ridge Associated Universities
- Oak Ridge National Laboratory, USA

The following individuals are also acknowledged:

- All co-authors in the papers list at the end of this preface
- Dr. Rivera-Díaz-del-Castillo in the Department of Materials Science and Metallurgy, University of Cambridge, for useful communications
- Dr. William Warke
- Dr. M. K. Miller and Ms. K. F. Russell
- Dr. Kazutoshi Ichikawa of Nippon Steel Corporation for supplying Nippon fire resistant steels
- Mr. G. M. Newman and Professor C. G. Bailey, for useful discussions
- Professor P. A. M. Basheer for help with mix design, discussion, and provision of materials and concrete laboratory facilities

Wei Sha

Research Papers Used in the Preparation of this Book

(Chap. 2)

1. Kinetics of ferrite to Widmanstätten austenite transformation in a high-strength low-alloy steel revisited. Zhanli Guo, Wei Sha, *Zeitschrift für Metallkunde*, **95**, 2004, 718–723.
2. Modeling the diffusion-controlled growth of needle and plate-shaped precipitates. Z. Guo, W. Sha, *Modeling and Numerical Simulation of Materials Behavior and Evolution*, Materials Research Society (MRS) Symposium Proceedings, vol. 731, Symposium on Modeling and Numerical Simulation of Materials Behavior and Evolution held at the 2002 MRS Spring Meeting (Symposium W), 2–5 April 2002, San Francisco, CA, ed: Antonios Zavaliangos, Veena Tikare, Eugene A. Olevsky, Materials Research Society, Warrendale, PA, paper W7.5, pp. 215–220.
3. Determination of activation energy of phase transformation and recrystallization using a modified Kissinger method. W. Sha, *Metallurgical and Materials Transactions A*, **32A**, 2001, 2903–2904.
4. Change of tensile behavior of a high-strength low-alloy steel with tempering temperature. Wei Yan, Lin Zhu, Wei Sha, Yi-yin Shan, Ke Yang, *Materials Science and Engineering A*, **517**, 2009, 369–374.
5. Delamination fracture related to tempering in a high-strength low-alloy steel. Wei Yan, Wei Sha, Lin Zhu, Wei Wang, Yi-yin Shan, Ke Yang, *Metallurgical and Materials Transactions A*, **41**, 2010, 159–171.

(Chap. 3)

6. Development of structural steels with fire resistant microstructures. W. Sha, F.S. Kelly, P. Browne, S.P.O. Blackmore, A.E. Long, *Materials Science and Technology*, **18**, 2002, 319–325.
7. Atom probe field ion microscopy study of commercial and experimental structural steels with fire resistant microstructures. W. Sha, F.S. Kelly, *Materials Science and Technology*, **20**, 2004, 449–457.
8. Design and characterisation of experimental fire resistant structural steels. W. Sha, F.S. Kelly, S.P.O. Blackmore, K.H.J. Leong, *Proceedings for the 4th International Conference on HSLA Steels* (*HSLA Steels '2000*), 30 October-2

November 2000, Xi'an, China, eds: Guoquan Liu, Fuming Wang, Zubin Wang, Hongtao Zhang, Metallurgical Industry Press, Beijing, pp. 578–583.

9. Mechanical properties of structural steels with fire resistance. Wei Sha, *PRICM4: Fourth Pacific Rim International Conference on Advanced Materials and Processing*, 11–15 December 2001, Honolulu, Hawaii, vol. II, eds: S. Hanada, Z. Zhong, S.W. Nam, R.N. Wright, The Japan Institute of Metals, Sendai, pp. 2707–2710.
10. High temperature transient tensile properties of fire resistant steels. W. Sha, T.M. Chan, *Advances in Steel Structures*, Proceedings of the Third International Conference on Advances in Steel Structures (ICASS '02), 9–11 December 2002, Hong Kong, vol. 2, eds: S.L. Chan, J.G. Teng, K.F. Chung, Elsevier Science, Oxford, pp. 1095–1102.

(Chap. 4)

11. Microstructure evolution of a 10Cr heat-resistant steel during high temperature creep. Ping Hu, Wei Yan, Wei Sha, Wei Wang, Yiyin Shan, Ke Yang, *Journal of Materials Science & Technology*, **27**, 2011, 344–351.
12. Study on Laves phase in an advanced heat-resistant steel. Ping Hu, Wei Yan, Wei Sha, Wei Wang, Zhan-li Guo, Yi-yin Shan, Ke Yang, *Frontiers of Materials Science in China*, **3**, 2009, 434–441.
13. Microstructural evolution and mechanical properties of short-term thermally exposed 9/12Cr heat-resistant steels. Wei Wang, Wei Yan, Wei Sha, Yiyin Shan, Ke Yang, *Metallurgical and Materials Transactions A*, **43**, 2012, 4113–4122.

(Chap. 5)

14. Microstructure and mechanical properties of a nitride-strengthened reduced activation ferritic/martensitic steel. Qiangguo Zhou, Wenfeng Zhang, Wei Yan, Wei Wang, Wei Sha, Yiyin Shan, Ke Yang, *Metallurgical and Materials Transactions A*, **43**, 2012, 5079–5087.
15. Nitride-strengthened reduced activation ferritic/martensitic steels. Ping Hu, Wei Yan, Li-fen Deng, Wei Sha, Yi-yin Shan, Ke Yang, *Fusion Engineering and Design*, **85**, 2010, 1632–1637.
16. Effect of carbon reduction on the toughness of 9CrWVTaN steels. Wei Yan, Ping Hu, Lifen Deng, Wei Wang, Wei Sha, Yiyin Shan, Ke Yang, *Metallurgical and Materials Transactions A*, **43**, 2012, 1921–1933.
17. The impact toughness of a nitride-strengthened martensitic heat resistant steel. Wenfeng Zhang, Wei Yan, Wei Sha, Wei Wang, Qiangguo Zhou, Yiyin Shan, Ke Yang, *Science China Technological Sciences*, **55**, 2012, 1858–1862.

(Chaps. 6 and 7)

18. Phase transformations in maraging steels. W. Sha, H. Leitner, Z. Guo, W. Xu, *Phase transformations in steels*. Volume 2: Diffusionless transformations, high strength steels, modelling and advanced analytical techniques, eds: Elena Pereloma, David V. Edmonds, Woodhead Publishing, Cambridge, UK, 2012, Chapter 11, 332–362.

(Chap. 7)

19. Precipitation, microstructure and mechanical properties of low nickel maraging steel. W. Sha, Q. Li, E.A. Wilson, *Materials Science and Technology*, **27**, 2011, 983–989.
20. Microstructure and mechanical properties of low nickel maraging steel. W. Sha, A. Ye, S. Malinov, E.A. Wilson, *Materials Science and Engineering A*, **536**, 2012, 129–135

(Chap. 8)

21. In stainless steel reinforcement a viable option? Wei Sha, Roslyn Kee, *Proceedings of Metal 2000: 9th International Metallurgical Conference* (CD-R), 16–18 May 2000, Ostrava, Tanger Ltd., Ostrava, Czech Republic.
22. Differential scanning calorimetry study of normal portland cement paste with 30% fly ash replacement and of the separate fly ash and ground granulated blast-furnace slag powders. W. Sha, G.B. Pereira, *Proceedings of the Seventh CANMET/ACI International Conference on Fly Ash, Silica Fume, Slag and Natural Pozzolans in Concrete*, 22–27 July 2001, Chennai (Madras), India, Supplementary volume, compiled: Maria Venturino, ACI, Detroit, pp. 295–309.
23. Differential scanning calorimetry study of hydrated ground granulated blast-furnace slag. W. Sha, G.B. Pereira, *Cement and Concrete Research*, **31**, 2001, 327–329

(Chap. 9)

24. Optimization of cold-formed steel portal frame topography using real-coded genetic algorithm. D.T. Phan, J.B.P. Lim, C.S.Y. Ming, T. Tanyimboh, H. Issa, W. Sha, *Procedia Engineering*, **14**, 2011, 724–733.
25. Design optimization of cold-formed steel portal frames taking into account the effect of building topology. Duoc T. Phan, James B.P. Lim, Wei Sha, Calvin Y.M. Siew, Tiku T. Tanyimboh, Honar K. Issa, Fouad A. Mohammad, *Engineering Optimization*, **45**, 2013.
26. An efficient genetic algorithm for the design optimization of cold-formed steel portal frame buildings. Duoc T. Phan, James B.P. Lim, Wei Sha, Tiku T. Tanyimboh, *Proceedings of the 21st International Specialty Conference on Cold-formed Steel Structures*, 24–25 October 2012, St. Louis, MO, eds: Roger A. LaBoube, Wei-Wen Yu, Missouri University of Science and Technology, pp. 473–483.
27. Full-scale fire tests on a cold-formed steel portal frame building. Yixiang Xu, James Lim, Wei Sha, Christine Switzer, Richard Hull, Andrew Taylor, Ross McKinstray. Institution of Structural Engineers Research Award 2011.
28. Large-scale fire tests on a cold-formed steel portal frame building. Yixiang Xu, James Lim, Wei Sha, Christine Switzer, Richard Hull, Andrew Taylor, Ross McKinstray. Institution of Structural Engineers Research Award Application, 2011.

(Chap. 10)

29. Fire safety design and recent developments in fire engineering. W. Sha, N.C. Lau, *Structural Engineering, Mechanics and Computation*, Proceedings of the International Conference on Structural Engineering, Mechanics and Computation, 2–4 April 2001, Cape Town, South Africa, vol. 2, ed: A. Zingoni, Elsevier Science, Oxford, pp. 1071–1078.
30. Heat transfer in fire across a wall in shallow floor structure. Wei Sha, *Journal of Structural Engineering*, **127**, 2001, 89–91.
31. Heat transfer in steel structures and their fire resistance. W. Sha, T.L. Ngu, *Structural Engineering, Mechanics and Computation*, Proceedings of the International Conference on Structural Engineering, Mechanics and Computation, 2–4 April 2001, Cape Town, South Africa, vol. 2, ed: A. Zingoni, Elsevier Science, Oxford, pp. 1095–102.
32. Intumescent fire protection coating thickness for shallow floor beams. W. Sha, T.M. Chan, *Advances in Building Technology*, Proceedings of the International Conference on Advances in Building Technology (ABT 2002), 4–6 December 2002, Hong Kong, vol. 2, eds: M. Anson, J.M. Ko, E.S.S. Lam, Elsevier Science, Oxford, pp. 1321–1328.

(Chap. 11)

33. Fire resistance of slim floors protected using intumescent coatings. W. Sha, *Proceedings of the Eighth International Conference on Civil and Structural Engineering Computing*, ed: B.H.V. Topping, Civil-Comp Press, Stirling, UK, 2001, Paper 65, pp. 163–164.
34. Fire resistance of protected asymmetric slim floors beams. W. Sha, *Proceedings of the Eighth International Conference on Civil and Structural Engineering Computing*, ed: B.H.V. Topping, Civil-Comp Press, Stirling, UK, 2001, Paper 67, pp. 167–168.

Contents

Chapter 1
Introduction

Abstract This chapter introduces the various topics included in the book, in subsequent chapters. As the research monograph assumes that the reader has graduate level materials science knowledge, this is not discussed in the introductory chapter here. Instead, the Introduction focuses on different types of steels, including high-strength low-alloy steel, fire-resistant structural steel, heat-resistant steel, nitride-strengthened ferritic/martensitic steel and low nickel maraging steel. Examples of application of steels in structural engineering follow, in cold-formed steel portal frame, and fire safety design and developments in fire engineering of structural steelwork. The chapter can be read independently of the rest of the book, but, with its wide referencing, is also useful in assisting in reading the in-depth chapters that follow and being referred back to. Directly referring to later chapters and sections helps with the latter.

1.1 High-Strength Low-Alloy Steel

In order to reduce the materials cost and improve the transportation efficiency, high-strength low-alloy (HSLA) steels are widely employed in modern car manufacturing due to their excellent strength-toughness combination and weldability (Wang et al. 2009a; Kim et al. 2002; Zhao et al. 2002). HSLA plate steels are widely used throughout the world for various structural applications. These steels combine particularly high strengths with good toughness, especially developed by the controlled hot rolling process.

1.1.1 Kinetics of Ferrite to Widmanstätten Austenite Transformation

Various models have been developed to describe the diffusion-controlled growth of precipitates with shapes approximating to needles or plates. In the most comprehensive theory (Guo and Sha 2004a), the needle is assumed to be in the form

W. Sha, *Steels*, DOI: 10.1007/978-1-4471-4872-2_1,

of a paraboloid of revolution and the plate a parabolic cylinder, Fig. 1.1a and b, respectively. The solutions obtained for specified conditions are shape-preserving when the radius of the advancing tip for either needle or plate, ρ, is several times the critical nucleus size (radius value) for plate, ρ_c, or for needle (which is twice that for plate), ρ_c', and in this context they allow rigorously for changes in capillarity and interface kinetics as the curvature of the interface varies along the parabolic surfaces. The theory predicts a constant lengthening rate in one direction when assuming that the needle or plate tip advances into fresh parent phase.

This theory is popular because many factors that influence precipitate growth, i.e. diffusion, interface kinetics and capillarity, are accounted for within one equation, which can be solved mathematically strictly. However, the theory was developed based on many assumptions. For modelling the growth of precipitate needles or plates by diffusion of solute in solid–solid phase transformations, it assumes that interphase energy and kinetic coefficient are independent of crystallographic orientation, and neglects elastic strain energy and the anisotropy of surface properties. Nevertheless, in most solid–solid phase transformations the migration of atoms across the interface is believed to be quite rapid, so the interface kinetics effect can be neglected. Interphase energy is related to the degree of mismatch between precipitates and parent phase, which is usually anisotropic. Anisotropy of the interphase boundary structure leads to faster growth in certain directions than in others, which was not considered in the theory. In addition, the existence of the

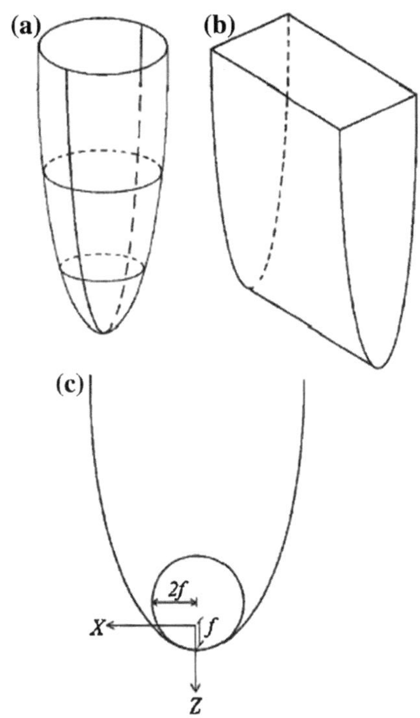

Fig. 1.1 Shapes used to represent needle and plate-shaped precipitates. **a** Paraboloid of revolution for needle; **b** parabolic cylinder for plate; **c** radius of the parabola tip

stress and the strain caused by the transformation will also affect the shape, or the aspect ratio, of precipitates.

In Chap. 2, the effects of transformation stress/strain and anisotropic interphase energy on precipitate morphology are incorporated into the theory by using experimental values of the radius of the advancing tip, to study the growth kinetics of Widmanstätten austenite in ferrite in HSLA Fe–C–Mn–Nb steel. The growth theory is summarised in Sect. 2.1.1.

1.1.2 Change of Tensile Behaviour with Tempering Temperature

Thermomechanical control process (TMCP) substituting for the traditional rolling process has effectively promoted the development of HSLA steels. Therefore, in recent decades the line-pipe steels have been developed from grade X60 to the current X80 and X100 grades (Koo et al. 2004; Asahi et al. 2004a, b; Fairchild et al. 2004).

To achieve a combination of high strength and toughness, the microstructure of lower bainite or ferrite plus martensite is designed for HSLA steels. Hence, in spite of additions of alloying elements such as molybdenum and boron to enhance lower bainite and martensite in these steels, rapid cooling after finishing rolling is introduced. Generally, there are three cooling treatments after rolling. The first treatment is direct quenching followed by tempering (DQT). The second is accelerated continuous cooling (ACC), in which the as-rolled steel is cooled down to room temperature at a given cooling rate. The third is interrupted accelerated cooling (IAC), in which the steel is subject to water-cooling in the phase transformation temperature region, and then air-cooled to room temperature. Self-tempering could happen during air-cooling because of the slower cooling rate in the component interior. A critical point in these three cooling treatments is that the rapidly cooled steel should be tempered, which is necessary to achieve a good strength/toughness combination. However, as the lower bainite microstructure is only recently introduced in HSLA steels, studies on the mechanical properties have rarely been related to the tensile behaviour after tempering.

Tensile tests carried out on structural steels may provide valuable information related to the microstructure. In a typical tensile curve of annealed low carbon steel, the upper and lower yield points are well related to the interactions between dislocations and carbon as well as nitrogen atoms. This theory may not be able to explain the yield behaviour in other metals with fcc or hcp lattice structures. However, a convincing explanation should involve two aspects, the density of mobile dislocations and the rate of dislocation glide. The strain rate of metals is related to the value of the Burger vector, b, the mobile dislocations density, ρ, and the rate of dislocation glide, $\bar{v}$, as given in Eq. 1.1:

$$\dot{\varepsilon} = b \cdot \rho \cdot \bar{v} \tag{1.1}$$

where the rate of dislocation glide $\bar{v}$ depends on the applied stress, as shown in Eq. 1.2:

$$\bar{v} = k \cdot \bar{v}_0 \left(\frac{\tau}{\tau_0}\right)^m \tag{1.2}$$

where τ is the shear stress in the sliding plane; τ_0 is the shear stress for dislocation glide of a unit speed; m is the stress exponent for dislocation glide, where the rate is thermally activated. Equation 1.2 illustrates that a higher stress will produce a faster dislocation glide rate.

In the as-recrystallized state and prior to tensile loading, the mobile dislocations density may be relatively low, so a high dislocation rate is necessary to meet the demand of plastic deformation. Correspondingly, a stress peak will occur at the upper yield point on the tensile curve. Once moving, the mobile dislocations density increases quickly. Hence, a lower dislocation rate may be possible to meet the demand of plastic deformation and the stress will correspondingly decrease. Consequently, the lower yield point appears on the tensile curve. When the moving dislocations are blocked (or impeded) or re-pinned and the mobile dislocations density decreases, the same cycle described above happens again. This explanation is in principle reasonable and suitable for the plastic deformation of most metals.

Another important aspect revealed in tensile tests (and the recorded stress–strain curves) is the strain-hardening exponent. The true stress and true strain relations of homogeneous plastic deformation can be described by Eqs. 1.3–1.5:

$$s = ke^n \tag{1.3}$$

$$s = \sigma(1 + \varepsilon) \tag{1.4}$$

$$e = \ln(1 + \varepsilon) \tag{1.5}$$

where s is the true stress that can be calculated by Eq. 1.4, e is the true strain and can be calculated by Eq. 1.5, σ is the engineering stress, ε is the engineering strain, k is the hardening coefficient, and n is the strain-hardening exponent, which demonstrates the amount of work hardening at an incremental deformation strain. If $n = 1$, this shows that the material exhibits a linear work hardening characteristic. If $n = 0$, this indicates that the material has no strain hardening ability and behaves ideally plastic.

Section 2.3 is concerned with the tensile behaviour of HSLA steel with lower bainite microstructure after tempering at different temperatures.

1.1.3 Delamination Fracture Related to Tempering

Due to possible applications in severe service environments, the low temperature impact toughness is being given more attention and becomes significant to the application of HSLA steels. Therefore, the toughness at a certain low temperature is always one of the most important properties specified for HSLA steels. In particular, their impact toughness at −30 °C should be high enough to meet service demands. However, when the strength is very high, it becomes more difficult to achieve excellent toughness, especially at low temperatures, due to the conflict of the commonly used mechanisms for improving strength and toughness and the limited capability of manufacturing facilities. In addition, delamination, i.e. splitting in the form of single or multiple secondary cracks perpendicular to the main crack and parallel to the plate surface, is often encountered during impact fracture of hot-rolled high-strength steels (Yang et al. 2008a, b). The density of delamination usually initially increases with the decreasing temperature (Tsuji et al. 2004; Song et al. 2005, 2006), and then goes through a maximum and decreases.

The effect of delamination on impact toughness has been widely investigated on many HSLA steels such as X60, X70 and X80 pipeline steels (Wallin 2001; Silva et al. 2005, Guo et al. 2002). It has often been reported that the delamination reduced the notch impact value in the upper shelf region (Tsuji et al. 2004; Song et al. 2005, 2006; Otárola et al. 2005; Verdeja et al. 2003). On the other hand, Kimura et al. (2007), Zhao et al. (2005), and Pozuelo et al. (2006) showed that the delamination can improve toughness due to the delamination toughening effect. It will be shown in Sect. 2.4 that delamination does not seem to have much influence on the low temperature toughness at −30 °C (Yang et al. 2008b). This point should be the subject of further study.

The initiation of the delamination in either dynamic (e.g. impact) or static conditions has not yet been sufficiently explained. From previous studies, it seems that such features as bent ferrite-pearlite microstructure, elongated grain shape, strip microstructure (Yang et al. 2008a), certain texture characteristics (Tsuji et al. 2004; Verdeja et al. 2003; Zhao et al. 2005), decohesion of grain boundaries, segregation of impurity atoms, and aligned particles (Otárola et al. 2005) and inclusions (Yang et al. 2008a) could lead to delamination, either individually (Tsuji et al. 2004) or cooperatively (Yang et al. 2008a). Although these proposed mechanisms have widely different aspects, they do have the common characteristic that the microstructures are anisotropic. Hence, it is quite natural that frequent delamination was not thought to be the result of insufficient roll-bonding after ARB (accumulative roll bonding), but rather a characteristic feature of the ultrafine elongated grain structures produced by heavy deformation (Tsuji et al. 2004).

The delamination on the fracture surfaces in Sect. 2.4 is observed on steel reheated in a certain range of temperature after rolling. In Sect. 2.4, the delamination is described in detail and their effect on the low temperature toughness at −30 °C is evaluated. The relationship between delamination and the microstructural changes resulting from the reheating is described.

1.2 Fire-Resistant Structural Steel

Steel is the most important metallurgical material available to civil engineers worldwide. Interest in the area of fire resistance of steel for safety and economy has grown in the last two decades. The development of fire-resistant steels is one aspect of the effort to enhance building safety in fire. This effort has been mainly from Nippon Steel, where some fire-resistant steels have been developed containing niobium and/or molybdenum. These steels have increased yield strength at elevated temperatures, believed to be especially for meeting the requirements of Japanese building regulations.

Traditionally, fire resistance of steel building structures is achieved by applying fire protection to steel columns and beams. However, fire engineering has been rapidly developed and one aim has been to design buildings with significant built-in fire resistance via the use of reduced or no traditional fire protection.

A further approach towards ensuring fire safety of steel structures is the use of a fire resistant-type structural steel in construction. This term 'fire resistant steel' simply refers to a structural steel that has better strength performance at elevated temperatures, typical of those that a steel section in a building would experience in a fire. Using such steel would not complicate current design routines significantly, as only new strength reduction factors or limiting temperature tables would need to be used in fire design calculations. Microalloyed steel of this nature, with elevated temperature properties, is well established in the boiler and superheated market areas. At present, however, these steels while technically possible, are not available in the general construction market, nor would they be economically competitive with established and readily available systems.

The driving force for the development in Japan of structural steels with fire resistant microstructures was the Japanese construction standards. They are stricter than Western standards and restrict the maximum steel temperature to the temperature at which the steel retains two-thirds of its room temperature yield strength. A number of fire-resistant steels so named by the Japanese steel manufacturers have found use in the Japanese market. These, and other developments of such steels, were reviewed in an overview paper (Sha et al. 2001) that includes extensive references.

In Japan, a significant amount of research has been carried out by Nippon Steel, where several fire-resistant steels have been developed. These steels represent a notable improvement over conventional steels, judging by the Japanese 'New fire safety design system', which allows the maximum permissible steel temperature to be set in terms of elevated temperature yield strength. However, their advantage is limited, as Eurocodes, and other building standards in the West do not use this parameter in design. Broken Hill Proprietary Co. (BHP) in Australia carried out earlier research on fire-resistant steels.

The author has carried out a programme of research to develop fire-resistant steels for building construction. Comprehensive microstructural characterisation and mechanical testing have been carried out on two Japanese fire-resistant steels made by Nippon Steel. The chemical compositions of these steels are given in Table 3.1. The research has shown that there is a strong relationship between the relatively high elevated temperature tensile strength and fine precipitation as well as coarse inclusion particles. The good high temperature strength and creep properties of Nippon steels are owing to the high lattice friction stresses. These are a result of the very fine distribution of precipitates with limiting theoretical carbide having a composition of MX where X is nitrogen and/or carbon, molybdenum in solid solution and a strong secondary wave of precipitation at approximately 550 °C. Here, the term 'lattice friction stress' should be regarded as the inherent resistance of an iron crystal structure to dislocation motion plus the hardening effect of solutes, precipitates and dislocations. This lattice friction stress maintains strength up to 600 °C when grain boundary sliding begins.

The effectiveness of using these fire-resistant steels in building floor construction has been evaluated, based on established fire test data and the current codes of standard. It has been found that they can achieve a relatively small gain in 'hot' moment resistance and the load ratio that a steel structure can withstand in fire. It has also been found through the calculations that, in order to achieve a 30-min increase in fire resistance in conventional composite floor structures, the steel is required to retain its strength at 150 °C higher than conventional commercial structural steels.

Several experimental fire-resistant steels have been designed (Chap. 3), to test the effectiveness of two methods devised by the author for achieving better fire performance. The first is to react to the high temperature reached in a fire by precipitation, which leads to in situ strengthening. The second is to rely on the steel's thermal stability to provide good elevated temperature strength. These steels can provide better fire resistance than conventional structural steels.

Chapter 3 describes the microstructural characterisation and the mechanical properties of the fire-resistant steels. The control materials are the two Nippon steels and the conventional S275 steel (designation used in Eurocodes) formerly known as grade 43 steel (designation used in British Standards). Detailed analyses on these steels are presented.

Further to the literature reviewed by Sha et al. (2001), additional new experimental fire-resistant steels were developed (Chap. 3). Sections 3.3 and 3.4 describe and discuss atom probe field ion microscopy work on two examples of Japanese commercial steels and two experimental steels developed at Queen's University Belfast. The research will be compared with that obtained with a multiple of other, more conventional techniques Sha et al. (2001) and the references quoted in it and in Chap. 3. An overview of the behaviour of structural steels at elevated temperatures and the design of fire resistant steels, which forms a general presentation of the background of Chap. 3, are given elsewhere by Sha et al. (2001) and will not be repeated here.

1.3 Heat-Resistant Steel

To save the non-recycled energy source and reduce the CO_2 emission requires improving the efficiency of fossil power plants, which can be achieved by developing super critical power plants. The 9–12 %Cr ferritic/martensitic heat-resistant steels have been widely used in ultra super critical (USC) power plants because of the high thermal conductivity, low thermal expansion coefficient and low susceptibility to thermal fatigue (Masuyama 2001).

The high chromium ferritic/martensitic heat-resistant steels are replacing the austenitic stainless steels as an alternative structural material for USC power plants over recent years due to their excellent mechanical properties and low cost. The main advantages of these kinds of steels include improvement of oxidation and corrosion resistance as well as long-term creep rupture strength, which mainly relies on the innovation and optimisation of the chemical composition. The chromium content was increased to enhance the oxidation and corrosion resistance and to meet the demand of the increasing operating steam temperature (Wang et al. 2009b). Addition of tungsten or substitution of tungsten for molybdenum (Miyata and Sawaragi 2001) was found beneficial to improve the creep rupture strength. Cobalt was added to balance the chromium equivalent and suppress the δ-ferrite which is detrimental to creep rupture strength during the high temperature normalising process (Yamada et al. 2003). Over these years, steels with C–Cr–W–Co as main alloying elements have been paid increasing attention for their excellent creep resistance performance (Abe et al. 2007a; Toda et al. 2003). It was found that the creep resistance is proportional to the tungsten content in the range of 1–4 wt%. In addition, cobalt has been found to make a great contribution to creep rupture strength due to its three effects: (1) suppressing δ-ferrite (Yoshizawa and Igarashi 2007; Yamada et al. 2003); (2) possible effect of inhabiting the coarsening of $M_{23}C_6$ carbides (Gustafson and Ågren 2001); and (3) increasing the atomic binding forces (Li et al. 2003).

However, recent research shows that the microstructure stability of the 9–12 % Cr ferritic/martensitic heat-resistant steels is the most critical issue to the creep rupture strength in the long time service. How to improve the microstructure stability has been recognised as a key issue in developing steels with high creep rupture strength (Kimura et al. 2010). On the one hand, tungsten was found to prevent the migration of martensitic laths (Abe 2004). On the other hand, tungsten was discovered to facilitate the formation of large-size Laves phase and harmful to the microstructure stability (Abe 2001). In addition, cobalt seemed to accelerate the precipitation of tungsten, resulting in the quick coarsening of Laves phase, which will initiate creep cavities (Cui et al. 2001; Lee et al. 2006).

In addition to creep resistance, the oxidation resistance at elevated temperature is an important issue for heat-resistant steel and it can be improved by raising the chromium content in the steel. However, chromium is the main element in forming Z phase (Danielsen and Hald 2009; Golpayegani et al. 2008), which would result in deterioration of the creep rupture strength (Sawada et al. 2006b, 2007). Moreover, the driving force of Z phase formation has already been reported to depend on the chromium content in high chromium ferritic steels (Danielsen and Hald 2004). Considering these aspects, appropriate chromium content is needed to achieve a balance between oxidation-resistance and creep-resistance properties.

In Chap. 4, based on the chemical composition of ASME-P92, through tungsten and cobalt addition, a 10Cr ferritic/martensitic steel is designed. The creep-resistance property of the steel and its microstructure evolution during high temperature creep are shown. The aim of this chapter is to develop some basic idea of improving microstructure stability from the view of chemical composition and contribute to the development of advanced high chromium heat-resistant steels superior to ASME-P92 steel in both oxidation-resistance and creep-resistance properties.

1.4 Nitride-Strengthened Ferritic/Martensitic Steel

1.4.1 Reduced Activation Ferritic/Martensitic Steels and Effect of Carbon Reduction

Reduced activation ferritic/martensitic (RAFM) steels have been regarded as candidate structural materials of future fusion and fission power reactors. Compared to austenitic stainless steels, the RAFM steels have not only good mechanical properties and thermal conductivity, but also excellent resistance to void swelling. The present RAFM steels, Eurofer 97 (the European reference material), JLF-1 (among the Japanese Low activation Ferritic steel series), F82H, 9Cr-2WVTa (Fe-9Cr-2W-0.25V-0.12Ta-0.1C) and CLAM (Chinese Low Activation Martensite), are alloyed with chromium, tungsten, manganese, vanadium, tantalum, carbon and nitrogen (Baluc et al. 2007b), and developed in Europe, Japan, US and China. India also has fabricated RAFM steels and oxide dispersion strengthened (ODS) (Wong et al. 2008; Saroja et al. 2011). All these steels are strengthened by both solution hardening and precipitation hardening. Solution hardening of these steels is dependent on the addition of chromium and tungsten whereas precipitation hardening relies on precipitates such as $M_{23}C_6$, where M is mainly Cr with substitution of Fe and MX (M = V, Ta; X = C, N) carbonitride. It is well known that coarsening of precipitates during creep results in an increase in inter-particle distance, but the precipitation strengthening effect is inversely proportional to this distance; therefore, the creep rupture strength would decrease eventually. Different precipitates have different coarsening rates. Sawada et al. (2001) found that the coarsening rate of $M_{23}C_6$ carbide is much higher than that of the MX type precipitates. Therefore, optimal control of the carbon content may be an effective method to improve the creep rupture strength of RAFM steels through changing carbides to nitrides or carbonitrides. Taneike et al. (2003) studied the creep-resistance property of the steels with different carbon contents and found that the time to rupture could be significantly increased by reducing the carbon content to a very low level, which was attributed to the elimination of carbides and the formation of fine, thermally stable and homogeneously distributed carbonitrides or nitrides in the microstructure. Much work has been carried out on the nitride-strengthened high chromium ferritic/martensitic heat-resistant steels used as structural materials for USC power generation (Taneike et al. 2004; Yin et al. 2007; Yin and Jung 2009; Abe et al. 2007b; Toda et al. 2005; Sawada et al. 2004). However, the development of nitride-strengthened RAFM steels has been limited. In Chap. 5, the microstructure and mechanical properties of the nitride-strengthened RAFM steels with higher creep rupture strength are discussed.

In these steels, both molybdenum and niobium are removed and replaced by tungsten and tantalum in order to obtain the low activation property. The typical microstructure of these heat-resistant steels is composed of tempered lathy martensite with precipitates dispersed in the matrix (Ghassemi-Armaki et al. 2009). These precipitates are of great importance to the microstructure stability. However, some work (Maruyama et al. 2001; Gustafson and Ågren 2001) (Sects. 5.1 and 5.2) has shown that with increasing service time, the $M_{23}C_6$ carbide grows too fast to pin the dislocation movement and cannot prevent grain boundaries or lath boundaries from migrating, resulting in premature fracture. Seeking thermally stable particles to achieve a highly stable microstructure has always been a goal in developing heat-resistant steels. Oxides such as fine Y_2O_3 and $YTiO_3$ have impressive thermal stability and could prevent the microstructure from degradation. Thus, ODS steels such as Eurofer 97-ODS and CLAM-ODS steels have been developed (de Castro et al. 2007; Olier et al. 2009; Klimenkov et al. 2009). However, the microstructures of ODS steels are usually anisotropic due to the manufacturing process and their ductile–brittle transition temperatures (DBTT) are very high (Lindau et al. 2005; Kurtz et al. 2009), though the steels showed better thermal stability (Schaeublin et al. 2002; Yu et al. 2005). Meanwhile, the fabrication involves the complicated and expensive process of powder alloying. It will be very difficult to put the ODS steels into industrial-scale practice.

Besides oxides, nitrides are also thermally stable. For a metal element, its nitride has a slower growth rate than its carbides under the same condition (Yong 2006). It was found by Taneike et al. (2003) that $M_{23}C_6$ carbide was suppressed in the 9 %Cr martensitic steel when carbon content was reduced to as low as 0.018 % (wt%) whereas the nitrides were stimulated to homogeneously disperse, which greatly improved the microstructure stability and enhanced creep strength. In order to eliminate the $M_{23}C_6$ carbide, Taneike et al. decreased the carbon content to even below 20 ppm and developed the totally nitride-strengthened martensitic steel. In this steel, no $M_{23}C_6$ was precipitated, and MX nitrides distributed along the grain boundaries and lath boundaries (Abe et al. 2007b; Taneike et al. 2004). The nitride-strengthened martensitic steel showed excellent creep property at elevated temperature because of the high thermal stability of nitrides.

It is well known that a single-phase microstructure should be beneficial for steels to achieve high creep strength. However, when the carbon content is reduced to an extremely low level, if no other composition change is made, δ ferrite will inevitably form (Sects. 5.1, 5.2, 5.3), which is harmful to both toughness and creep resistance by acting as the weak part during creep (Ryu et al. 2006; Yoshizawa and Igarashi 2007). Therefore, composition design should consider δ ferrite elimination. Taneike et al. added 3 % cobalt to suppress δ ferrite. Nevertheless, unlike the ordinary martensitic heat-resistant steels, the reduced activation martensitic steels cannot contain cobalt due to its negative effect on the reduced activation property. However, it is interesting to notice that manganese and cobalt are both adjacent to iron in the periodic table of elements and it is possible that manganese can be used to suppress δ ferrite in RAFM steel. Therefore, the nitride-strengthened reduced activation martensitic steel may be developed by reduction of carbon and proper addition of manganese.

The nitride-strengthened reduced activation martensitic steel is novel and literature reports have been scarce. In Sects. 5.4, 5.5, 5.6, the possible changes of microstructure and mechanical properties with decreasing carbon content are investigated. Some special characteristics caused by carbon reduction are demonstrated.

The eventual target of the alloy development in terms of property levels is defined in the published work in this area (van der Schaaf et al. 2000; Klueh et al. 2000, 2002; Jitsukawa et al. 2002; Baluc et al. 2007a). In most of the papers, the target mechanical properties were expected to be comparable or better than 9Cr-1Mo tempered martensitic steels and the target for induced activity level was indicated. In addition, expected service condition of RAFM steels was described (Baluc 2009). The target level of toughness after irradiation was indicated (Jitsukawa et al. 2009). 9CrWVTaN martensitic steels were introduced (Klueh 2008).

In relation to the target performance of the development of the steels, the expected operation temperature under service condition of most 7-12Cr martensitic steels ranges from 300 to 550 °C. Toughness degradation by irradiation at temperatures below 400 °C would be one of the biggest issues for the development of the steels. Therefore, higher toughness is required even before irradiation.

Improvement of high temperature strength is another direction of the development. Increasing the upper temperature limit above 600 °C (up to 700 °C) has been attempted (Kurtz et al. 2009; Klueh 2008; Klueh et al. 2007; de Carlan et al. 2004). Tempering temperature of those steels was 750 °C or even higher. The tempering temperature being comparable or lower than the expected operation temperatures can lead to unstable microstructure during service.

Information about steel competitors is given above. The new steels described in Sects. 5.4, 5.5, 5.6 however, are experimental steels, as parts of a large programme of steel development. As such, further refinement in composition and processing is expected. The aim of Sects. 5.4, 5.5, 5.6 is to provide a basis for future development of steels, and to contribute to our materials science understanding of the materials, by using 9Cr base steels with two different carbon contents.

1.4.2 Impact Toughness

The efficiency of power plants could be improved by enhancing the steam parameter. At present, heat-resistant steels for the high steam parameter of 650 °C are being developed. This has put heat-resistant steels such as T/P91, T/P92 and E211 out of consideration because of the loss of the microstructure stability during service at the high temperature (Weisenburger et al. 2008). More advanced steels should be developed to meet this requirement.

It is well accepted in heat-resistant steels that highly stable microstructure will produce excellent creep strength. The precipitates are basically $M_{23}C_6$ and MX, the carbonitride of Nb, V or Ti. The MX-type carbonitrides show much better stability than the $M_{23}C_6$ type carbide. In order to achieve microstructure with high stability, stable precipitates such as MX-type carbonitrides are expected in heat-resistant steels.

In addition to this initial tempered martensitic microstructure, long-term microstructure stability requires attention. Such coarse precipitates as Laves phase

(Fe_2W or Fe_2Mo) and Z phase ((Cr,Nb)N) should be delayed, although they could only form after a long service time (Sawada et al. 2006a). The formation of Laves phase and Z phase is a thermally automatic process which cannot be avoided (Shen et al. 2009). However, this process can be delayed by reducing the content of tungsten, molybdenum and nitrogen.

Nitride-strengthened martensitic heat-resistant steel is developed, based on the above ideas. Following the alloy design and the mechanical properties of the nitride-strengthened martensitic steels, Sect. 5.7 will present the excellent impact toughness of the steel after tempering.

1.5 Low Nickel Maraging Steel

Maraging steels, a special class of ultrahigh strength steels simultaneously with good toughness, are martensitic hardenable alloys (Guo et al. 2004). Maraging steels are hardened by precipitates and they differ from other steels hardening by carbon (Sha and Guo 2009). The term 'maraging' refers to age hardening in a low carbon, iron-nickel lath martensite matrix. The precipitation hardening is due to the formation of precipitates referring to the hardening phases of intermetallic compounds between nickel and molybdenum, nickel and titanium or other additional alloying elements with nickel, or other element combinations. After precipitation hardening, the steel properties are improved significantly, characterised by high tensile strength combined with high toughness and good weldability and malleability.

Maraging steels were developed for special purposes, where the combination of high strength and good toughness was required. Maraging steels have long been regarded as excellent materials and have been applied widely for decades, in industries, e.g. aircraft, aerospace and tooling applications, due to their good machining properties (Sha and Guo 2009; Guo and Sha 2004b).

Commercial maraging steels have sufficiently high nickel content (mostly 18 wt%) to produce martensite following air cooling to room temperature upon solution treatment (Sha and Guo 2009). Nickel has been widely used in maraging steels. The advantages are that nickel raises the yield strength of iron, but also lowers the DBTT (cleavage) of iron, so it is an alloying method of raising strength and increasing toughness. However, the high nickel content has cost implications, leading to the application of these steels being largely restricted to specialised sections such as aerospace. This has generated attempts to develop alternative compositions giving equivalent properties at reduced cost.

Numerous studies have focused on developing steels with low nickel and cobalt contents, as nickel and cobalt are costly and highlighted as strategic alloying elements. Developing novel maraging steels with low nickel and cobalt contents is necessary. There was a tremendous increase in cobalt price in the 1970s, which promoted the development of cobalt-free maraging steels (Leitner et al. 2010). The development of maraging steel with 18 wt% nickel and cobalt-free was a remarkable achievement (Teledyne Vasco and Inco, USA). Subsequently, the content of nickel was reduced to 14 wt%, cobalt-free. Another example of these further developments was the PH 13-8 Mo types (Guo et al. 2003).

Chapter 7 looks into further developing a low nickel content maraging steel, which should have reduced cost. The high cost of nickel demands second thoughts about the actual amounts required of this element in these steels and restricts the areas of application as well, since high cost materials are not suitable for general application, like in civil construction.

Table 1.1 shows the compositions of existing and developmental maraging steels and their approximate alloy costs per tonne. Alloy savings of about £2400 per tonne are possible on 12 % nickel steels compared with standard 18 % nickel steel, and about £800 per tonne on 12 % nickel steels compared with standard cobalt-free 18 % nickel steel. The alloy costs shown in Table 1.1 are based on recent prices in the UK for pure metals and therefore are for illustrative purpose only, because, obviously, metal prices change with time and place.

In order to achieve a better understanding of the function of nickel in maraging steels, the Fe-12.94Ni-1.61Al-1.01Mo-0.23Nb (wt%, for all compositions in this book unless otherwise specified) maraging steel is investigated in Chap. 7. The eventual aim of the research is to develop novel grain refined maraging steels with reduced nickel content, for high strength applications with good toughness at a reduced steel cost. The objectives are to complete mechanical testing and microstructural characterisation of the maraging steel. The main contribution would relate to commercially viable steels. Table 1.2 highlights the difference between the philosophy and the previous research (Howe 2000; Morris et al. 2000; Leinonen 2001; Priestner and Ibraheem 2000) in developing high strength steels.

X-ray diffraction analysis will be used to evaluate the formation of retained or reverted austenite. The hardness will be shown both before and after ageing treatment and then the hardness curve is determined. The toughness is measured by Charpy impact testing. One primary aim is improving toughness of this kind of maraging steel through a lower austenitisation temperature and intercritical annealing.

1.6 Cold-Formed Steel Portal Frame

The majority of portal frames use conventional hot-rolled steel sections for the primary load carrying members (i.e. columns and rafters) and cold-formed steel for the secondary members (i.e. purlins, side rails and cladding). Using hot-rolled steel, spans

Table 1.1 Estimation of alloy cost made with pure metal prices

Composition	Approximate cost per tonne
Fe-18Ni-3.3Mo-8.5Co-0.2Ti-0.1Al	£3807
Fe-18.5Ni-3Mo-0.7Ti-0.1Al	£2212
Fe-12.94Ni-1.61Al-1.01Mo-0.23Nb	£1400

Table 1.2 The novelty and contribution of the research

This research	Previous research
Improve toughness but with fair strength	Improve strength but retain fair toughness
Refine grain size → ageing treatment → use	Refine grain size → use (Howe 2000)
Achieve fine (<5 μm) grain	Achieve ultrafine (~1 μm) grain (Howe 2000)
Reduce cost by using cheaper alloying elements	Reduce cost by choosing the most economical processing routes (Morris et al. 2000; Leinonen 2001; Priestner and Ibraheem 2000)

of up to 60 m can be achieved. On the other hand, for frames of more modest spans, the use of cold-formed steel for the primary load carrying members (i.e. columns and rafters) should be a viable alternative to conventional hot-rolled steel.

For buildings of modest span of up to 30 m, cold-formed steel portal frames (see Fig. 1.2) are an increasingly popular form of construction, particularly in Australia and the UK, and are typically used for low-rise commercial, light industrial and agricultural buildings. Such buildings use cold-formed steel channel-sections for columns and rafters, with joints formed through back-to-back gusset plates bolted to the webs of the channel-sections (see Fig. 1.3).

In the practical design of cold-formed steel portal frames, involving use of sections from one cold-formed steel manufacturer, an economical design can be achieved by selecting the appropriate sections for column and rafter through a number of different permutations of channel-sections. This is feasible since each cold-formed steel manufacturer would only have a discrete number of section sizes in their catalogue. More importantly, the parameters relating to configuration and topography of frame (frame pitch and frame spacing) are usually considered to be fixed.

However, the topography of the frame (frame pitch and frame spacing) plays a key role in the overall structural response. Therefore, the determination of these parameters, in conjunction with the selection of the cold-formed steel sections as described above, can achieve an efficient, cost-wise, building. The research

Fig. 1.2 Cold-formed steel portal framing system

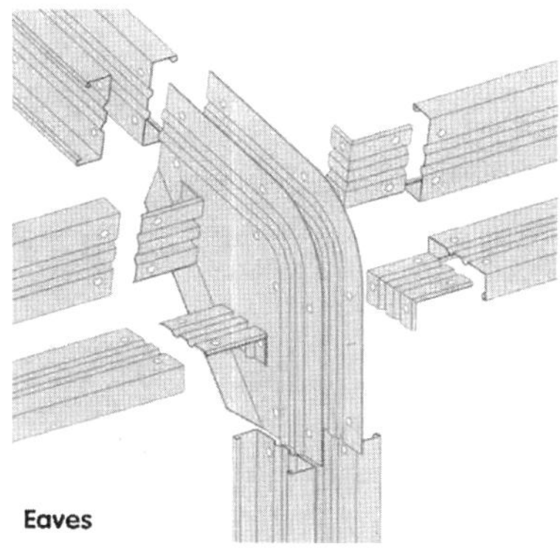

Fig. 1.3 Swagebeam eaves joint

described in Chap. 9 will concentrate on optimising the portal frame topography, with the objective function being the minimum cost per metre length of building.

Because fabrication and erection costs for cold-formed steel are much lower than for hot-rolled steel, there is scope to vary the frame spacing and pitch. Other advantages of cold-formed steel portal frame systems compared to hot-rolled steel portal frames are as follows (Lim and Nethercot 2002). Pre-galvanised cold-formed steel sections that do not require painting to prevent rusting are maintenance-free. The transportation costs are lower due to efficient stacking of cold-formed steel sections. Also, the acquisition costs are lower as the cold-formed steel used for the secondary members can be purchased from the same manufacturer/supplier.

For such frames, moment-resisting joints at the eaves and apex can be formed through mechanical interlock (see Fig. 1.4a). As can be seen, the joints are formed through brackets that are bolted to the webs of the channel-sections. Under moment, the matching swages in both the brackets and the webs of the channel-sections interlock, thus forming a rigid joint. With longer span frames, to reduce the section sizes of both the column and rafter members, a knee brace is often included at the eaves (see Fig. 1.4b). The effect of including a knee brace is that the bending moment that needs to be resisted by both the column and rafter around the joint is reduced with the axial load carried through the knee brace (Rhodes and Burns 2006).

Over the past three decades, the design optimisation of structures has attracted much attention from researchers (Gero et al. 2005). Many structural engineering design problems have discrete decision variables. For instance, in the design of steel frames, the section sizes of the columns and rafters are selected from standard tables. One of the most efficient methods for solving complex combinatorial optimisation problems such as the design of steel portal frames is genetic algorithms (GAs), based on the Darwinian principle of survival of the fittest and adaptation.

Binary-coded GAs have been applied to the design of hot-rolled steel frames to find the discrete section sizes for the members that minimise the weight of the structure (Kameshki and Saka 2001; Toropov and Mahfouz 2001; Gero et al. 2006). However, one of the limitations of binary-coded GAs is the extra computational complexity of the algorithm associated with continuous decision variables

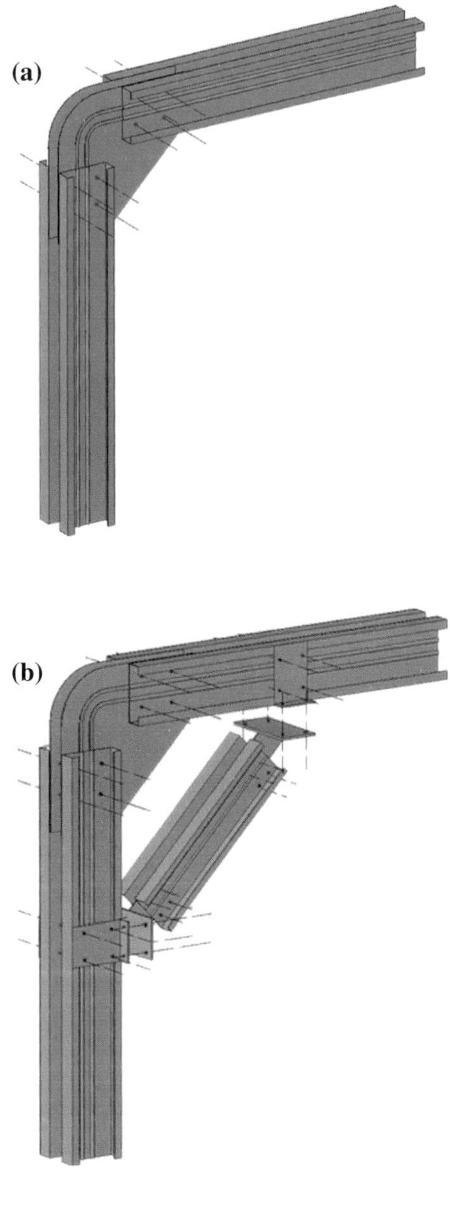

Fig. 1.4 Details of eaves joint arrangements. **a** Mechanical interlock; **b** knee brace

(Deb 2001). Therefore, real-coded GAs (Deb 2001; Deb and Gulati 2001) were proposed to resolve the drawback of binary coded GAs.

In design optimisation of hot-rolled steel portal frames, Saka (2003) described a binary-coded GA to minimise the weight of a portal frame through selecting the most appropriate hot-rolled steel section sizes for the columns and rafters, from a catalogue of available standard sections, based on elastic analysis and design as described in the British Standards (BS 5950-1). More recently, Issa and Mohammad (2010) described a GA using binary strings, to study the same problem. They varied the length and depth of the haunched part of the rafter in a specified range using fixed intervals to determine the optimum size of the haunched member. Hernandez et al. (2005) proposed an optimum design software named PADO, based on mathematical programming, to optimise the design of a hot-rolled steel portal frame in accordance with the Spanish code of practice (EA-95). Chen and Hu (2008) used GAs to optimise hot-rolled steel portal frames having tapered members, according to the Chinese specification for portal frames (CECS-102).

Previous research focused on the design of hot-rolled steel portal frames with fixed topology where the pitch and frame spacing were not optimised and were specified a *priori*. In Chap. 9, a GA is proposed to minimise the cost of cold-formed steel portal frame buildings by minimising the cost of the main structural elements per unit length of the building. Although any code of practice can be used the Australian code was adopted, since in Australia the spans of the frames can be larger as there is less snow.

The optimisation method addresses all the relevant combinations of the permanent and imposed loads, incorporates the full range of design constraints and considers all feasible wind load combinations. It is assumed here that full lateral restraint is applied to columns and rafters. Also, it is assumed that the cost of the purlins, side rails, and sheeting is independent of frame spacing. The research in Chap. 9 differs from the previous work on hot-rolled steel portal frames in that the section sizes of the columns and rafters and the topology of the building, including the pitch and frame spacing, are all optimised simultaneously. The decision variables used in the design optimisation are the spacing of the frames, the pitch of the roof and the section sizes of the main structural elements. Self-evidently, the solution space has both discrete and continuous variables. Also, unlike previous research on hot-rolled steel frames that used binary coding for the GA, real coding is used in Chap. 9.

1.7 Fire Engineering

To comply with the building regulations, fire safety design is a necessary measure. In general, fire protection of buildings can be divided into two types of measures, active and passive. The active measure is concerned with the detection and extinction of fire at its early stage, achieved by introducing alarm, control of smoke and other hazardous elements, in-built fire fighting or control and other fire safety management systems. The passive measure of fire safety design is mainly concerned with the structural fire protection and means of escape in case of fire. This can be achieved by means of enhancing structural performance including the use of fire-protected beams and columns, compartmentation, control of flammability of the structural fabric and provision of fixed escape routes.

Fire engineering is a passive measure towards fire safety for building structures. Traditionally, provision of protective material to steel columns and beams has been used to achieve fire resistance in steel structures. Fire protection is prescriptive in that the engineer designs the steel structure based on its full strength, and then determines the fire protection required by using pre-defined charts and tables depending on the specified fire resistance. The alternative passive measure, fire engineering, is to analytically design the steel structures by using the properties and behaviour of steel at high temperature. This type of passive measure is more economical in that the needs for fire protection may be minimised or eliminated.

The development of fire engineering based on the analytical approach has resulted in steel structures that have built-in fire resistance such as slim, also called shallow, floor beam construction. Two types of slim floor beams have been developed by British Steel (now Tata Steel) and The Steel Construction Institute (SCI), UK, the fabricated Slimflor beam and the asymmetric Slimflor beam, the Slimdek. In addition, Top-Hat beam was developed in Scandinavia (Chaps. 10 and 11). The Slimflor system consists of a Slimflor beam and either pre-cast floor slab with in-fill concrete or composite construction with deep decking. The composite Slimdek system is constructed using an asymmetric Slimflor beam (ASB) with deep decking. Another option of fire engineering in steel structures is to use fire-resistant steels, steels that have better high temperature strength than conventional steels.

Chapter 10 is concerned with the development and use of fire engineering computer software, including two types of software originally developed by SCI to model respectively the moment capacity and temperature development in fire of floor beams. The first type of programme is used to calculate the moment capacity at elevated temperatures for slim floor beams and conventional composite I-beam floor with fire protection, which can then be used to calculate the fire resistance. The second type of programme, TFIRE, is used to model the heat transfer and the temperature development in Slimflor beams with intumescent coating protection.

References

Abe F (2001) Creep rates and strengthening mechanisms in tungsten-strengthened 9Cr steels. Mater Sci Eng A 319–321:770–773. doi:10.1016/S0921-5093(00)02002-5

Abe F (2004) Coarsening behavior of lath and its effect on creep rates in tempered martensitic 9Cr-W steels. Mater Sci Eng A 387–389:565–569. doi:10.1016/j.msea.2004.01.057

Abe F, Semba H, Sakuraya T (2007a) Effect of boron on microstructure and creep deformation behavior of tempered martensitic 9Cr steel. Mater Sci Forum 539–543:2982–2987

Abe F, Taneike M, Sawada K (2007b) Alloy design of creep resistant 9Cr steel using a dispersion of nano-sized carbonitrides. Int J Press Vessels Pip 84:3–12. doi:10.1016/j.ijpvp.2006.09.003

Asahi H, Hara T, Sugiyama M, Maruyama N, Terada Y, Tamehiro H, Koyama K, Ohkita S, Morimoto H (2004a) Development of plate and seam welding technology for X120 linepipe. Int J Offshore Polar Eng 14:11–17

Asahi H, Tsuru E, Hara T, Sugiyama M, Terada Y, Shinada H, Ohkita S, Morimoto H, Doi N, Murata M, Miyazaki H, Yamashita E, Yoshida T, Ayukawa N, Akasaki H, Macia ML, Petersen CW, Koo JY (2004b) Pipe production technology and properties of X120 linepipe. Int J Offshore Polar Eng 14:36–41

Baluc N (2009) Material degradation under DEMO relevant neutron fluences. Phys Scr 2009:014004. doi:10.1088/0031-8949/2009/T138/014004

Baluc N, Abe K, Boutard JL, Chernov VM, Diegele E, Jitsukawa S, Kimura A, Klueh RL, Kohyama A, Kurtz RJ, Lässer R, Matsui H, Möslang A, Muroga T, Odette GR, Tran MQ, van der Schaaf B, Wu Y, Yu J, Zinkle SJ (2007a) Status of R&D activities on materials for fusion power reactors. Nucl Fusion 47:S696–S717. doi:10.1088/0029-5515/47/10/S18

Baluc N, Gelles DS, Jitsukawa S, Kimura A, Klueh RL, Odette GR, van der Schaaf B, Yu J (2007b) Status of reduced activation ferritic/martensitic steel development. J Nucl Mater 367–370:33–41. doi:10.1016/j.jnucmat.2007.03.036

Chen Y, Hu K (2008) Optimal design of steel portal frames based on genetic algorithms. Front Archit Civ Eng China 2:318–322. doi:10.1007/s11709-008-0055-1

Cui J, Kim IS, Kang CY, Miyahara K (2001) Creep stress effect on the precipitation behavior of Laves phase in Fe-10% Cr-6% W alloys. ISIJ Int 41:368–371. doi:10.2355/isijinternational.41.368

Danielsen HK, Hald J (2004) Z-phase in 9–12%Cr steels. In: Viswanathan R, Gandy D, Coleman K (eds) Proceedings of the 4th International Conference on Advances in Materials Technology for Fossil Power Plants. ASM International, Materials Park, OH, pp 999–1012

Danielsen HK, Hald J (2009) Tantalum-containing Z-phase in 12%Cr martensitic steels. Scr Mater 60:811–813. doi:10.1016/j.scriptamat.2009.01.025

Deb K (2001) Multi-objective optimization using evolutionary algorithms. Wiley, Chichester

Deb K, Gulati S (2001) Design of truss-structures for minimum weight using genetic algorithms. Finite Elem Anal Des 37:447–465. doi:10.1016/S0168-874X(00)00057-3

de Carlan Y, Murugananth M, Sourmail T, Bhadeshia HKDH (2004) Design of new Fe-9CrWV reduced-activation martensitic steels for creep properties at 650 °C. J Nucl Mater 329–333:238–242. doi:10.1016/j.jnucmat.2004.04.017

de Castro V, Leguey T, Muñoz A, Monge MA, Fernández P, Lancha AM, Pareja R (2007) Mechanical and microstructural behaviour of Y_2O_3 ODS EUROFER 97. J Nucl Mater 367–370:196–201. doi:10.1016/j.jnucmat.2007.03.146

Fairchild DP, Macia ML, Bangaru NV, Koo JY (2004) Girth welding development for X 120 linepipe. Int J Offshore Polar Eng 14:18–28

Gero MBP, García AB, del Coz Díaz JJ (2005) A modified elitist genetic algorithm applied to the design optimization of complex steel structures. J Constr Steel Res 61:265–280. doi:10.1016/j.jcsr.2004.07.007

Gero MBP, García AB, del Coz Díaz JJ (2006) Design optimization of 3D steel structures: genetic algorithms vs. classical techniques. J Constr Steel Res 62:1303–1309. doi:10.1016/j.jcsr.2006.02.005

Ghassemi-Armaki H, Chen RP, Maruyama K, Yoshizawa M, Igarashi M (2009) Static recovery of tempered lath martensite microstructures during long-term aging in 9–12% Cr heat resistant steels. Mater Lett 63:2423–2425. doi:10.1016/j.matlet.2009.08.024

Golpayegani A, Andrén HO, Danielsen H, Hald J (2008) A study on Z-phase nucleation in martensitic chromium steels. Mater Sci Eng A 489:310–318. doi:10.1016/j.msea.2007.12.022

Guo W, Dong H, Lu M, Zhao X (2002) The coupled effects of thickness and delamination on cracking resistance of X70 pipeline steel. Int J Pressure Vessels Pip 79:403–412. doi:10.1016/S0308-0161(02)00039-X

Guo Z, Sha W, Vaumousse D (2003) Microstructural evolution in a PH13-8 stainless steel after ageing. Acta Mater 51:101–116. doi:10.1016/S1359-6454(02)00353-1

Guo Z, Sha W (2004a) Kinetics of ferrite to Widmanstätten austenite transformation in a high-strength low-alloy steel revisited. Z Metallkd 95:718–723

Guo Z, Sha W (2004b) Comments on small-angle neutron scattering analysis of the precipitation behaviour in a maraging steel by Staron, Jamnig, Leitner, Ebner & Clemens (2003). J Appl Crystallogr 37:325–326. doi:10.1107/S0021889803028127

Guo Z, Sha W, Li D (2004) Quantification of phase transformation kinetics of 18 wt.% Ni C250 maraging steel. Mater Sci Eng A 373:10–20. doi:10.1016/j.msea.2004.01.040

Gustafson Å, Ågren J (2001) Possible effect of Co on coarsening of $M_{23}C_6$ carbide and Orowan stress in a 9% Cr steel. ISIJ Int 41:356–360. doi:10.2355/isijinternational.41.356

Hernández S, Fontán AN, Perezzán JC, Loscos P (2005) Design optimization of steel portal frames. Adv Eng Softw 36:626–633. doi:10.1016/j.advengsoft.2005.03.006

Howe AA (2000) Ultrafine grained steels: industrial prospects. Mater Sci Technol 16:1264–1266. doi:10.1179/026708300101507488

Issa HK, Mohammad FA (2010) Effect of mutation schemes on convergence to optimum design of steel frames. J Constr Steel Res 66:954–961. doi:10.1016/j.jcsr.2010.02.002

Jitsukawa S, Tamura M, van der Schaaf B, Klueh RL, Alamo A, Petersen C, Schirra M, Spaetig P, Odette GR, Tavassoli AA, Shiba K, Kohyama A, Kimura A (2002) Development of an extensive database of mechanical and physical properties for reduced-activation martensitic steel F82H. J Nucl Mater 307–311:179–186. doi:10.1016/S0022-3115(02)01075-9

Jitsukawa S, Suzuki K, Okubo N, Ando M, Shiba K (2009) Irradiation effects on reduced activation ferritic/martensitic steels-tensile, impact, fatigue properties and modelling. Nucl Fusion 49:115006. doi:10.1088/0029-5515/49/11/115006

Kameshki E, Saka MP (2001) Optimum design of nonlinear steel frames with semi-rigid connections using a genetic algorithm. Comput Struct 79:1593–1604. doi:10.1016/S0045-7949(01)00035-9

Kim YM, Kim SK, Lim YJ, Kim NJ (2002) Effect of microstructure on the yield ratio and low temperature toughness of linepipe steels. ISIJ Int 42:1571–1577. doi:10.2355/isijinternational.42.1571

Kimura Y, Inoue T, Yin F, Sitdikov O, Tsuzaki K (2007) Toughening of a 1500 MPa class steel through formation of an ultrafine fibrous grain structure. Scr Mater 57:465–468. doi:10.1016/j.scriptamat.2007.05.039

Kimura K, Toda Y, Kushima H, Sawada K (2010) Creep strength of high chromium steel with ferrite matrix. Int J Press Vessels Pip 87:282–288. doi:10.1016/j.ijpvp.2010.03.016

Klimenkov M, Lindau R, Möslang A (2009) New insights into the structure of ODS particles in the ODS-Eurofer alloy. J Nucl Mater 386–388:553–556. doi:10.1016/j.jnucmat.2008.12.174

Klueh RL (2008) Reduced-activation steels: future development for improved creep strength. J Nucl Mater 378:159–166. doi:10.1016/j.jnucmat.2008.05.010

Klueh RL, Cheng ET, Grossbeck ML, Bloom EE (2000) Impurity effects on reduced-activation ferritic steels developed for fusion applications. J Nucl Mater 280:353–359. doi:10.1016/S0022-3115(00)00060-X

Klueh RL, Gelles DS, Jitsukawa S, Kimura A, Odette GR, van der Schaaf B, Victoria M (2002) Ferritic/martensitic steels—overview of recent results. J Nucl Mater 307–311:455–465. doi:10.1016/S0022-3115(02)01082-6

Klueh RL, Hashimoto N, Maziasz PJ (2007) New nano-particle-strengthened ferritic/martensitic steels by conventional thermo-mechanical treatment. J Nucl Mater 367–370:48–53. doi:10.1016/j.jnucmat.2007.03.001

Koo JY, Luton MJ, Bangaru NV, Petkovic RA, Fairchild DP, Petersen CW, Asahi H, Hara T, Terada Y, Sugiyama M, Tamehiro H, Komizo Y, Okaguchi S, Hamada M, Yamamoto A, Takeuchi I (2004) Metallurgical design of ultra high-strength steels for gas pipelines. Int J Offshore Polar Eng 14:2–10

Kurtz RJ, Alamo A, Lucon E, Huang Q, Jitsukawa S, Kimura A, Klueh RL, Odette GR, Petersen C, Sokolov MA, Spätig P, Rensman JW (2009) Recent progress toward development of reduced activation ferritic/martensitic steels for fusion structural applications. J Nucl Mater 386–388:411–417. doi:10.1016/j.jnucmat.2008.12.323

Lee JS, Armaki HG, Maruyama K, Maruki T, Asahi H (2006) Causes of breakdown of creep strength in 9Cr-1.8W-0.5Mo-VNb steel. Mater Sci Eng A 428:270–275. doi:10.1016/j.msea.2006.05.010

Leinonen JI (2001) Processing steel for higher strength. Adv Mater Process 159(11):31–33

Leitner H, Schober M, Schnitzer R (2010) Splitting phenomenon in the precipitation evolution in an Fe-Ni-Al-Ti-Cr stainless steel. Acta Mater 58:1261–1269. doi:10.1016/j.actamat.2009.10.030

Li PJ, Xiong YH, Liu SX, Zeng DB (2003) Electron theory study on mechanism of action of cobalt in Fe-Co-Cr based high-alloy steel. Chin Sci Bull 48:208–210

Lim JBP, Nethercot DA (2002) Design and development of a general cold-formed steel portal framing system. Struct Eng 80(21):31–40

Lindau R, Möslang A, Rieth M, Klimiankou M, Materna-Morris E, Alamo A, Tavassoli AAF, Cayron C, Lancha AM, Fernandez P, Baluc N, Schäublin R, Diegele E, Filacchioni G, Rensman JW, van der Schaaf B, Lucon E, Dietz W (2005) Present development status of

EUROFER and ODS-EUROFER for application in blanket concepts. Fusion Eng Des 75–79:989–996. doi:10.1016/j.fusengdes.2005.06.186

Maruyama K, Sawada K, Koike J (2001) Strengthening mechanisms of creep resistant tempered martensitic steel. ISIJ Int 41:641–653. doi:10.2355/isijinternational.41.641

Masuyama F (2001) History of power plants and progress in heat resistant steels. ISIJ Int 41:612–625. doi:10.2355/isijinternational.41.612

Miyata K, Sawaragi Y (2001) Effect of Mo and W on the phase stability of precipitates in low Cr heat resistant steels. ISIJ Int 41:281–289. doi:10.2355/isijinternational.41.281

Morris JW, Krenn CR, Guo Z (2000) 19th ASM Heat Treating Society conference and exposition including steel heat treating in the new millenium: an international symposium in honor of Professor George Krauss. ASM International, Materials Park, pp 526–535

Olier P, Bougault A, Alamo A, de Carlan Y (2009) Effects of the forming processes and Y_2O_3 content on ODS-Eurofer mechanical properties. J Nucl Mater 386–388:561–563. doi:10.1016/j.jnucmat.2008.12.177

Otárola T, Hollner S, Bonnefois B, Anglada M, Coudreuse L, Mateo A (2005) Embrittlement of a superduplex stainless steel in the range of 550–700 °C. Eng Fail Anal 12:930–941. doi:10.1016/j.engfailanal.2004.12.022

Pozuelo M, Carreño F, Ruano O (2006) A delamination effect on the impact toughness of an ultrahigh carbon-mild steel laminate composite. Compos Sci Technol 66:2671–2676. doi:10.1016/j.compscitech.2006.03.018

Priestner R, Ibraheem AK (2000) Processing of steel for ultrafine ferrite grain structures. Mater Sci Technol 16:1267–1272. doi:10.1179/026708300101507497

Rhodes J, Burns R (2006) Development of a portal frame system on the basis of component testing. In: Proceedings of the 18th international specialty conference on cold-formed steel structures, University of Missouri-Rolla, Missouri, pp 367–385

Ryu SH, Lee YS, Kong BO, Kim JT, Kwak DH, Nam SW et al (2006) In: Proceedings of the 3rd international conference on advanced structural steels. The Korean Institute of Metals and Materials, pp 563–569

Saka MP (2003) Optimum design of pitched roof steel frames with haunched rafters by genetic algorithm. Comput Struct 81:1967–1978. doi:10.1016/S0045-7949(03)00216-5

Saroja S, Dasgupta A, Divakar R, Raju S, Mohandas E, Vijayalakshmi M, Rao KBS, Raj B (2011) Development and characterization of advanced 9Cr ferritic/martensitic steels for fission and fusion reactors. J Nucl Mater 409:131–139. doi:10.1016/j.jnucmat.2010.09.022

Sawada K, Kubo K, Abe F (2001) Creep behavior and stability of MX precipitates at high temperature in 9Cr-0.5Mo-1.8W-VNb steel. Mater Sci Eng A 319–321:784–787. doi:10.1016/S0921-5093(01)00973-X

Sawada K, Taneike M, Kimura K, Abe F (2004) Effect of nitrogen content on microstructural aspects and creep behavior in extremely low carbon 9Cr heat-resistant steel. ISIJ Int 44:1243–1249. doi:10.2355/isijinternational.44.1243

Sawada K, Kushima H, Kimura K (2006a) Z-phase formation during creep and aging in 9–12% Cr heat resistant steels. ISIJ Int 46:769–775. doi:10.2355/isijinternational.46.769

Sawada K, Kushima H, Kimura K, Tabuchi M (2006b) Creep strength degradation by Z phase formation in 9-12%Cr heat resistant steels. In: Proceedings of the 3rd international conference on advanced structural steels. The Korean Institute of Metals and Materials, pp 532–537

Sawada K, Kushima H, Kimura K, Tabuchi M (2007) TTP diagrams of Z phase in 9–12% Cr heat-resistant steels. ISIJ Int 47:733–739. doi:10.2355/isijinternational.47.733

Schaeublin R, Leguey T, Spätig P, Baluc N, Victoria M (2002) Microstructure and mechanical properties of two ODS ferritic/martensitic steels. J Nucl Mater 307–311:778–782. doi:10.1016/S0022-3115(02)01193-5

Sha W, Guo Z (2009) Maraging steels: modelling of microstructure, properties and applications. Woodhead Publishing, Cambridge. doi:10.1533/9781845696931

Sha W, Kirby BR, Kelly FS (2001) The behaviour of structural steels at elevated temperatures and the design of fire resistant steels. Mater Trans 42:1913–1927

Shen YZ, Kim SH, Cho HD, Han CH, Ryu WS (2009) Precipitate phases of a ferritic/martensitic 9% Cr steel for nuclear power reactors. Nucl Eng Des 239:648–654. doi:10.1016/j.nucengdes.2008.12.018

Silva MC, Hippert Jr E, Ruggieri C (2005) In: Proceedings of ASME pressure vessels and piping conference. ASME, Denver, pp 87–94

Song R, Ponge D, Raabe D (2005) Mechanical properties of an ultrafine grained C-Mn steel processed by warm deformation and annealing. Acta Mater 53:4881–4892. doi:10.1016/j.actamat.2005.07.009

Song R, Ponge D, Raabe D, Speer JG, Matlock DK (2006) Overview of processing, microstructure and mechanical properties of ultrafine grained bcc steels. Mater Sci Eng A 441:1–17. doi:10.1016/j.msea.2006.08.095

Taneike M, Abe F, Sawada K (2003) Creep-strengthening of steel at high temperatures using nano-sized carbonitride dispersions. Nature 424:294–296. doi:10.1038/nature01740

Taneike M, Sawada K, Abe F (2004) Effect of carbon concentration on precipitation behavior of $M_{23}C_6$ carbides and MX carbonitrides in martensitic 9Cr steel during heat treatment. Metall Mater Trans A 35A:1255–1262. doi:10.1007/s11661-004-0299-x

Toda Y, Seki K, Kimura K, Abe F (2003) Effects of W and Co on long-term creep strength of precipitation strengthened 15Cr ferritic heat resistant steels. ISIJ Int 43:112–118. doi:10.2355/isijinternational.43.112

Toda Y, Tohyama H, Kushima H, Kimura K, Abe F (2005) Improvement in creep strength of precipitation strengthened 15Cr ferritic steel by controlling carbon and nitrogen contents. JSME Int J Ser A 48:35–40. doi:10.1299/jsmea.48.35

Toropov VV, Mahfouz SY (2001) Design optimization of structural steelwork using a genetic algorithm, FEM and a system of design rules. Eng Comput 18:437–460. doi:10.1108/02644400110387118

Tsuji N, Okuno S, Koizumi Y, Minamino Y (2004) Toughness of ultrafine grained ferritic steels fabricated by ARB and annealing process. Mater Trans 45:2272–2281. doi:10.2320/matertrans.45.2272

van der Schaaf B, Gelles DS, Jitsukawa S, Kimura A, Klueh RL, Möslang A, Odette GR (2000) Progress and critical issues of reduced activation ferritic/martensitic steel development. J Nucl Mater 283–287:52–59. doi:10.1016/S0022-3115(00)00220-8

Verdeja JI, Asensio J, Pero-Sanz JA (2003) Texture, formability, lamellar tearing and HIC susceptibility of ferritic and low-carbon HSLA steels. Mater Charact 50:81–86. doi:10.1016/S1044-5803(03)00106-2

Wallin K (2001) Upper shelf energy normalisation for sub-sized Charpy-V specimens. Int J Pressure Vessels Pip 78:463–470. doi:10.1016/S0308-0161(01)00063-1

Wang W, Shan Y, Yang K (2009a) Study of high strength pipeline steels with different microstructures. Mater Sci Eng A 502:38–44. doi:10.1016/j.msea.2008.10.042

Wang Y, Mayer KH, Scholz A, Berger C, Chilukuru H, Durst K, Blum W (2009b) Development of new 11%Cr heat resistant ferritic steels with enhanced creep resistance for steam power plants with operating steam temperatures up to 650 °C. Mater Sci Eng A 510–511:180–184. doi:10.1016/j.msea.2008.04.116

Weisenburger A, Heinzel A, Müller G, Muscher H, Rousanov A (2008) T91 cladding tubes with and without modified FeCrAlY coatings exposed in LBE at different flow, stress and temperature conditions. J Nucl Mater 376:274–281. doi:10.1016/j.jnucmat.2008.02.026

Wong CPC, Salavy JF, Kim Y, Kirillov I, Kumar ER, Morley NB, Tanaka S, Wu YC (2008) Overview of liquid metal TBM concepts and programs. Fusion Eng Des 83:850–857. doi:10.1016/j.fusengdes.2008.06.040

Yamada K, Igarashi M, Muneki S, Abe F (2003) Effect of Co addition on microstructure in high Cr ferritic steels. ISIJ Int 43:1438–1443. doi:10.2355/isijinternational.43.1438

Yang M, Chao YJ, Li X, Tan J (2008a) Splitting in dual-phase 590 high strength steel plates: Part I. Mechanisms. Mater Sci Eng A 497:451–461. doi:10.1016/j.msea.2008.07.067

Yang M, Chao YJ, Li X, Immel D, Tan J (2008b) Splitting in dual-phase 590 high strength steel plates: Part II. Quantitative analysis and its effect on Charpy impact energy. Mater Sci Eng A 497:462–470. doi:10.1016/j.msea.2008.07.066

Yin F, Jung W (2009) Nanosized MX precipitates in ultra-low-carbon ferritic/martensitic heat-resistant steels. Metall Mater Trans A 40A:302–309. doi:10.1007/s11661-008-9716-x

Yin F, Jung W, Chung S (2007) Microstructure and creep rupture characteristics of an ultra-low carbon ferritic/martensitic heat-resistant steel. Scr Mater 57:469–472. doi:10.1016/j.scriptamat.2007.05.034

Yong Q (2006) The second phase in steels. Metallurgical Industry Press, Beijing

Yoshizawa M, Igarashi M (2007) Long-term creep deformation characteristics of advanced ferritic steels for USC power plants. Int J Press Vessels Pip 84:37–43. doi:10.1016/j.ijpvp.2006.09.005

Yu G, Nita N, Baluc N (2005) Thermal creep behaviour of the EUROFER 97 RAFM steel and two European ODS EUROFER 97 steels. Fusion Eng Des 75–79:1037–1041. doi:10.1016/j.fusengdes.2005.06.311

Zhao MC, Shan YY, Xiao FR, Yang K, Li YH (2002) Investigation on the H_2S-resistant behaviors of acicular ferrite and ultrafine ferrite. Mater Lett 57:141–145. doi:10.1016/S0167-577X(02)00720-6

Zhao X, Jing TF, Gao YW, Qiao GY, Zhou JF, Wang W (2005) Annealing behavior of nano-layered steel produced by heavy cold-rolling of lath martensite. Mater Sci Eng A 397:117–121. doi:10.1016/j.msea.2005.02.007

Part I
Materials Science of Alloy Steels

Chapter 2
High-Strength Low-Alloy Steel

Abstract Growth kinetics of Widmanstätten austenite in ferrite in high-strength low-alloy steel is based on a model that describes diffusion controlled growth of precipitates with shapes approximating to needles or plates, where all the factors that may influence the precipitate growth, i.e. diffusion, interface kinetics and capillarity, are accounted for within one equation. The ratio between calculated and experimental values of the radius of the advancing tip is inversely proportional to the degree of supersaturation. Following this theoretical work, the tensile behaviour of high-strength low-alloy steel after tempering is discussed, and well explained in view of the interactions of mobile dislocations and dissolved carbon and nitrogen atoms and their effects on the strain hardening exponent. In the final section, splitting during fracture of tensile and impact loading is examined. Delamination does not occur in the as-rolled condition, but is severe in steel tempered in the temperature range of 500–650 °C. Steel that has been triple quench-and-tempered to produce a fine equiaxed grain-size also does not exhibit splitting. It is concluded that the elongated as-rolled grains and grain boundary embrittlement resulting from precipitates (carbides and nitrides) formed during reheating are responsible for the delamination.

2.1 Kinetics of the Diffusion-Controlled Ferrite to Widmanstätten Austenite Transformation

2.1.1 Growth Theory

2.1.1.1 Precipitate Plates and Needles

The equation relating the Peclet number (a dimensionless velocity parameter) $p = V\rho/2D$ to the dimensionless degree of supersaturation Ω_0 for the growth of plates is

$$\Omega_0 = \sqrt{\pi/p}\,\exp(p)\mathrm{erfc}(\sqrt{p})\left[1 + \frac{V}{V_c}\Omega_0 S_1(p) + \frac{\rho_c}{\rho}\Omega_0 S_2(p)\right] \quad (2.1)$$

W. Sha, *Steels*, DOI: 10.1007/978-1-4471-4872-2_2,

where V is the lengthening rate, D is the diffusion coefficient of solute in the matrix phase, erfc is the complementary error function, S_1 and S_2 are functions involved in the growth of plate. Other parameters can be calculated as below

$$\rho_c = \frac{\sigma \nu}{RT} \frac{1 - c_\alpha}{c_\beta - c_\alpha} \frac{c_\alpha}{c_0 - c_\alpha} \tag{2.2}$$

where σ is the interfacial free energy per unit area between precipitate and matrix, ν is the molar volume of the precipitate, R is the gas constant, T is the temperature (in Kelvin), c_α is the solid solubility of the controlling element in the parent phase α, c_β is the concentration of the controlling element in the new phase β, c_0 is the concentration in the matrix before precipitation. The radius of curvature $\rho = 2f$, where f is the focal distance of the parabola of either plate or needle (Fig. 1.1c), defined uniquely for a parabola lengthening along the Z direction and thickening along the X direction. In Eq. 2.1, V_c is the velocity or the lengthening rate of a flat interface during interface controlled growth (i.e. when almost all the free energy is dissipated in the transfer of atoms across the interface so that the concentration difference in the matrix vanishes) and is given by

$$V_c = \mu_0 (c_0 - c_\alpha) \tag{2.3}$$

For curved interfaces, the growth rate is a function of the interface curvature via the Gibbs–Thomson effect. The curvature at which the growth rate becomes zero is $1/\rho_c$. The concentration parameters c_0, c_α, and c_β are illustrated in Fig. 2.1. The values of functions $S_1(p)$ and $S_2(p)$ are given by Rivera-Díaz-del-Castillo and Bhadeshia (2001).

Equation 2.1 gives the general solution for the growth of precipitate plates. Note that the right-hand side is a sum of three terms of which the first term is for the case of the iso-concentration boundary. The second and the third terms are corrections to that solution due to the interface kinetics and capillarity effects respectively. The value of Ω_0 can be obtained from the phase diagram when the average composition of the alloy is known:

$$\Omega_o = \frac{c_0 - c_\alpha}{c_\beta - c_\alpha} \tag{2.4}$$

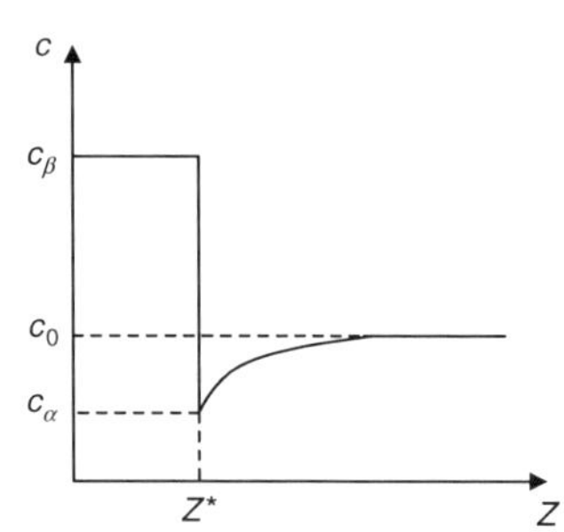

Fig. 2.1 Concentrations at precipitate boundary ($Z = Z^*$) for flat interface (From Guo and Sha (2004) © Carl Hanser Verlag, Muenchen)

Equation 2.1 reveals that, for a given value of Ω_0, multitudes of exact solutions are possible for the growth rate depending upon the value of the radius of curvature. The experimental observation that a plate grows at a constant rate indicates that a steady-state diffusion field exists with respect to the advancing interface and that the tip radius ρ remains unchanged during the growth. This implies that of many solutions available only one solution is stable with respect to small perturbation in the radius of curvature of the tip of the plate. This solution corresponds to the maximum growth rate. This maximum growth rate can be obtained by differentiating Eq. 2.1 with respect to ρ and setting $\partial V/\partial \rho = 0$, which gives us another relationship between Ω_0, V, and ρ. The simultaneous solution of this equation with Eq. 2.1 then gives us unique values for V and ρ for a given value of Ω_0.

The equation relating the Peclet number to the dimensionless supersaturation Ω_0 for the growth of needles is

$$\Omega_0 = p\exp(p)\mathrm{E}_1(p)\left[1 + \frac{V}{V_c}\Omega_0\mathrm{R}_1(p) + \frac{\rho_c'}{\rho}\Omega_0\mathrm{R}_2(p)\right] \tag{2.5}$$

where E_1 is the exponential integral function, R_1 and R_2 are functions involved in the growth of needle, and $\rho_c{}'$ is the critical radius of curvature for nucleation for a needle which is equal to twice that for a plate. The values of functions $\mathrm{R}_1(p)$ and $\mathrm{R}_2(p)$ have been given by Rivera-Díaz-del-Castillo and Bhadeshia (2001). Equation 2.5 gives the general solution for the growth of needles. The right-hand side is a sum of three terms of which the first is where interface kinetics and capillarity are neglected. The second and third terms account respectively for those effects. For a given value of Ω_0, though many solutions are available only one is stable. This solution corresponds to the maximum growth rate. This maximum growth rate can be obtained by differentiating Eq. 2.5 with respect to ρ and setting $\partial V/\partial \rho = 0$, which gives us another relationship between Ω_0, V, and ρ. The simultaneous solution of the resulting equation with Eq. 2.5 then gives unique values for V and ρ for a given value of Ω_0. Growth theory for plate growth follows similar formulas and can be referred to from the earlier part of this section.

2.1.1.2 Approximation for the Growth of Needles or Plates

At a given supersaturation degree, the values of p and $\rho/\rho_c{}'$ were given by Rivera-Díaz-del-Castillo and Bhadeshia (2001). Especially when the supersaturation is small, the values of p and $\rho/\rho_c{}'$ approach asymptotically to simple relationships between p and Ω_0. For needles, one has

$$\Omega_0 = \frac{2p[\ln(\kappa p)]^2}{1 - \ln(\kappa p)} \tag{2.6a}$$

$$\frac{\rho}{\rho_c'} = -\frac{3.841}{4}\frac{\Omega_0 \ln(\kappa p)}{\Omega_0 + p\ln(\kappa p)} \tag{2.6b}$$

where $\kappa = \exp(-\gamma)$, γ being the Euler constant ($\gamma = 0.57722\ldots$).

Simplified kinetic expressions for the growth of precipitate plates are:

$$p = \frac{9}{16\pi}\Omega_*^2 \tag{2.7a}$$

$$\frac{\rho}{\rho_c} = \frac{32}{3}\frac{1}{\Omega_*} \tag{2.7b}$$

where the corrected degree of supersaturation (dimensionless)

$$\Omega_* = \frac{\Omega_0}{1 - \frac{2}{\pi}\Omega_0 - \frac{1}{2\pi}\Omega_0^2} \tag{2.8}$$

The above-simplified expressions are valid even when p is up to 1 or Ω_0 to 0.85. The final lengthening rate expression can be obtained by using the definition of the Peclet number, which results in:

$$V = \frac{9}{8\pi}\frac{D}{\rho}\Omega_*^2 = \frac{27}{256\pi}\frac{D}{\rho_c}\Omega_*^3 \tag{2.9}$$

where D is the diffusion rate of alloying elements in the ferrite matrix. These expressions are simple and show clearly the dependence of radius of curvature and growth rate on the degree of supersaturation.

2.1.2 Parameter Determination and Calculation

The ferrite-to-Widmanstätten austenite transformation in high-strength low-alloy (HSLA) Fe–C–Mn–Nb steel is studied using this theory with experimental values of the radius of the advancing tip. The chemical composition of the steel is Fe-0.693C-1.514Mn-0.022Nb-0.028N-0.178Si-0.047Al (at.%), referred to as Nb-steel.

The specimens were encapsulated in quartz tubes under vacuum and solution treated at 1500 K for 2 h. They were then isothermally transformed at 923 K for 30 min and quenched to obtain a structure consisting of ferrite and pearlite. Ferrite to austenite transformation took place when the specimens were reheated to 1003–1073 K and held. During the reverse transformation from ferrite to austenite, the pearlitic structure transforms at first to austenite. The concentrations of manganese in austenite and ferrite are almost the same because the austenite to ferrite transformation in the initial structure occurs by the paraequilibrium mode. The initial growth of austenite into ferrite in the intercritical annealing range occurs by diffusion of carbon in austenite and across the interface between austenite and ferrite. This constitutes the second stage of the ferrite to austenite transformation. However, in low carbon steels such as HSLA steels, the relative amount of pearlite is small and the extent of the second stage of ferrite to austenite transformation is limited. As soon as the diffusion field of carbon in austenite reaches the original grain boundary position, further transformation of ferrite to austenite can occur only under a truncated diffusion field. The situation is ideal for further growth of austenite to occur by the diffusion of manganese in ferrite. This is because the diffusivity of manganese in ferrite is two or three orders of magnitude higher than that in austenite. The growth of austenite by the diffusion of manganese in ferrite constitutes the third stage of the transformation, observed as the formation of Widmanstätten austenite.

The first step is to determine the necessary parameters involved in the calculation, which include supersaturation degree, critical nucleus size, interfacial free energy, diffusion coefficient and the experimental values of the radius of the advancing tip. The equilibrium concentrations of manganese in ferrite and austenite in the Nb-steel as functions of temperature are shown in Fig. 2.2. The temperature for ferrite to disappear in the Nb-steel is 1084 K. This section concentrates on the temperature range below this temperature, where precipitate grows into needles. Temperature increase affects the degree of supersaturation Ω_0 and the size of critical radius. Ω_0 and $\rho_c{}'$ as functions of temperature are shown in Figs. 2.3 and 2.4.

The molar volume of the austenite phase is 7.304×10^{-6} m^3/mol, calculated using the weighted average of the molar volumes of pure iron and manganese

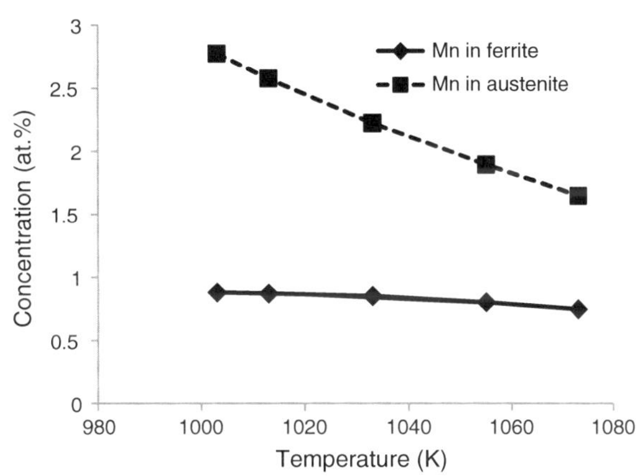

Fig. 2.2 Concentration of manganese in ferrite and austenite as functions of temperature in the Nb-steel (From Guo and Sha (2004) © Carl Hanser Verlag, Muenchen)

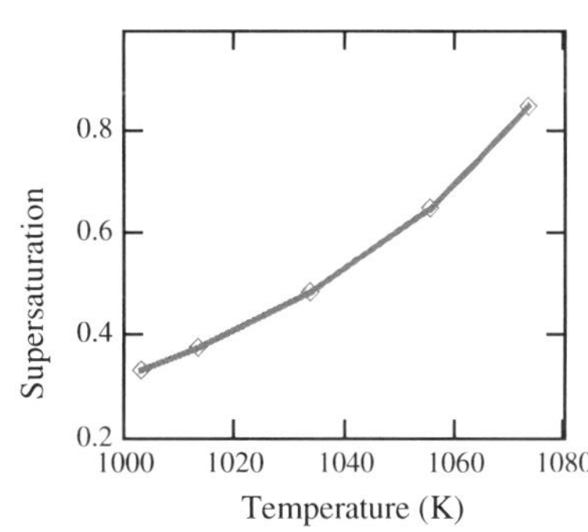

Fig. 2.3 Degree of supersaturation as a function of temperature in the Nb-steel (From Guo and Sha (2004) © Carl Hanser Verlag, Muenchen)

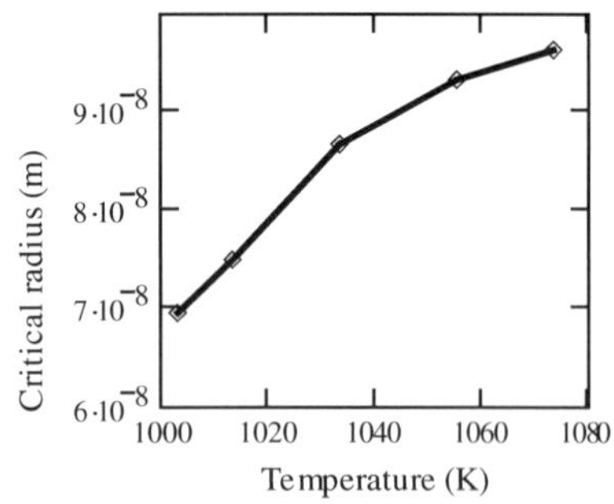

Fig. 2.4 Critical radius size as a function of temperature in the Nb-steel (From Guo and Sha (2004) © Carl Hanser Verlag, Muenchen)

considering Fe-1.5 wt% Mn alloy. The interphase energy of the interface between austenite and ferrite is taken as the value for the interface between α-Fe and γ-Fe, 560 mJ/m^2, at the advancing tip. The diffusion coefficient of alloying elements in the ferrite matrix is $D = D_0\exp(-Q/RT)$, where D_0 is the pre-exponential term in Arrhenius expression for the diffusion coefficient and Q is the diffusion activation energy. Parameters are $D_0 = 1.49 \times 10^{-4}$ m^2/s and $Q = 233.25$ kJ/mol for ferromagnetic iron in the temperature range 973–1033 K. Between 1073–1173 K, the values for D in paramagnetic iron are used, $D_0 = 3.5 \times 10^{-4}$ m^2/s and $Q = 219.45$ kJ/mol. Since the difference in diffusivity values calculated from the two sets of D_0 and Q is small for temperature 1033–1073 K, the ferromagnetic D values are used in this temperature interval.

The values of Peclet number p and ρ/ρ_c' ratio at different degrees of supersaturation from the original theory can be obtained from the work by Rivera-Díaz-del-Castillo and Bhadeshia (2001). The experimental ρ values are used. The experimental values of ρ and calculated ρ/ρ_c' ratios are listed in Table 2.1. At the two temperature points where the experimentally observed radius of the advancing tip for either needle or plate (in this case needle) ρ_e is not known, ρ_e/ρ_c' is interpolated by assuming that there is a linear relationship between ρ_e/ρ_c' and Ω_0. The ratio between calculated ρ and ρ_e as a function of Ω_0 is shown in Fig. 2.5. The calculated lengthening rates of the Nb-steel using calculated ρ and ρ_e are shown

Table 2.1 Comparison between the calculated ρ/ρ_c' and experimental values for the Nb-steel

Temperature (K)	ρ_c' (nm)	ρ_e (nm)	ρ_e/ρ_c'	ρ/ρ_c' (calculated)	ρ (calculated, nm)	ρ (calculated)/ρ_e
1003	69	111.2	1.61	6.0	414	3.7
1013	75	-	1.64[a]	5.7	428	3.5
1033	87	-	1.71[a]	4.9	426	2.9
1055	93	169.0	1.81	3.9	363	2.2
1073	96	184.4	1.92	2.9	278	1.5

[a]Estimated based on the values at 1003, 1055 and 1073 K, by assuming a linear relationship between ρ_e/ρ_c' and the supersaturation degree Ω_0

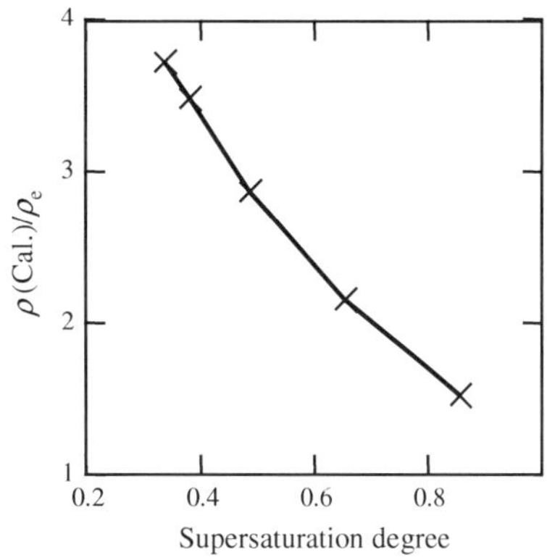

Fig. 2.5 Ratio between calculated ρ and ρ_e as a function of supersaturation degree for the Nb-steel (From Guo and Sha (2004) © Carl Hanser Verlag, Muenchen)

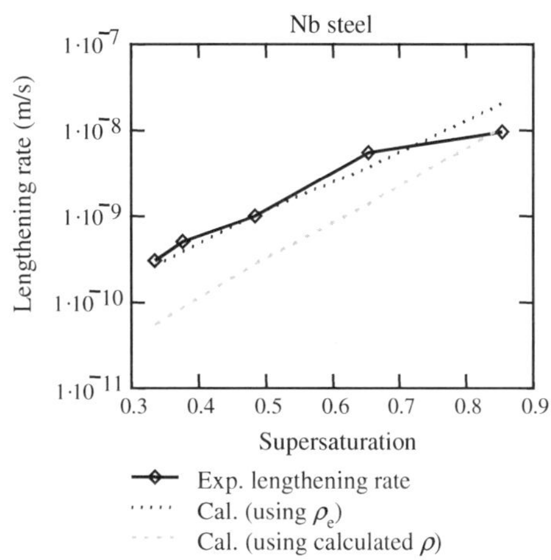

Fig. 2.6 Comparison between experimental and calculated results using calculated ρ and ρ_e, for the lengthening of austenite in the Nb-steel (From Guo and Sha (2004) © Carl Hanser Verlag, Muenchen)

in Fig. 2.6, together with experimental data for comparison. It can be seen that the values using ρ_e fit the experimental data much better than the values calculated using calculated ρ.

The development of the original theory neglects the transformation stress and strain and the anisotropic nature of interphase energy. Consequently, the shape of the precipitate may not be accurately predicted. In order to obtain lengthening rates in agreement with experimental observations, previous studies on precipitate lengthening usually included optimisation procedures on such as interphase energy, interface kinetic coefficients or diffusion coefficient.

The effects of stress and strain caused by precipitate formation and anisotropic interphase energy are reflected in the aspect ratio of the newly formed phase, which can be in turn reflected in the change of the ρ/ρ_c or ρ/ρ_c' ratio. Such effects can be compensated by using the original theory with experimental values of the radius of the advancing tip. The ratio between the theoretically calculated ρ/ρ_c or ρ/ρ_c' with experimental values ρ_e/ρ_c or ρ_e/ρ_c' reflects the compression effects on precipitate morphology due to transformation stress and strain and anisotropic interphase energy.

The focal distance of the advancing tip changes to ρ_e/ρ(calculated) of the value obtained from the original calculation, as illustrated in Fig. 2.7. The relationships between Ω_0 and the Peclet number p are not affected by such revisions. The ρ/ρ_c and $\rho/\rho_c{}'$ values are calculated using Eqs. 2.7b and 2.6b, respectively. The lengthening rate will be ρ(calculated)/ρ_e times faster if one uses $p = V\rho_e/2D$ in place of $p = V\rho$(calculated)/$2D$. For instance, the growth rate for precipitate plates will be

$$V = \frac{9}{8\pi}\frac{D}{\rho_e}\Omega_*^2 \tag{2.10}$$

That the original theory is not able to predict the $\rho/\rho_c{}'$ ratio well is not new. In relation to the growth of Widmanstätten needles in Fe–Mn–C steels, the experimentally observed $\rho/\rho_c{}'$ ratios for needle growth are about 1.6–1.9. The calculated results using the original model, Eq. 2.5, range from 2.9 to 6, giving ρ(calculated)/ρ_e of 1.5–3.7 (Table 2.1). When studying the lengthening kinetics of ferrite and bainite side plates in Fe–C steels (C from 0.24 to 0.45 wt%), the theoretically calculated radius of curvature is larger than observed experimentally. Ascribing different values to interface kinetics coefficient μ_0 could achieve fair agreement between calculation and experimental observation.

The development of the original theory ignores the existence of transformation stress and strain and the anisotropic nature of interphase energy. Consequently, the calculated radius of the advancing tip (Fig. 1.1) is larger than its real size and the precipitate has a different shape. When quantification of precipitate lengthening using the original theory includes optimisation procedures, errors that are caused by using the wrong calculated tip radius would be transferred to other parameters to be optimised. The geometry of the precipitate will be incorrect, as shown in Fig. 2.7. Using the original theory with experimental values of the radius of the advancing tip compensates the neglected effects from transformation stress and strain and interphase energy on precipitate morphology. It makes the original theory physically more accurate. Purely from a theoretical point of view, when

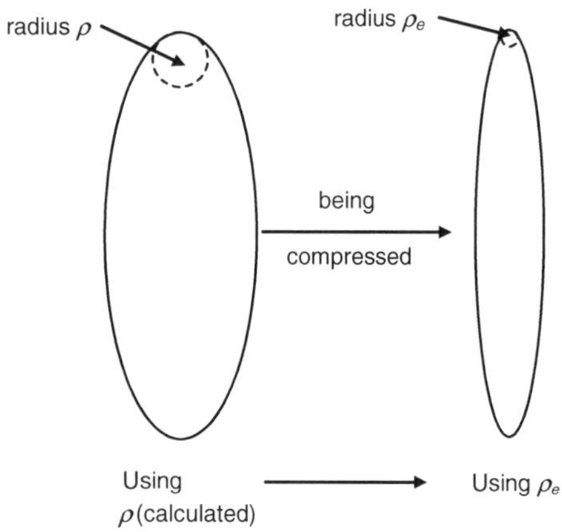

Fig. 2.7 Illustration of the compression effect on precipitate morphology and tip radius with ρ(calculated)/ρ_e. The precipitate geometry from using ρ(calculated) shown here is what one should expect when optimisation procedures were applied to meet the experimental lengthening rate (From Guo and Sha (2004) © Carl Hanser Verlag, Muenchen)

plate growth at small degrees of supersaturation is quantified using the original theory with calculated ρ, the ρ/ρ_c ratio calculated from Eq. 2.7b approaches infinity, when virtually there is no lengthening, Eq. 2.7a. The geometric requirement of the original theory no longer exists. The use of experimental values of the radius of the advancing tip is however of rather phenomenological or empirical nature.

Theoretical determination of the effects from transformation stress and strain and interphase energy remains difficult, which is justified and reflected by the difficulties in theoretical estimation of the aspect ratio of precipitates. Figure 2.5 shows that the value of ρ(calculated)/ρ_e decreases with increasing degree of supersaturation. In fact, the product of ρ(calculated)/$\rho_e \cdot \Omega_0$ is between 1.25 and 1.40 (1.33 $\pm$ 0.07), close to a constant, i.e. ρ(calculated)/ρ_e seems to be inversely proportional to the supersaturation degree. If the tip radius of the precipitate is measured at one temperature, ρ(calculated)/ρ_e and the supersaturation degree at that temperature can be calculated. Given another temperature, ρ(calculated)/ρ_e can be estimated since the supersaturation degree is known. The lengthening rate of precipitates at that temperature can therefore be calculated.

As aforementioned, two factors that influence the aspect ratio are transformation stress and strain and the anisotropy of interphase energy. Usually, one or the other dominates the precipitate growth process. When a precipitate is first formed, it is coherent with the matrix. The contribution from the transformation stress and strain is little due to the small size (volume). Interphase energy is the dominant factor. When the precipitate grows into larger size whereas the interface remains coherent, the interphase energy of the coherent interface is small while compared with the strain energy and therefore it can be neglected. When the precipitate grows to a certain size, it will lose coherency with the matrix through the introduction of dislocations into the interface. The energy of the incoherent interface becomes the dominant factor, whose value is much larger than the value of a coherent interface. At a certain stage, it is possible that the compression effect on the advancing tip radius is mainly due to one factor, either transformation strain/stress or the anisotropy of interphase energy.

2.1.3 Summary

A growth theory of precipitates of needle and plate shapes is used in a revised form. The experimental values of the radius of the advancing tip are used to compensate the influences from transformation stress and strain and anisotropic interphase energy on the precipitate morphology, which are ignored in the initial development of the theory. The theory, with the adaptation, is applied to the growth kinetics of Widmanstätten austenite in ferrite in HSLA Fe–C–Mn–Nb steel. Results agree well with the experimental observations. The ratio between theoretical and experimental values of the radius of the advancing tip for the needlelike precipitate growth is inversely proportional to the degree of supersaturation.

2.2 Determination of Activation Energy of Recrystallisation Using a Kissinger Method

Recrystallisation is commonly studied under continuous heating conditions, using differential thermal analysis or differential scanning calorimetry, to measure the heat evolution during the process, and thermal and mechanical simulation, to measure the dimension change during the process. Experiments can be conducted under different heating rates, with higher heating rates increasing the recrystallisation temperature, as it is a thermally activated process. A modified Kissinger method has been developed to evaluate the activation energy of such processes, based on the following equation:

$$\ln\frac{T_f^2}{Q} = \frac{E}{RT_f} + \ln\frac{E}{RK_0} + \ln\beta_f^* \tag{2.11}$$

where T_f is the characteristic temperature for the given process corresponding to a fixed stage of recrystallisation measured for a number of heating rates, Q is the heating rate, E is the activation energy, R is the gas constant and K_0 and $\beta_f{}^*$ are constants (Sha 2001). The term T_f can be recrystallisation start, 50 %, or finish temperature. The original Kissinger analysis and variants thereof for determination of activation energies have been derived on the basis of homogeneous reaction kinetics.

Recrystallisation in steels is a process extensively studied in the past and present. The processes under continuous heating have been studied by many workers, using mainly thermal analysis. In this section, the activation energy of recrystallisation in low and ultralow carbon steels (Muljono et al. 2001) is shown.

The heating rates used are 50, 200, 500 and 1000 °C/s. The intention of the original work was to understand the effect of high heating rates on recrystallisation behaviour of cold-rolled steel strip. The transverse flux induction heating may offer advantages over the more conventional annealing processes by reducing annealing time, allowing a more compact annealing line and providing greater control of microstructure and properties. This process is capable of continuously annealing steel strip at heating rates up to 1000 °C/s. The curves of $\ln(T_f^2/Q)$ versus $1/RT_f$ are shown in Figs. 2.8 and 2.9, and the activation energy values calculated from the linear regression shown in the figures are given in Table 2.2. The ultralow carbon steel has higher recrystallisation activation energy compared to the two low-carbon steels. The role of carbon on the activation energy is complicated. The lowest carbon steel (0.003C) has a larger initial grain size, resulting in a small number of nucleation sites. The high recrystallisation activation energy also explains why this steel has much higher recrystallisation temperatures compared to the other two steels.

In Fig. 2.9, the activation energies for recrystallisation in the 0.05C steel after different heat treatments are evaluated in comparison. The hot-rolled sample is the same sample used in Fig. 2.8, but the data were taken from a different diagram

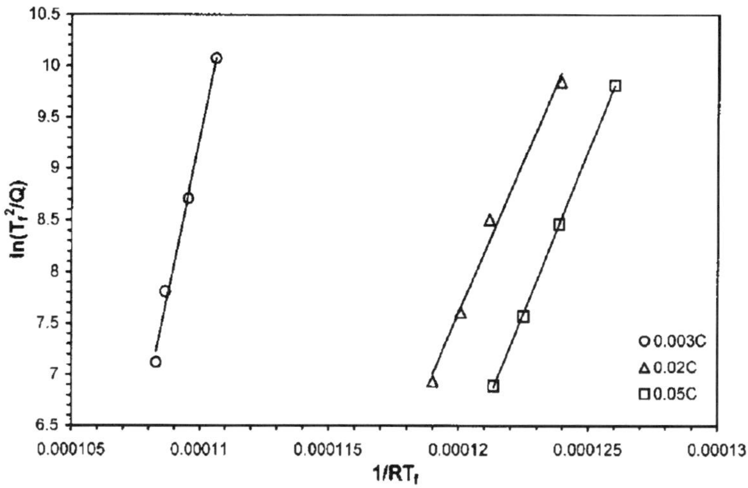

Fig. 2.8 $\ln(T_f^2/Q)$ versus $1/RT_f$ plots for recrystallisation in hot-rolled low and ultralow carbon steels. With kind permission from Springer Science + Business Media: Metallurgical and Materials Transactions A, Determination of activation energy of phase transformation and recrystallization using a modified Kissinger method, 32, 2001, 2903–2904, W. Sha, Fig. 2

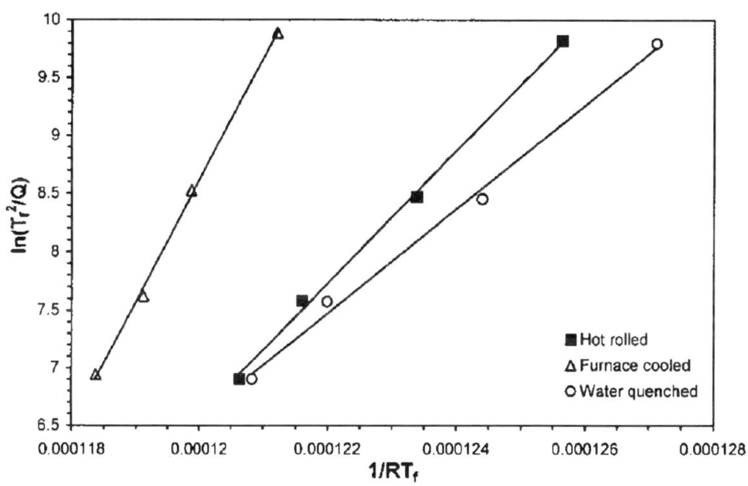

Fig. 2.9 $\ln(T_f^2/Q)$ versus $1/RT_f$ plots for recrystallisation in 0.05C steel after different treatments. With kind permission from Springer Science + Business Media: Metallurgical and Materials Transactions A, Determination of activation energy of phase transformation and recrystallization using a modified Kissinger method, 32, 2001, 2903–2904, W. Sha, Fig. 3

from Muljono et al. (2001), so the difference in the calculated activation energy values gives a measure of the error in the data read from the diagrams in Muljono et al. (2001). This is found to be 59 kJ/mol and is larger than the errors estimated using standard statistical theory for linear regression, for the 0.02 and 0.05C steels under all conditions. Therefore, this is taken as the error of the activation energy

Table 2.2 Recrystallisation activation energy

Steel	Condition	Activation energy (kJ/mol)[a]
0.003C	Hot rolled	1217
0.02C	Hot rolled	591
0.05C	Hot rolled	631
0.05C	Hot rolled	572
	Furnace cooled	1050
	Water quenched	448

[a]The errors are ± 77 kJ/mol for the 0.003C steel, and ± 59 kJ/mol for the other two steels

values obtained. The furnace-cooled sample has a much larger grain size than the water-quenched sample, which results in the higher recrystallisation temperatures as the driving force is reduced. This is also associated with a much higher activation energy for recrystallisation.

2.3 Change of Tensile Behaviour with Tempering Temperature

2.3.1 Microstructure and Tensile Properties

Lower bainite microstructures with pancake-shaped grains are obtained for the steel of a composition as listed in Table 2.3 with the employed thermomechanical control process (TMCP, Yan et al. 2009) (Fig. 2.10). Each pancake grain is around 30 μm thick, 100 μm wide and several hundred microns long. The bainite ferrite laths are characterised by the carbides distributed along the lath boundaries. It is worth noticing that the laths in each grain mainly show only one direction. With increase in tempering temperature, the appearance of ferrite lath boundaries becomes less clear. When the tempering temperature is high up to 700 °C, small recrystallised grains appears, as shown in Fig. 2.10e.

The tensile stress–strain curves (Fig. 2.11) of the steel treated below 400 °C does not show obvious yield phenomenon. However, when the tempering temperature is 500 °C and higher, the upper yield point appears gradually. This phenomenon becomes obvious when the tempering temperature increases from 600 to 650 °C. Another point worth considering is that the tensile curves of the steel tempered at 500–650 °C exhibit plateaus after their yield point. When the tempering temperature is as high as 700 °C, the upper yield point disappears and the tensile curve is characterised by a round roof shape, indicating good formability.

Table 2.3 Chemical composition of the steel (wt%)

C	Mn	Nb	V	Ti	Mo	Cr	Cu	Ni	Si	P	Al	N
0.046	1.79	0.049	0.03	0.023	0.31	0.31	0.2	0.77	0.1	0.0061	0.1	0.003

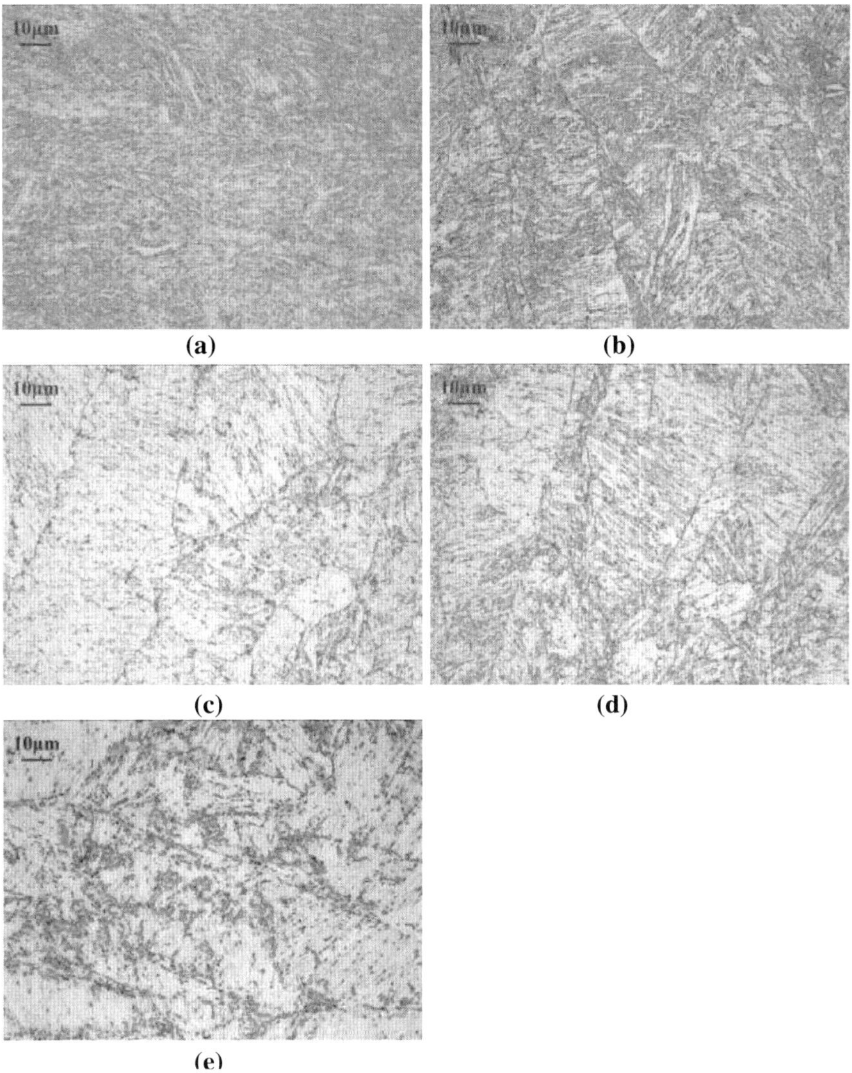

Fig. 2.10 Microstructures of the as-processed steel in the as-rolled and tempered state. **a** As-rolled; **b** tempered at 200 °C; **c** tempered at 400 °C; **d** tempered at 600 °C; **e** tempered at 700 °C. Reprinted from Materials Science and Engineering: A, Vol 517, Wei Yan, Lin Zhu, Wei Sha, Yi-yin Shan, Ke Yang, Change of tensile behaviour of a high-strength low-alloy steel with tempering temperature, Pages 369–374, 2009, with permission from Elsevier

In order to present the change in strength, both the yield strengths (YS) and ultimate tensile strengths (UTS) of the steel tempered at different temperatures are shown in Fig. 2.12. This reveals that the yield strength does not decrease with increasing tempering temperatures below 650 °C, even possessing

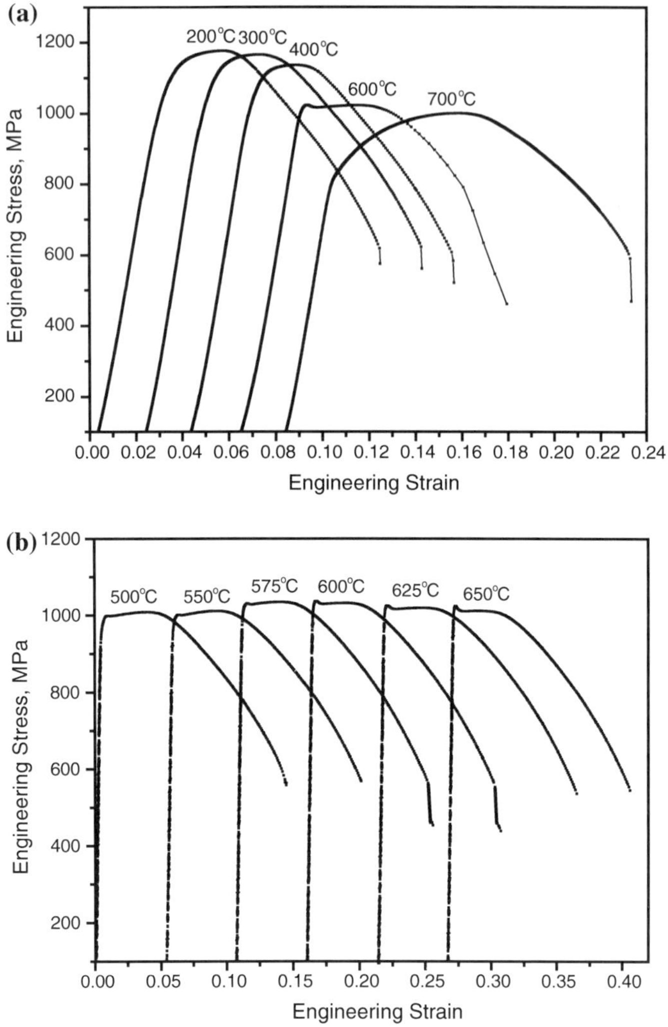

Fig. 2.11 Tensile curves of the steel tempered at different temperature. Reprinted from Materials Science and Engineering: A, Vol 517, Wei Yan, Lin Zhu, Wei Sha, Yi-yin Shan, Ke Yang, Change of tensile behaviour of a high-strength low-alloy steel with tempering temperature, Pages 369–374, 2009, with permission from Elsevier

a small peak at 600 °C. Nevertheless, the tensile strength remarkably decreases to about 1000 MPa when tempered at 500 °C and is close to the yield strength when the tempering temperature is up to 650 °C. At the highest tempering temperature of 700 °C, the yield strength decreases more steeply than the tensile strength. Therefore, with increasing temperature the difference between the tensile strength and the yield strength becomes smaller and smaller, and it almost

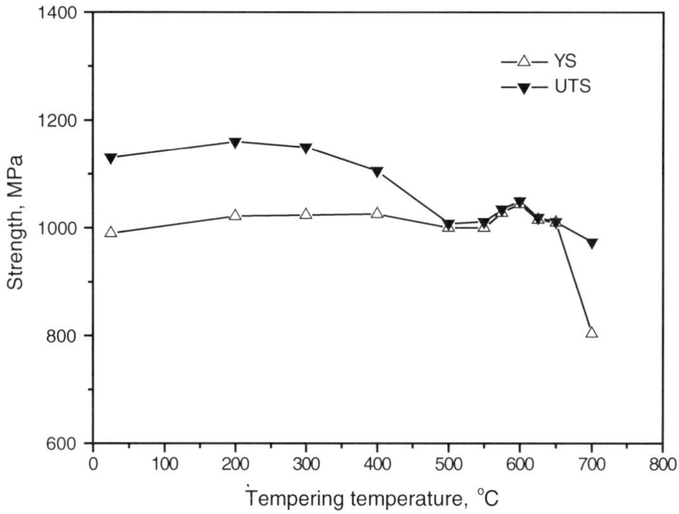

Fig. 2.12 Ultimate tensile strength (UTS) and yield strength (YS) of the HSLA steel as functions of the tempering temperature. Reprinted from Materials Science and Engineering: A, Vol 517, Wei Yan, Lin Zhu, Wei Sha, Yi-yin Shan, Ke Yang, Change of tensile behaviour of a high-strength low-alloy steel with tempering temperature, Pages 369–374, 2009, with permission from Elsevier

Table 2.4 Strain hardening exponent *n* of the steel tempered at 500–650 °C

Tempering temperature (°C)	strain hardening exponent *n*
500	0.036
550	0.036
575	0.029
600	0.076
625	0.029
650	0.021

vanishes in the temperature range between 500 and 650 °C. When the tempering temperature reaches 700 °C, the difference in the stress values increases.

The strain hardening exponents for differently heat-treated steel according to Eqs. 1.3–1.5 are given in Table 2.4. The strain hardening exponent decreases slightly when the tempering temperature is below 400 °C, but is subjected to an abrupt decrease to as low as 0.021, when the tempering temperature is 650 °C. At the tempering temperature of 700 °C, the *n* value increases to 0.258. The strain hardening exponent also displays a peak at the tempering temperature of 600 °C.

2.3.2 The Upper Yield Point

For tempering at 500 °C, dislocations in the steel should have enough thermal energy to move and interact with each other. Thus, many dislocations with opposite Burgers vectors, i.e. the positive and negative dislocations, would interact and will annihilate. Consequently, the dislocations density will considerably decrease. The ferrite lath boundaries would also begin to disappear due to the movement of dislocations, as shown in Fig. 2.10c. Simultaneously, iron carbide precipitates should form during treatment at this temperature, since the dissolved carbon atoms have been activated, resulting in quite fast diffusion. Then, the precipitates act as strong pinning obstacles for dislocations glide. Therefore, the mobile dislocations density will be strongly reduced. On the other hand, the reduced dislocations density and disappearing lath boundaries in turn afford much space for moving dislocations. A higher stress will be needed in order to drive the dislocations off the pinning to move. Thus, the upper yield strength occurs. When the dislocations break away from the pinning precipitates, the mobile dislocations density will increase again. Hence, the stress needed for gliding dislocations decreases. Consequently, it is reasonable that the upper yield point appears on the stress–strain curves of the steel when the tempering temperature is above 500 °C, according to the above-described mechanisms. With increasing tempering temperature, more ferrite laths disappear and the dislocations density will be reduced, while the amount of precipitation increases. From this, it is deduced that the upper yield point will become more pronounced with increasing tempering temperature. The nucleation temperature for large amount of precipitations in HSLA steels is about 600 °C (Chap. 3), which is related to the upper yield point and the yield strength peak. However, it is noteworthy that there is only one peak in the tensile curves, not like the typical oscillating yield points (Portevin-LeChatelier effect) on stress–strain curves of mild steels, because the mobile dislocations density cannot be reduced effectively in this steel.

However, tempering at 200 °C will lead to the moving dislocations in the steel interacting with point defects such as vacancies and interstitial atoms. The mobile dislocations density may be much lower and will hardly increase. If the tempering temperature increases up to 300 and 400 °C, the dislocations will be able to move and interact with each other, but the density might not be strongly reduced, which indicates the small change in the strength. Additionally, the ferrite lath boundaries are still stable, so the mean free path of gliding dislocations is restricted. Even if they will move, other immobile dislocations and the ferrite lath boundaries would immediately block them. Therefore, the amount of mobile dislocations will hardly increase. Hence, the upper yield points do not appear in the steel tempered at 300 and 400 °C.

When the tempering temperature increases up to 700 °C, recrystallisation occurs. Coarsening of precipitates in the steel takes place and the effective pinning of dislocations diminishes. Therefore, the yield strength of the steel shows a strong decrease. The mobile dislocations density may be quite low and dislocation glide is inhibited by a large number of newly formed grain boundaries in the recrystallised grains. Thus, the density of mobile dislocations cannot significantly increase. Therefore, the upper yield points do not appear on tensile curves.

2.3.3 Strain Hardening Exponent

The strain hardening exponents are calculated for the uniform plastic deformation range. A lower strain hardening exponent indicates that the material possesses a lower strain hardening ability. Generally, the as-recrystallised metals and alloys and the severely strain hardened material show this feature of a low *n* value. The primary reason in the case of the as-recrystallised metals and alloys is that the matrix has not enough barriers for the moving dislocations, and the reason in the case of the severely strain hardened materials is that the dislocations cannot move any more in the matrix. In either case, the yield strength is nearly equal to the tensile strength, for the described tempered steel at 500–650 °C, as shown in Fig. 2.12.

The phenomenon of low strain hardening exponents of HSLA steels is unexpected. The steel tempered at 500–650 °C is obviously not so strongly strain hardened, as described as the second reason above. No recrystallisation occurs at these temperatures; see the primary reason as aforementioned. Therefore, there must be a third reason. Lowering of *n* value for continuously annealed cold-rolled steel has been observed because of boron additions (Funakawa et al. 2001). The *n* value decreases significantly when the carbon content in steel is lower than 0.002 %. The reason for this is explained in terms of morphological changes of the carbide precipitates in matrix and at grain boundaries. Antoine et al. (2005) analysed the relationship between *n* and the yield strength of titanium containing IF steel. The *n* value is controlled by dislocation-precipitate and dislocation-grain boundary interactions. Smaller amount of dissolved carbon and nitrogen atoms reduces the efficiency of grain boundaries to block the dislocations movement and thus a decrease in the *n* value. Therefore, tempering at 500–650 °C can result in a depletion of the dissolved carbon and nitrogen atoms due to the precipitation of ε-carbides (FeC_2) or carbonitrides, which might be responsible for the low strain hardening exponent. In addition, the decrease in the dislocation density as well as the disappearing of the ferrite lath during tempering would contribute to the low *n* value.

The precipitates could also pin gliding dislocations, and hence increase the strain hardening rate. From the above-described results, it is concluded that the dissolved carbon and nitrogen atoms and their interactions with dislocations have a much stronger effect on the strain hardening exponent than the precipitates. The *n* value will increase gradually with an increase in the amount of precipitates, but *n* decreases steeply with the decrease in the content of dissolved carbon and nitrogen atoms. The formation of the precipitates would consume the dissolved carbon and nitrogen atoms, which can be demonstrated from the relation between the strain hardening exponent and the content of the dissolved carbon and nitrogen atoms as shown in Fig. 2.13. The increasing amount of precipitates is consistent with the decreasing content of dissolved carbon and nitrogen atoms. Therefore, the *n* value shows a considerable change as illustrated by the bold line in Fig. 2.13, i.e. the *n* value reaches a minimum at 575 °C. When the tempering temperature is above 600 °C and increases up to 650 °C, the precipitation kinetics is saturated

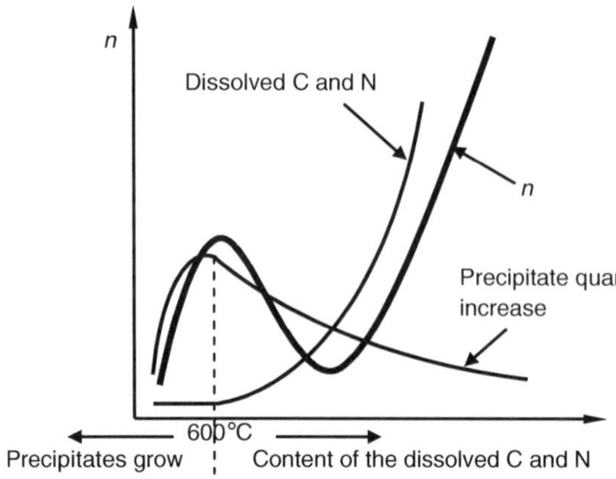

Fig. 2.13 Schematic representation of the strain hardening exponent variation as a function of the dissolved carbon and nitrogen atoms contents and the quantity and size of the precipitates. Reprinted from Materials Science and Engineering: A, Vol 517, Wei Yan, Lin Zhu, Wei Sha, Yi-yin Shan, Ke Yang, Change of tensile behaviour of a high-strength low-alloy steel with tempering temperature, Pages 369–374, 2009, with permission from Elsevier

and the growth rate of the precipitates decreases. Hence, the n value decreases again. Therefore, the n value shows a small peak at 600 °C since the 600 °C tempering can produce the largest number of fine precipitates. This explanation agrees well with the results shown in Table 2.4 and Fig. 2.12.

When tempering at 700 °C, the newly formed, small, recrystallised grains not only increase the amount of grain boundaries, but also behave as the second phase, resulting in high strain hardening ability of this steel.

2.3.4 Summary

Lower bainite microstructure causing high strength of the HSLA steel is achieved by proper thermomechanical control process. The yield strength of the steel shows a slight increase up to the tempering temperature of 550 °C, then passing a maximum at 600 °C. Beyond 650 °C, the yield strength decreases steeply. The tensile strength decreases beyond 200 °C and becomes close to the yield strength when the tempering temperature ranges between 500 and 650 °C.

There is no pronounced yielding point appearing on the tensile curves of the steel tempered below 400 °C. However, the steel gradually develops upper yield points with relatively low strain hardening exponents when the tempering temperature is increased from 500 to 650 °C. The low strain hardening exponent reaches a peak at 600 °C. This can be explained in view of the interactions of the mobile dislocations with the dissolved carbon and nitrogen atoms and their effect on the strain hardening exponent.

The steel tempered at 700 °C shows 'round roof' shaped tensile curves possessing high strain hardening exponents due to the fine-grained microstructure in the as-recrystallised state. The governing mechanism is the strong interactions of dislocations and interstitials with grain boundaries and the fine-dispersed precipitates of carbides and carbonitrides.

2.4 Delamination Fracture Related to Tempering

2.4.1 Microstructures

The as-rolled steel (composition in Table 2.3) has a lower bainite microstructure. The microstructure of a transverse section is shown in Sect. 2.3. The microstructures of the tempered steel in the longitudinal or rolling direction are shown in Fig. 2.14. The length of the pancake grains along the rolling direction is much larger than the width, which is an obvious characteristic of the anisotropic microstructure. Before rolling, the grains are equiaxed and the size is 50–100 μm when soaked at 1200 °C. After rolling at austenite recrystallisation zone, the grain size could be refined to 20 μm. With increase of the tempering temperature, the ferrite laths in the grains disappear and the prior austenite grain boundaries become more evident, which should indicate a change in the matrix precipitates. There is little change after tempering at 200–600 °C. During tempering at 700 °C, reaustenitisation begins, so many small ferrite grains are observed in the microstructure.

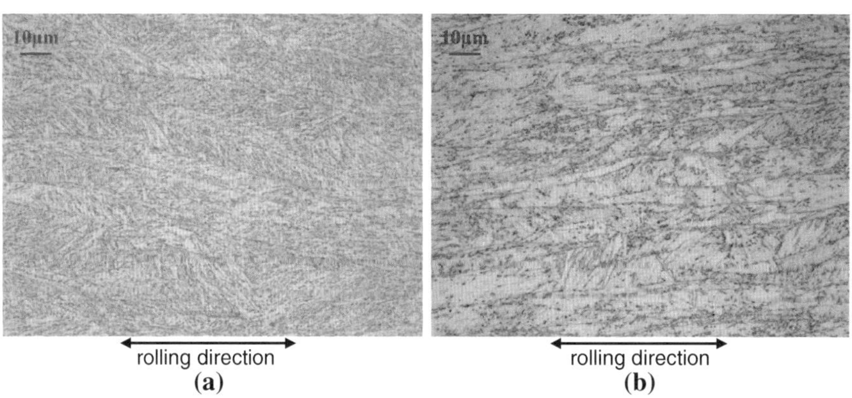

Fig. 2.14 Optical micrographs showing the microstructures of the as-rolled and of the tempered steel along the rolling direction (longitudinal sections). **a** As-rolled; **b** tempered at 700 °C. The etchant was 4 % nital. With kind permission from Springer Science + Business Media: Metallurgical and Materials Transactions A, Delamination fracture related to tempering in a high-strength low-alloy steel, 41, 2010, 159–171, Wei Yan, Wei Sha, Lin Zhu, Wei Wang, Yi-Yin Shan, Ke Yang, Fig. 3

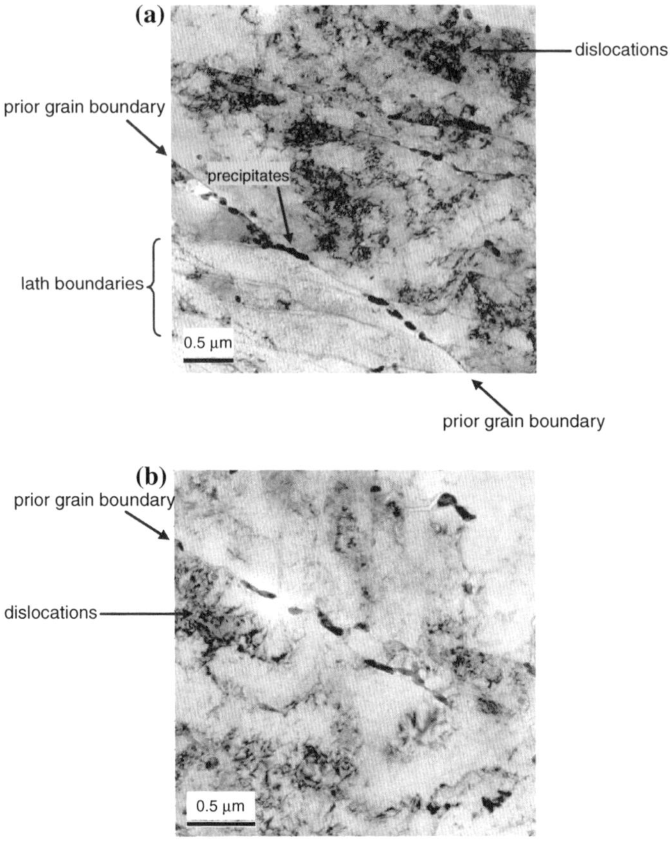

Fig. 2.15 TEM micrographs showing precipitates, probably alloy carbonitrides, in the boundaries of the elongated grains in the steel tempered at 600 °C. With kind permission from Springer Science + Business Media: Metallurgical and Materials Transactions A, Delamination fracture related to tempering in a high-strength low-alloy steel, 41, 2010, 159–171, Wei Yan, Wei Sha, Lin Zhu, Wei Wang, Yi-Yin Shan, Ke Yang, Fig. 4

Figure 2.15 shows two TEM images of material tempered at 600 °C. Precipitates, most likely carbides and nitrides, are densely distributed along the boundaries of the elongated grains in this steel, proving that tempering at 600 °C leads to grain boundary precipitation. The dense dislocations are a result of the heavy deformation during rolling. The mottled structure inside the grains in the TEM micrographs may be due to surface oxidation of the TEM thin foil between the times when the specimen was made and the microscopy imaging. Such contrast, however, does not interfere with the microstructural features being discussed here, viz. precipitates and grain boundaries. Mottled contrast also arises in some optical and scanning electron microscopy images, for similar reasons and possible surface contamination and inhomogeneous etching.

The precipitates seen in optical micrographs are not the same precipitates observed in TEM, as the magnifications are at very different levels. The precipitates

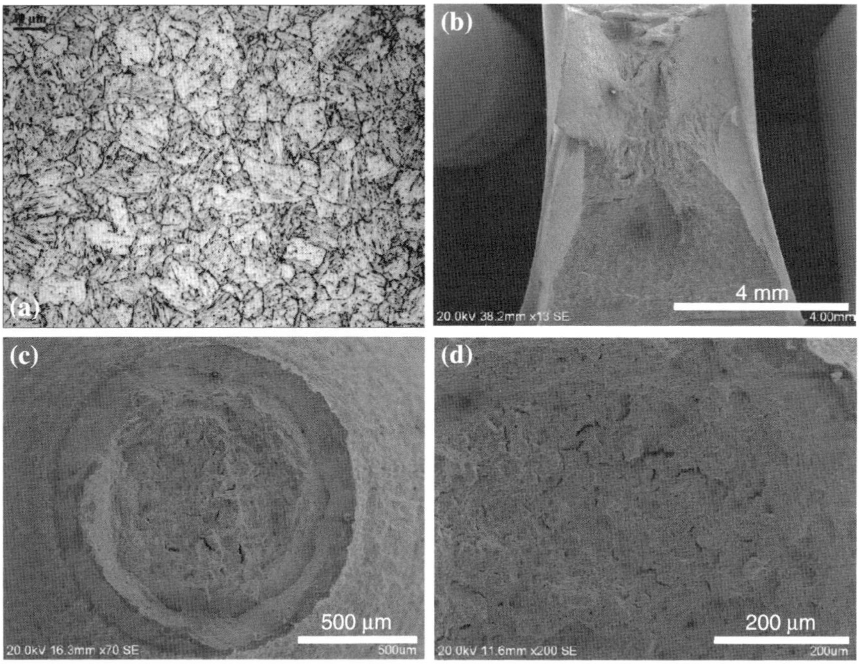

Fig. 2.16 Optical photomicrograph and SEM fractographs of the triple quenched and tempered, equiaxed grain steel. **a** Optical image showing equiaxed grains; **b** fractograph of a sample after impact at –30 °C; **c** fractograph of the room temperature tensile sample; **d** magnified from (**c**) showing small cracks. With kind permission from Springer Science + Business Media: Metallurgical and Materials Transactions A, Delamination fracture related to tempering in a high-strength low-alloy steel, 41, 2010, 159–171, Wei Yan, Wei Sha, Lin Zhu, Wei Wang, Yi-Yin Shan, Ke Yang, Fig. 5

apparent in optical micrographs such as those in Figs. 2.14 and 2.16a are much larger precipitates, which may be inclusions, made to appear bigger due to the etching effect. Their number density is actually very low, orders of magnitude lower than the type of precipitates in the TEM scale. The precipitates in the optical scale, though apparently evenly distribution throughout the entire microstructure, are not relevant to the discussion in this section, because of their very small numbers (though appearing numerous on the optical micrographs due to their low magnification).

After triple quenching and tempering heat treatment, small equiaxed grains of around 25 μm form in the steel (Fig. 2.16a). There are precipitates distributed along the grain boundaries that accentuated the outline of the grains.

2.4.2 Delamination on Tensile and Impact Fracture Surfaces

The steel tempered at 200, 400 and 700 °C shows no delamination on tensile fracture surfaces, while remarkable delamination occurs on the fracture surfaces of the tensile specimens tempered in the range of 500–650 °C (Yan et al. 2010).

The tensile fracture surfaces with delamination no longer display typical cup-and-cone fracture. This may indicate poor ductility. The fracture surfaces are separated into two halves approximately in the middle of the fracture, which would correspond to the mid-thickness of the rolled plate. Many local undeveloped splits lay on the surface, parallel to the main one. Although there is no delamination in the fracture surfaces of the steel tempered at 200, 400 and 700 °C, small split-like cracks are distributed along the same direction. This type of small cracks is also in the fracture surface of the steel tempered at 500 °C, parallel to the main split.

The details of delamination on each impact fracture surfaces at −30 °C are summarised in Table 2.5. The impact fracture surfaces at room temperature have no delamination, with impact energy 50 J for as-rolled steel, and 48 J for the steel tempered at 500, 600, and 650 °C. The −30 °C impact fracture surfaces of the steel tempered at different temperatures are shown in Fig. 2.17. With increase of the tempering temperature, delamination in these fracture surfaces first increases in both number and length, and then decreases in both. A quantitative measure of degree of splitting could be made by measuring the total length of split per unit area of fracture surface. Multiple continuous splitting occurs in the fracture surface of the steel tempered in the range of 500–650 °C, while single local splitting is in the top (near the notch) area of the fracture surfaces of the steel tempered at 400 and 700 °C and no splitting is on the fracture surface of the steel tempered at 200 °C. Continuous splitting does not extend all the way to the back of the fracture surface when the tempering temperature is 650 °C, and is limited to the area near the notch when the tempering temperature reaches 700 °C.

The continuous, deep splitting segments the delaminated fracture surface, with what appears to be 45° shear fracture between the splits. Small local splits are in the fracture surfaces, as well as large delamination. The details of the surfaces are displayed in Fig. 2.18. Smooth fracture surfaces mixed with ductile fracture areas are illustrated. The smooth fracture surface likely indicates grain boundary decohesion, as can be clearly seen in Fig. 2.18c marked with a black arrow. Song et al. (2006) explain this phenomenon in more detail.

Table 2.5 Impact toughness and fracture splitting of the tempered steel at −30 °C on half-size specimens (5 mm thick)

Tempering temperature (°C)	Impact energy (J)	Splitting
As-rolled	56	No
200	60	No
300	50	No
400	51	Local
500	47	Throughout
550	45	Throughout
575	49	Throughout
600	50	Throughout
625	53	Throughout
650	49	Throughout
700	55	Local

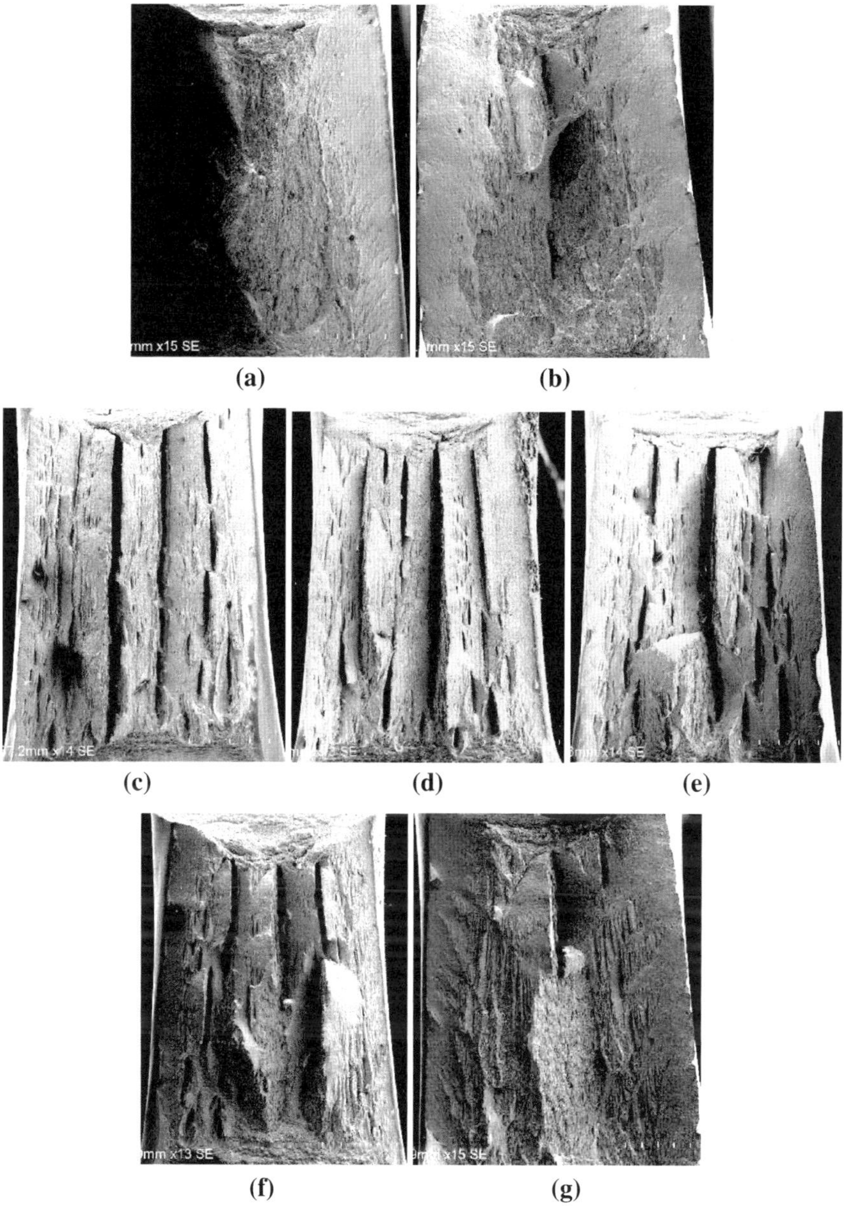

Fig. 2.17 Macrofractographs of the impact fracture surfaces at −30 °C, showing the development of delamination with tempering temperature from none to severe to mild splitting. Tempered at **a** 200 °C; **b** 400 °C; **c** 500 °C; **d** 550 °C; **e** 600 °C; **f** 650 °C; **g** 700 °C. The entire width of the impact specimens, 5 mm, is shown in each image. With kind permission from Springer Science + Business Media: Metallurgical and Materials Transactions A, Delamination fracture related to tempering in a high-strength low-alloy steel, 41, 2010, 159–171, Wei Yan, Wei Sha, Lin Zhu, Wei Wang, Yi-Yin Shan, Ke Yang, Fig. 11

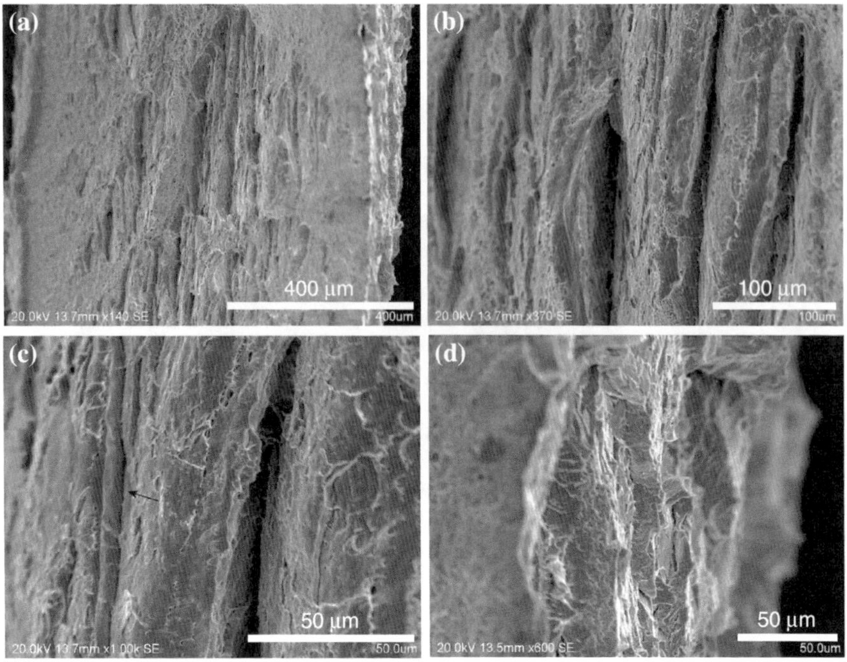

Fig. 2.18 SEM fractographs showing the details of impact fracture surfaces of the steel tempered at 600 °C. **a** Delaminated fracture surface; **b** magnified; **c** magnified from (**b**) showing decohesion; **d** edge of the piece. Since the steps are not in the same plane, the area near the left and right edges of the images is faint. With kind permission from Springer Science + Business Media: Metallurgical and Materials Transactions A, Delamination fracture related to tempering in a high-strength low-alloy steel, 41, 2010, 159–171, Wei Yan, Wei Sha, Lin Zhu, Wei Wang, Yi-Yin Shan, Ke Yang, Fig. 12

2.4.3 Fractography of the Equiaxed Grain Sample and Toughness

When the triple quench heat treatment changes the elongated grains into the equiaxed morphology, delamination disappears from the fracture surfaces after the −30 °C impact and the room temperature tension even after tempering at 600 °C (Fig. 2.16b, c). However, instead of the general delamination seen previously, many small secondary cracks are seen on the tensile fracture surface (Fig. 2.16d).

The Charpy V-notch impact energy is generally employed to characterise the toughness of steel. The toughness values, at −30 °C, of this HSLA steel, with elongated grains, tempered at different temperatures are given in Table 2.5. The steel has a similar level of toughness for all the tempering temperatures. The steel as-rolled and tempered at 200 °C has higher toughness values of 56 and 60 J, respectively, and the other tempering temperatures all lead to toughness around 50 J. When the elongated grains are changed into equiaxed grains, the steel has a little higher toughness of 65 J at −30 °C due to the small grain size.

2.4.4 *Splitting Tip Metallography and XRD of the Tempered Steel*

The tip of the split in the impact fracture surface from the steel tempered at 500 °C is shown in Fig. 2.19. The crack path is clearly revealed by both optical microscope and SEM to propagate along the prior austenite grain boundaries. The split can make minor adjustments in direction when crossing a grain boundary, as is clearly illustrated by the step-like morphology at the edge of the delaminated pieces shown in Fig. 2.18d. Since the steps are not in the same plane, the ambient area of the image is faint.

X-ray diffraction can qualitatively evaluate the presence of crystallographic texture in steel by the relative heights of the main diffraction peaks. Those planes with peaks showing much stronger intensity than from an entirely random sample would be qualified to be the orientation of texture. The main peaks of the body centred cubic structure show a somewhat higher than expected intensity for the (110) peak relative to the (200) and (211) peaks (Table 2.6), which indicates that some, limited texture may have existed in the tempered samples. The diffraction was from planes parallel to the rolling plane of the steel plate (Yan et al. 2010).

In the above sections, we have examined the phenomenon of delamination or splitting of tensile and impact fracture surfaces of a control-rolled and quenched, HSLA steel when loaded after tempering in the temperature range of 200–700 °C. In Sect. 2.3, the tensile and strain hardening behaviour is discussed. A summary of some of the important observations of Sect. 2.3 is appropriate:

(1) The microstructure in the as-rolled condition consists of lower bainite (ferrite laths containing elongated cementite precipitates and a high dislocation density) in highly elongated, flat prior austenite grains.
(2) After tempering at 200–400 °C, the steel exhibits normal stress-strain behaviour, strain hardening exponents of 0.11–0.13, and yield-tensile ratios of approximately 0.85.
(3) Upon tempering in the range of 500–650 °C, a yield point behaviour is observed, with strain hardening exponents of only 0.02–0.08 (i.e. essentially no strain hardening) and yield-tensile ratios of essentially unity (i.e. the maximum stress occurs at or near yield).
(4) There is a modest increase in the yield and tensile strengths across this tempering temperature range, with a maximum strength at 600 °C, similar to secondary hardening in the tempering of martensite.

Table 2.6 Relative intensities of the XRD peaks of the steel tempered at different temperatures

Peak index	Standard (no texture)	200 °C	400 °C	600 °C	700 °C
110	100	348	284	262	335
200	20	14	18	13	12
211	30	30	30	30	30

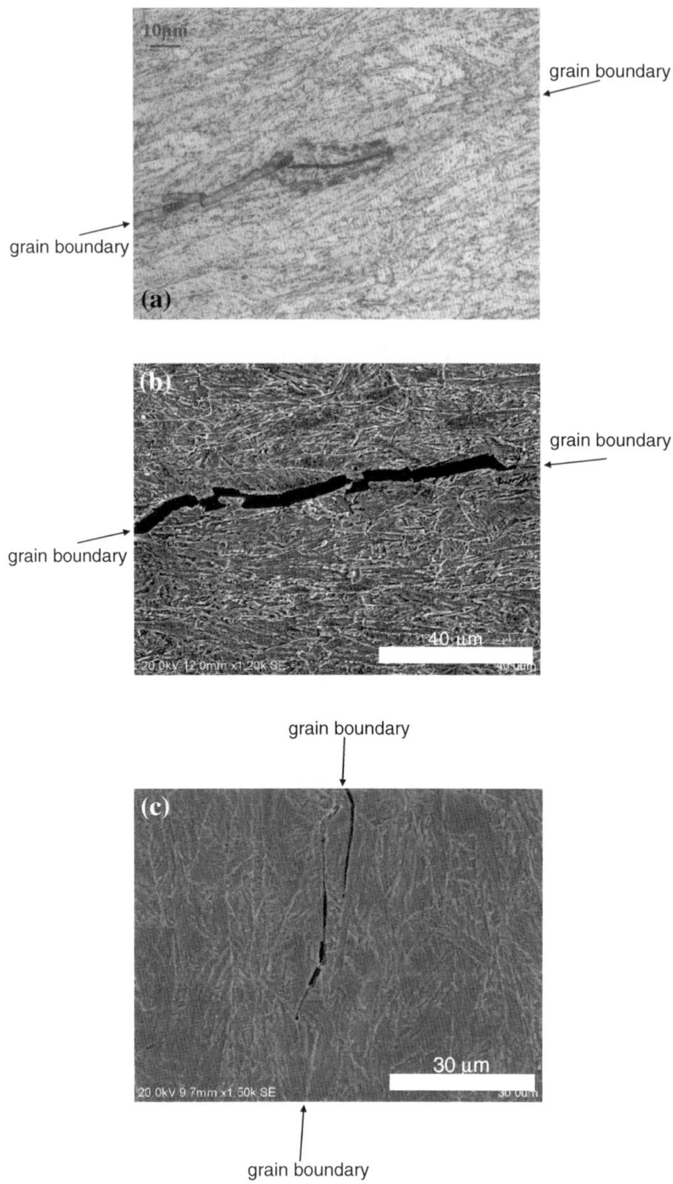

Fig. 2.19 Metallographic section through the tip of a split showing that crack propagation is along the grain boundaries. **a** Optical image showing the splitting tip. The etchant was 4 % nital. **b** SEM image showing the crack propagation path. **c** SEM image showing the splitting tip. With kind permission from Springer Science + Business Media: Metallurgical and Materials Transactions A, Delamination fracture related to tempering in a high-strength low-alloy steel, 41, 2010, 159–171, Wei Yan, Wei Sha, Lin Zhu, Wei Wang, Yi-Yin Shan, Ke Yang, Fig. 13

(5) Tempering at 700 °C produces a return to normal behaviour, with a strain hardening exponent of 0.25 and a yield-tensile ratio of about 0.8.

The important aspects of the present section (Sect. 2.4) so far that require discussion include:

(1) The microstructural changes occurring during tempering at temperatures in the range of 500–650 °C cause the steel to be susceptible to delamination during subsequent tensile and impact loading.
(2) The stress conditions lead to splitting of the room temperature tensile specimens, no splitting of room temperature impact specimens and severe splitting of the −30 °C impact specimens.
(3) The effect of delamination on the toughness values is obtained.

2.4.5 Delamination and the Anisotropic Microstructure

As mentioned above, delamination behaviour in hot-rolled steels is associated with many different phenomena, all related to anisotropic microstructure. Of those previously identified, the possible anisotropic microstructural features causing delamination could be texture, planar inclusion arrays, aligned precipitates and the elongated grain structure. In the present case, texture can be eliminated for the following reasons. First of all, the steel is not rolled within the two phase alpha-plus-gamma region that has been shown to favour the formation of strong textures (Yang et al. 2008). The strong (100) cleavage plane texture increases with increased annealing temperatures and the textures are retained during annealing up until full recrystallisation (Tsuji et al. 2004). Therefore, if texture played an important role in the observed delamination, the most serious cases should have occurred in the as-rolled or the low temperature tempered specimens, instead of those tempered at 500–650 °C. Furthermore, the XRD shows that there is no significant difference in texture among the tempered steel, confirming the above inference. The (110) peak relative to the (200) and (211) peaks is somewhat high, but (110) is not a normal cleavage plane and there is no trend with tempering temperature.

Precipitation is believed to become predominant when the steel is tempered in the temperature range of 500–650 °C. The precipitates have good opportunity to be one of the reasons for the delamination. With the purpose of distinguishing the roles of elongated grains and precipitates, it is necessary to evaluate the contribution of the elongated grain structure on the delamination. Hence, the as-rolled steel is subjected to the triple oil quench heat treatment that changes the elongated grains into an equiaxed grain structure. The steel, once having equiaxed grains, displays no delamination in the impact fracture surface at −30 °C, even after tempering at 600 °C. This finding indicates that elongated grains are a necessary condition for delamination. The steel even with the highly elongated grain structure shows no extensive delamination as long as the tempering temperature is not in the range of 500–650 °C. In this tempering temperature range, by analogy to martensite tempering, the cementite precipitates in the bainite most likely go into solution as carbide particles of the strong carbide forming elements in the steel (titanium, niobium, vanadium, molybdenum and chromium) precipitate within the grains and on the grain boundaries.

These two aspects suggest that the elongated grains resulting from hot rolling below the austenite recrystallisation temperature, and the densely aligned precipitates formed during tempering are likely responsible together for the severe delamination. The sole presence of either one would not lead to delamination. This observation also leads to the hypothesis that the precipitates embrittle the grain boundaries and make them the most probable propagation path of delamination. This speculation is actually demonstrated by the investigation of the splitting tip. In Fig. 2.19, the splitting crack propagation path was along the prior austenite grain boundaries. The smooth fracture surfaces on the pieces separated by splitting, in Figs. 2.18a–c, are another proof of the splitting propagation along grain boundary. Thus, it is clear that the continuous splits on the tensile and impact fracture surfaces are in the same direction and parallel to the plane of the flat, elongated grains, as depicted in Fig. 2.20.

The small cracks on the tensile fracture surface of the equiaxed grain steel, as shown in Fig. 2.16d, indicate that the precipitates formed during 600 °C tempering also make the equiaxed grain boundary brittle and easily separated, which further proves the hypothesis. However, due to the grain refinement, the grain boundary area per unit volume is enormously increased, and the precipitates per unit grain-boundary area would have been greatly reduced. Therefore, the embrittlement effect is diminished. Furthermore, the equiaxed grain structure would inhibit crack propagation, while the elongated grains can scarcely change the propagation direction due to their flat, elongated shape. Thus, delamination does not occur due to the absence of its favourable microstructural feature of the elongated grain structure.

Increasing the annealing temperature to beyond the reaustenitisation range, i.e. normalising, can eliminate the splitting phenomenon. Reaustenitisation replaces the anisotropic microstructure with an isotropic one. Therefore, delamination would be difficult and unlikely without such anisotropy. When the steel is tempered at 700 °C, reaustenitisation is not complete. Additionally, reaustenitisation starts primarily along the grain boundaries, which would have depleted the precipitates and erased the embrittlement effect. Hence, although significantly decreased, splitting is not eliminated.

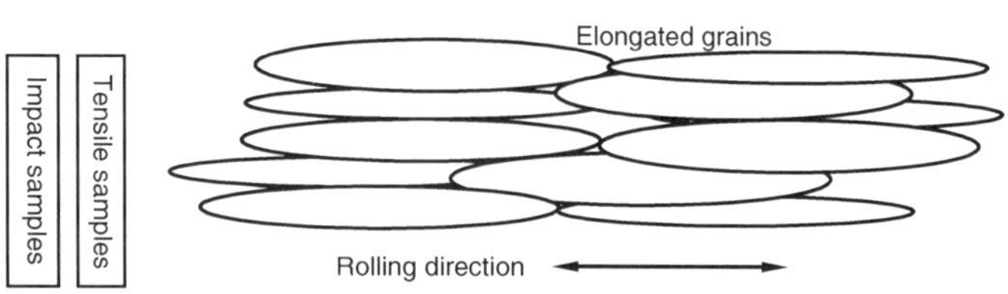

Fig. 2.20 Schematic plan view diagram of the rolled plate showing elongated grain structure and orientation of the tensile and impact test specimens. With kind permission from Springer Science + Business Media: Metallurgical and Materials Transactions A, Delamination fracture related to tempering in a high-strength low-alloy steel, 41, 2010, 159–171, Wei Yan, Wei Sha, Lin Zhu, Wei Wang, Yi-Yin Shan, Ke Yang, Fig. 1

2.4.6 Delamination and the Stress Conditions, and Effect of Delamination on Toughness

An anisotropic microstructure alone cannot lead to delamination; the stress conditions are a second important and, in fact, necessary factor. These stress conditions are actually dependent on the mechanical properties, especially the ductility, of the materials and the geometric dimension, especially the thickness, of the specimens. In both a Charpy impact test and a tensile test, a state of plane strain (i.e., a triaxial tensile stress) emerges. The deformation under plane strain conditions is so highly constrained that the stresses in this rigid zone can reach two to three times the yield stress (Yan et al. 2007).

This rigid plane strain zone is located in the centre of the neck in a cylindrical tensile specimen. The stress state in the neck of a tensile specimen consists of a uniform axial stress plus a hydrostatic tensile component. The hydrostatic component varies from zero at the neck surface to a maximum at the centre line, with the magnitude dependent on the degree of necking and the neck profile. Therefore, delamination in a tensile specimen can be fully developed in the middle, separating the fracture surface into half, provided there exists a plane of weakness. Necking generally begins and uniform strain ends at the maximum load or engineering stress. The split-like cracks in Yan et al. (2010) are undeveloped because of lack of the sufficiently strong stress condition due to the considerable uniform ductility of the steel. It is worth noting that the room temperature tensile specimens tempered at 500–650 °C all display the delamination phenomenon. The specimens should have experienced a certain amount of uniform plastic deformation before necking since the loading speed is low in tensile test. However, in the tensile tests, following tempering in the critical range, the maximum engineering stress occurs at or very near yield, so necking is premature and intense. Therefore, the radial and tangential stresses acting on the planes of weakness would be high leading to delamination.

Delamination is more extensive in the static tensile loading than in impact loading at room temperature. Generally, the stress condition in impact loading should be more severe than that in tensile loading, because of the notch and the high load ing speed. The possible explanation may be found in the conditions that exist during the plastic deformation before necking in tensile loading.

The stresses and the rigid zone location in a CVN specimen, however, are much more complex than in a tensile test specimen. As the specimen bends and a plastic zone is generated at the notch, contraction in the thickness direction is constrained, setting up a triaxial stress state. As the crack grows, the location of this zone will move from the area near the notch to spread across most of the fracture surface, which is the reason why delamination evolves as shown in Fig. 2.17 with increase of the tempering temperature. The steel tempered in the range of 500–650 °C has very low strain hardening ability, which leads to low tensile ductility (Song et al. 2006) (Sect. 2.3). Therefore, the stress conditions can easily cause delamination and the anisotropic microstructure (i.e., elongated grains in the triaxial tensile stress zone) would no doubt separate along the weak paths that are in the plane of hot rolling of the steel, leading to serious delamination. The steel as-rolled, and tempered at 200, 300, 400 and 700 °C, has better elongation-to-failure and the necking ability, resulting in no delamination or only minor local delamination.

The impact specimens tempered at 500–650 °C show no splitting at room temperature, but serious delamination at −30 °C (Table 2.5). In general, the number of delamination increases with a decrease of temperature (Song et al. 2005, 2006; Tsuji et al. 2004). It is well accepted that the yield strength will increase and the elongation-to-failure and the necking ability of steels will be reduced with decreasing temperature. This increase in yield strength can be as much as 10–15 % between room temperature and −30 °C. The stress in the thickness direction needs to be high enough to tear apart the anisotropic microstructure along its weak paths. This is believed to be the reason that delamination is more likely to happen at relatively low temperatures.

Prior plastic strain might induce some microcracks along planes of weakness, which would greatly reduce the required stress for cracks propagation and facilitate the delamination. However, in CVN test the steel has better elongation-to-failure and the necking ability at room temperature and the stress concentration cannot reach that high level.

According to the above analysis, it can also be concluded that the delamination can happen when plane strain condition is satisfied. Correspondingly, the delamination occurs after necking in tensile condition and before the main fracture in impact condition.

It is obvious that the splitting can produce several effects such as strong stress relaxation, reduction of the stress concentration at the main crack tip, promotion of plane stress conditions, i.e. biaxial stress state deep inside the specimen, which could make the delamination more difficult to continue, resulting in ductile fracture. The separated pieces will continue the fracture in the form of the thinner specimens. Local splitting is observed on the surfaces of the separated pieces, as shown in Figs. 2.18a–c. It is reasonable to suggest that this kind of shallow splitting on the primary crack surface should be formed at the same moment as the main splitting occurs due to the necessary strong stress conditions. The ductile fracture on the separated pieces with characteristic dimples may be the last fracture on the main fracture surface.

The steel tempered at 500–650 °C gives similar toughness values at room temperature and at −30 °C (Table 2.5). The fracture appearance and similar toughness values show that −30 °C is above the ductile brittle transition temperature (DBTT). Delamination has no significant influence on the upper shelf energy in the steel. Actually, the continuous splitting can only absorb a little energy, while most of the energy is absorbed by the ductile fracture. Therefore, it is natural that the impact energy is not changed as indicated by comparing the ductile percentages on the fracture surfaces. By comparing the room temperature toughness of the steel as-rolled and the steel tempered at 500–650 °C, the tempering treatment is responsible for the reduction of the toughness.

2.4.7 Summary

Regarding the delamination behaviour of hot-rolled and quenched steel when tensile and impact loaded following tempering at temperatures from 200 to 700 °C, tempering in the range of 500–650 °C leads to serious delamination or splitting during room temperature tensile and −30 °C impact loading. Elongated grains from hot rolling and grain boundary embrittlement resulted from precipitates (alloy carbides and nitrides) formed during tempering are responsible for the delamination. The delamination does not influence the upper shelf impact energy. The number and length of splitting on the low temperature impact fracture surface increase with increase of tempering temperature until 650 °C, and then decrease.

The reasons for the delamination are complex, but can be summarised as the combined effect of the anisotropic microstructure and triaxial stress conditions. The anisotropic microstructure will separate along planes of weakness when the proper plane strain state emerges during the loading. Therefore, the delamination happens after necking in tensile loading and before main fracture in impact loading. Due to the stress–strain behaviour of the steel after tempering in the critical range, delamination is easier under tensile than under impact loading. Decreasing temperature favours delamination, related to the corresponding increase in yield strength and decrease in elongation-to-failure.

References

Antoine P, Vandeputte S, Vogt JB (2005) Effect of microstructure on strain-hardening behaviour of a Ti-IF steel grade. ISIJ Int 45:399–404. doi:10.2355/isijinternational.45.399

Funakawa Y, Inazumi T, Hosoya Y (2001) Effect of morphological change of carbide on elongation of boron-bearing Al-killed steel sheets. ISIJ Int 41:900–907. doi:10.2355/isijinternational.41.900

Guo Z, Sha W (2004) Kinetics of ferrite to Widmanstätten austenite transformation in a high-strength low-alloy steel revisited. Z Metallkd 95:718–723

Muljono D, Ferry M, Dunne DP (2001) Influence of heating rate on anisothermal recrystallization in low and ultra-low carbon steels. Mater Sci Eng A 303:90–99. doi:10.1016/S0921-5093(00)01882-7

Rivera-Díaz-del-Castillo PEJ, Bhadeshia HKDH (2001) Growth of needle and plate shaped particles: theory for small supersaturations, maximum velocity hypothesis. Mater Sci Technol 17:25–29. doi:10.1179/026708301101509070

Song R, Ponge D, Raabe D (2005) Mechanical properties of an ultrafine grained C-Mn steel processed by warm deformation and annealing. Acta Mater 53:4881–4892. doi:10.1016/j.actamat.2005.07.009

Song R, Ponge D, Raabe D, Speer JG, Matlock DK (2006) Overview of processing, microstructure and mechanical properties of ultrafine grained bcc steels. Mater Sci Eng A 441:1–17. doi:10.1016/j.msea.2006.08.095

Sha W (2001) Crystallization and nematic-isotropic transition activation energies measured using the Kissinger method. J Appl Polym Sci 80:2535–2537. doi:10.1002/app.1362

Tsuji N, Okuno S, Koizumi Y, Minamino Y (2004) Toughness of ultrafine grained ferritic steels fabricated by ARB and annealing process. Mater Trans 45:2272–2281. doi:10.2320/matertrans.45.2272

Yan W, Shan YY, Yang K (2007) Influence of TiN inclusions on the cleavage fracture behavior of low-carbon microalloyed steels. Metall Mater Trans A 38A:1211–1222. doi:10.1007/s11661-007-9161-2

Yan W, Zhu L, Sha W, Shan YY, Yang K (2009) Change of tensile behavior of a high-strength low-alloy steel with tempering temperature. Mater Sci Eng A 517:369–374. doi:10.1016/j.msea.2009.03.085

Yan W, Sha W, Zhu L, Wang W, Shan YY, Yang K (2010) Delamination fracture related to tempering in a high-strength low-alloy steel. Metall Mater Trans A 41A:159–171. doi:10.1007/s11661-009-0068-y

Yang M, Chao YJ, Li X, Immel D, Tan J (2008) Splitting in dual-phase 590 high strength steel plates: Part II. Quantitative analysis and its effect on Charpy impact energy. Mater Sci Eng A 497:462–470. doi:10.1016/j.msea.2008.07.066

Chapter 3
Fire-Resistant Steel

Abstract This chapter is concerned with the design and characterisation of fire-resistant steels for building construction. Issues raised include effects of alloy additions, controlling the grain size, properties of substitutional elements, and processing, centred on the property requirements for fire-resistant steels. The new fire-resistant steels are microalloyed with molybdenum and niobium, or tungsten, titanium and boron, as the alloying additions for elevated temperature strength. The steels have satisfactory high temperature strength, owing partly to their relatively large grain sizes compared with conventional steels. The nature of equilibrium precipitation is calculated using thermodynamics. The high temperature strengthening mechanisms could be attributed to secondary formation of fine distribution of MC and Laves phase precipitates, molybdenum clusters, and molybdenum and niobium in solid solution. The atomic-scale microstructure of structural steels designed to have fire-resistant microstructures is characterised using atom probe field ion microscopy (APFIM). There is strong precipitation at approximately 650 °C. These probably coherent precipitates have the effect of producing a stable dispersion.

3.1 Design Against Fire

Steel design is an immensely complex area. A process of iteration is required to refine the properties until they are optimal. A structural steel has well-known requirements as far as mechanical properties are concerned. Fire-resistant steels have the additional constraint of requiring an adequate level of strength at elevated temperatures reached in a fire.

Fire protection is needed because conventional structural steel cannot withstand the temperatures experienced in real fires. Many tests were carried out on various large-scale constructions at Cardington Laboratory in England as part of a European collaboration and co-ordinated by the Building Research Establishment (BRE) and the formerly British Steel. Most of the steelwork involved was unprotected and the temperatures of the steel members were recorded. Maximum steel temperatures between 691 and 1060 °C were recorded. If these temperatures are reached, then the limiting stress of the steel falls below the working stress and failure ensues. Thus, it is necessary to protect steelwork and ensure that such temperatures are not reached within the specified fire resistance period.

W. Sha, *Steels*, DOI: 10.1007/978-1-4471-4872-2_3,

A structural member must be able to perform its function during the designated fire resistance time. The specified fire resistance period for a building depends on its size, location and use, with the required period being given in multiples of 30 min, e.g. 30, 60, 90 min, etc. Approximately 7 % of all UK multi-storey steelwork requires 30 min fire resistance. 60 % requires 60 min fire resistance, 10 % requires 90 min fire resistance and 15 % requires 120 min fire resistance. The fire resistance of a member is assessed from the results of fire tests, based on the response of an individual member to the standard fire curve, following the prescribed temperature path. This, along with natural fire curves, depends on fire loads, measured in kilograms of wood and the amount of ventilation. The standard fire curve is different in nature to natural fires, but remains a useful means of comparing the fire resistance of structural members.

Given the impressive performance of the fire-resistant steels manufactured by Nippon Steel, it would seem sensible to use their compositions as a base from which to improve the properties further. This may be achieved by using additions of alloying elements and manipulating the processing variables. The compositions of these steels are given in Table 3.1.

The design can be approached from two different angles, i.e. to design 'intelligent' steel, which would react to a fire by precipitation, and 'solution' steel, which would simply resist the effects of a fire. The composition of steel can improve strength at high temperature, by precipitation hardening and solid solution strengthening. Each of these properties is, by their very nature, suited to the task of creating intelligent steel and solution steel, respectively. Solid solution strengthening provides a constant source of strength, although static, and thus, is suitable for solution steel. Precipitation is a dynamic, reactive process and is thus more suitable for intelligent steel. For carbide precipitates such as MC and M_2C and carbonitrides such as (V,Ti,Nb)(C,N), the source of strength becomes ineffective above approximately 650 °C, due to their dissolving and coarsening. It makes sense to give both the steels a small measure of each property, i.e. for one steel to be mostly solid solution strengthened with some precipitation and the other steel to be mostly precipitation-based with some solid solution strengthening. This approach would also lend itself to the reuse of steel after a fire. A certain amount of precipitation in the matrix would prevent grain growth both in ferrite and in austenite during a fire and hence loss of strength after a fire. This effect has already manifested itself in the Nippon fire-resistant steels.

3.2 Steel Design Considerations

3.2.1 Controlling the Grain Size

There is a definite need for a certain amount of precipitation in the steel in order to control the grain size during exposure to high temperatures and during the rolling procedure. However, if the objective is to study the viability of a solid solution

Table 3.1 Chemical composition of manufactured fire-resistant steels and conventional S275 structural steel (wt%)

Steel	C	Si	Mn	Mo	Nb	N	Other elements	Yield strength (MPa)	Tensile strength (MPa)	Elongation (%)
S275	0.1	0.35	0.9	–	–	0.001	–	280	318	19.5
Nippon Nb-Mo steel	0.11	0.24	1.14	0.52	0.03	0.004	–	350	552	20
Nippon Mo steel	0.1	0.1	0.64	0.51	–	0.003	–	380	507	21
Nippon Mo steel (2)	0.11	1.13	0.23	0.56	–	–		448	562	24
P8123 (Intelligent Steel)	0.08	0.38	1.32	0.54	0.26	0.001	–	594	723	13.5
P8124 (Solution Steel)	0.02	0.36	0.87	0.16	0.63	0.001	–	411	538	15.5
P8240 (low-alloying Solution Steel)	0.014	0.28	0.28	0.21	0.58	–	–	214	389	29
P8241	0.004	0.19	0.22	–	–	–	0.5W, 0.04Ti, 0.013Al, 0.001B	200	346	25

type of steel, the volume fraction of precipitation should be kept to a minimum. Another condition of the chosen carbide type must be that precipitation must occur at a high enough temperature to refine the grain size, i.e. at around 1,200 °C, implying carbide of the MX type. As only the most stable carbides precipitate at such high temperatures and form MC type, the choice of carbide forming elements is limited to titanium, zirconium and niobium. These elements result in the formation of MC-type carbides that not surprisingly are among the most stable carbides available, owing to their extremely high affinity for carbon. It is also important that the precipitation does not increase the yield strength of the steel too dramatically, if the objective is to concentrate on solid solution strengthening. Therefore, it is important that the amount of precipitation that occurs during and after the $\gamma \rightarrow \alpha$ transformation is strictly limited or preferably eliminated. Ti and Nb precipitate almost exclusively in the γ phase owing to their very low solubility.

An arbitrary level of carbon of 0.01 wt% would require a level of 0.04 wt% titanium, or 0.08 wt% niobium, to obtain stoichiometric TiC, or NbC, carbide. (All compositions in this chapter are given in wt%, unless otherwise indicated.) For Fe-0.01C-0.04Ti and Fe-0.01C-0.08Nb steels, thermodynamic calculations predict that TiC and NbC precipitates have solubility temperatures of 957 and 1072 °C, respectively. The addition of 0.5 % tungsten changed these values slightly to 956 and 1012 °C, respectively. A 0.01C-0.04Ti steel would have all of the carbon in TiC and leave tungsten in solid solution. For totally ferritic steel with only a small volume fraction of precipitation, a carbon level of 0.01 % and a titanium level of 0.04 % should suffice.

3.2.2 Properties of Substitutional Elements

In Table 3.2 are the atomic sizes of various substitutional elements that have a suitable range of solid solution in iron along with their shear modulus and the calculated diffusion rate at 850 °C. Unfortunately, there is no available information concerning calculation of the actual strain energy created by the modulus interaction. However, it can be observed that the shear modulus of tungsten differs from that of iron by the greatest amount, creating the greatest modulus interaction. After tungsten, the elements decrease in the order of niobium, molybdenum, titanium, vanadium. A high modulus interaction is desirable. A solute atom can interact with screw dislocations only in this way.

There is a huge variation in the range of diffusion rates. Tungsten has the ability to retard the self-diffusion of iron in iron-tungsten alloys. This was observed in iron alloyed with tungsten up to 0.33 at. %, where the self-diffusion coefficient of iron was drastically reduced. This effect was also observed in the iron-molybdenum system, although to a much lesser degree.

The total available strain energy is limited by the solubility of each solute, i.e. the limit of solubility where the Laves phase will form. Therefore, calculations were carried out to assess the solubility limit of each element. It was assumed that

Table 3.2 Various physical parameters of relevant elements being considered for design

Element	Fe	Mn	Mo	Nb	W	Ti	Cr	V	Co	Ni
Atomic radius R_x (Å)	1.24	1.3	1.36	1.42	1.36	1.45	1.25	1.31	1.25	1.15
Ratio R_x/R_{Fe}	1	1.05	1.10	1.15	1.10	1.17	1.01	1.06	1.01	0.93
Electronegativity	1.83	1.55	2.16	1.60	–	1.54	1.66	1.63	1.88	1.91
Shear modulus G_x (GPa)	82	80	126	38	161	46	115	47	82	76
$(G_x - G_{Fe})/G_{Fe}$ (%)	0	−3	54	−54	97	−44	41	−43	0	−7
Diffusion D in Fe at 850 °C ($\times 10^{-2}$ cm^{2}s^{-1})	7.9	21	21	96	9	1791	3.4	22	6.5	0.0002
Diffusion ratio D_x/D_{Fe} in γ at 950 °C	1	5	125	16	–	612	1.3	6	69	4
Diffusion ratio D_x/D_{Fe} in α at 700 °C	1	7	1.7	10	1.1	10	0.2	0.05	2.3	9×10^{-6}

the Laves phase will not form below 600 °C owing to under cooling and the very slow diffusion of substitutional elements. The weight percentage of each element was then obtained for this Laves phase solubility limit at 600 °C. The results were 1.8, 0.7, 0.8 and 17 wt% for molybdenum, tungsten, titanium and vanadium, respectively. Laves phase can tie up some of the niobium.

There are many factors to be weighed against each other in the choice of a major substitutional element for the fire-resistant steel. However, it is possible to eliminate unsuitable elements. Vanadium can be eliminated owing to its small shear modulus interaction. Niobium is unsuitable owing to its very fast rate of diffusion. Of the remaining elements, tungsten has a very large modulus interaction, titanium and molybdenum having considerably smaller values. All have Laves phase solubility limits above 0.5 %. However, it must be taken into account that tungsten has a slow diffusion rate in ferrite. This would hinder dislocation climb, as it is a diffusion-controlled event. Tungsten would also interact strongly with screw dislocations owing to its modulus interaction, whereas titanium has a weak modulus interaction. On this basis, tungsten can be the major substitutional element. Given that the Laves phase forms at 0.7 % at 600 °C, a level of 0.5 % should be appropriate for tungsten.

3.2.3 Processing, Steel Composition and Manufacturing

Consideration needs to be given to the effects of alloying elements and processing parameters. Tungsten is a ferrite stabiliser of approximately equal strength to niobium. Therefore, for a tungsten content of 0.5 %, it is certain that air cooled steel shall be entirely ferritic. Figure 3.1 is the phase diagram for

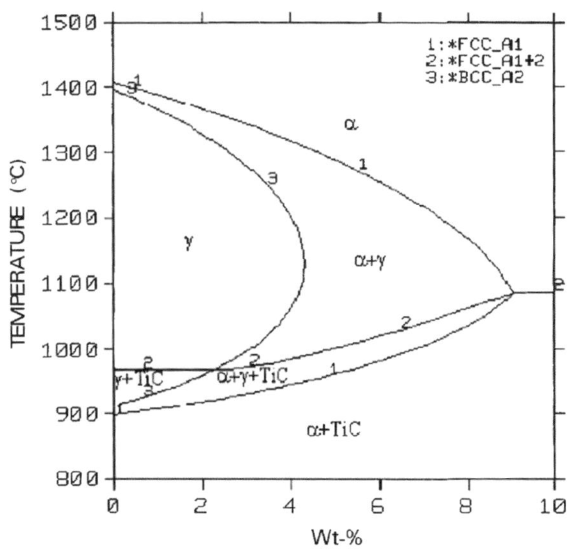

Fig. 3.1 Phase diagram for 0.01C-0.04Ti-0.1Si-0.2Mn (wt%) steel with varying temperature and W content. (From Sha et al. (2002), www.maney.co.uk/journals/mst and www.ingentaconnect.com/content/maney/mst.)

Table 3.3 Processing variables for experimental fire-resistant steels

Variable	Value
Cooling after rolling	Air cooled
Slab reheating temperature	1250 °C
Finish rolling temperature	1050 °C for P8123, P8124, P8240; 1000 °C for P8241
Thickness reduction factor	25 %
Number of passes	5
Initial plate thickness	75 mm
Final plate thickness	18 mm

0.01C-0.04Ti-0.1Si-0.2Mn steel alloyed with variable amount of W. The shape of the phase diagram is similar to those for stainless steels when significant levels of tungsten additions are employed. It can be seen that for a tungsten content of 0.5 %, the $\alpha + \gamma \rightarrow \gamma$ transformation temperature A_{e3} is ~920 °C and that the fcc TiC phase (line 2) should begin to precipitate at around 970 °C under equilibrium conditions. Owing to strain-induced precipitation after rolling, the actual precipitation temperature is likely to be considerably higher than that predicted.

The bulk compositions of the four experimental fire-resistant steels are given in Table 3.1, and the rolling parameters in Table 3.3.

3.3 Microstructure

3.3.1 Grain Structure

The microstructures of the steels are shown in Figs. 3.2 and 3.3 in the as-rolled condition and after heat treatment for 30 min at 650 °C (P8123 and P8124). These steels have similar grain structures.

The measured grain sizes in the as-rolled condition are 13 μm for the P8123 steel and 14 μm for the P8124 steel, compared with 6 μm for the Nippon fire-resistant steels. The grain size referred to here is the mean linear intercept, so the grains look larger in the photographs. The diameter of the grain is larger than the mean linear intercept by some 50 %. After the heat treatment, the grain sizes remain the same, showing no grain growth, which is as the case for the Nippon steels. These steels have polygonal ferrite microstructures, the P8123 steel also containing a proportion of acicular structure. There are large particles of the Laves phase, Fe_2Nb, in both P8123 and P8124 steels, particularly the P8124 (Fig. 3.4). This is expected for the P8124 steel, but is surprising in the case of the P8123 steel considering its lower niobium content.

The grain sizes are a little too large. A reduction in the finish rolling temperature from 1,050 to 1,000 °C results in a slightly smaller grain size.

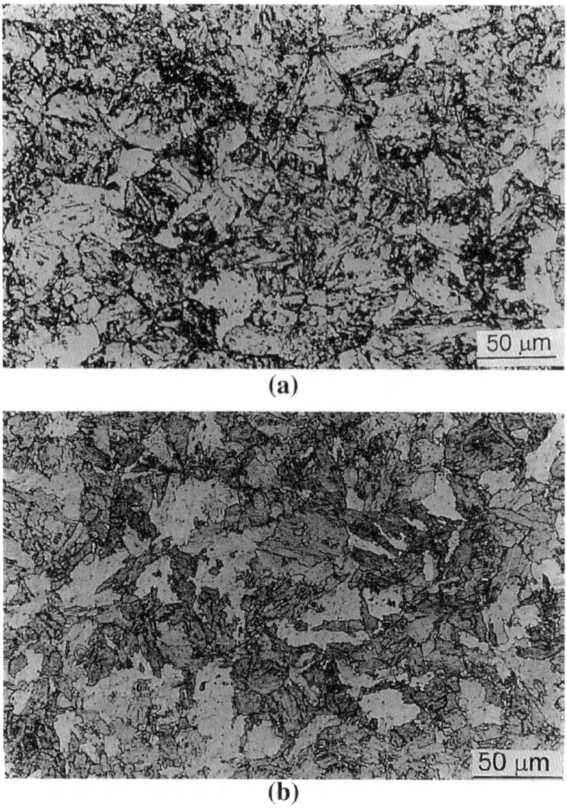

Fig. 3.2 Optical micrographs of P8123 steel. **a** As-rolled; **b** heat treated for 30 min at 650 °C. (From Sha et al. (2002), www.maney.co.uk/journals/mst and www.ingentaconnect.com/content/maney/mst.)

3.3.2 Ferrite–Austenite Transformation and Precipitation

The α–γ phase transformation (Fig. 3.5) temperatures are summarised in Table 3.4. The effect of the alloying elements is to push the ferrite to austenite transformation to higher temperature. This has benefits in strength retention at high temperatures, as the austenite phase is much softer than ferrite. Calculated using thermodynamics, A_{e1} is 679 and 877 °C for the P8123 and the P8124 steels, respectively. For P8240 and P8241, ferrite starts to transform to austenite (A_{e1}) at 946 and 934 °C, respectively.

Precipitates in the as-rolled P8123 steel are niobium rich compounds. The equilibrium ferrite matrix composition, from thermodynamic calculations, at

Table 3.4 α–γ transformation temperatures (°C)

Steel	A_{c1}	A_{c3}	A_{r1}	A_{r3}
P8124	927	939	853	863
P8240	944	952	901	913
P8241	968	989	909	915

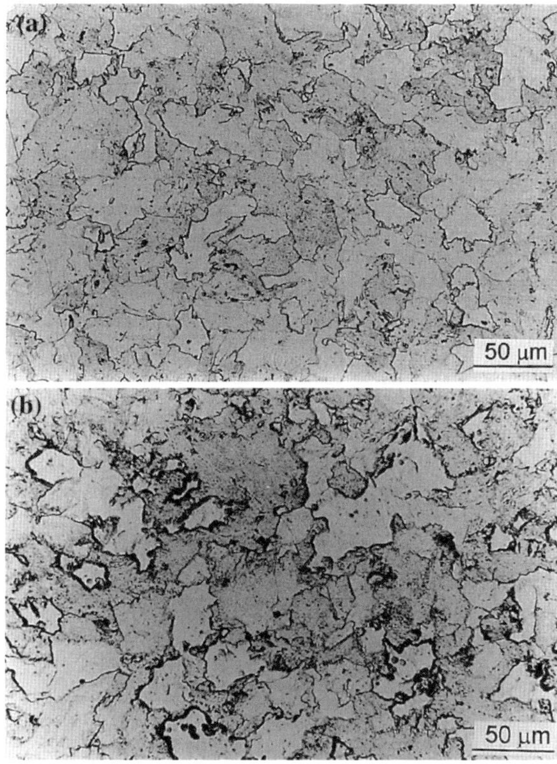

Fig. 3.3 Optical micrographs of P8124 steel. **a** As-rolled; **b** heat treated for 30 min at 650 °C. (From Sha et al. (2002), www.maney.co.uk/journals/mst and www.ingentaconnect.com/content/maney/mst.)

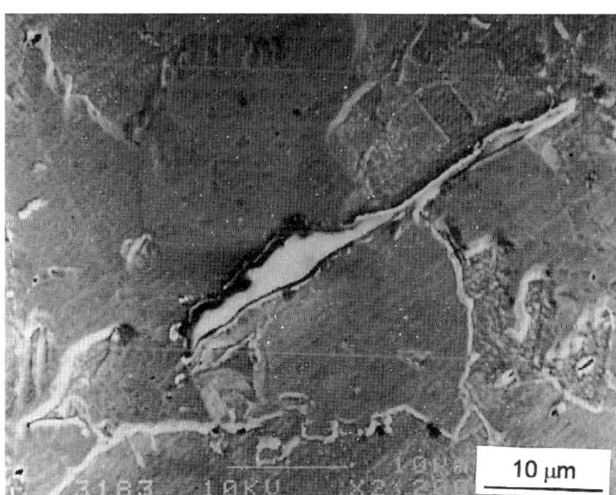

Fig. 3.4 Micrograph of large Fe_2Nb particle in P8124 steel in as-rolled condition (SEM). (From Sha et al. (2002), www.maney.co.uk/journals/mst and www.ingentaconnect.com/content/maney/mst.)

20 °C is Fe-0.39Si-0.72Mn. The calculated compositions of the precipitates at 20 and 650 °C are given in Table 3.5. Comparing the phase constitutions in addition to ferrite, at the two temperatures, cementite has disappeared at the higher

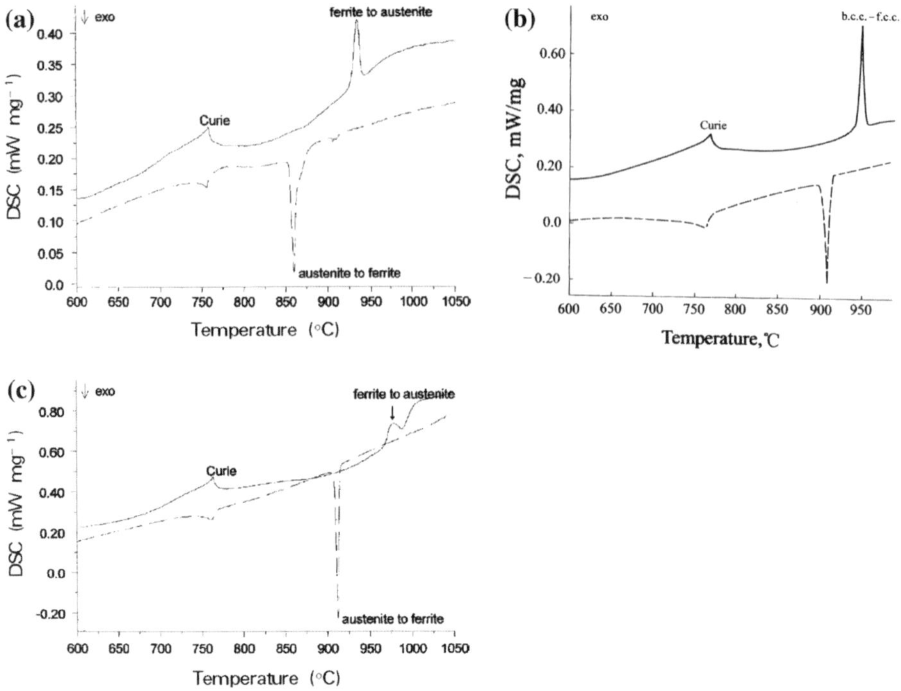

Fig. 3.5 Differential scanning calorimetry (DSC) curves: solid line is from heating and dashed line cooling. **a** P8124; **b** P8240; **c** P8241. ((**a**) and (**c**) from Sha et al. (2002), www.maney.co.uk/journals/mst and www.ingentaconnect.com/content/maney/mst.)

Table 3.5 Precipitate compositions (at.%) and mole fractions for P8123 and P8124 steels calculated

Steel	Temperature	Precipitate	C	Mn	Mo	Nb	Fe	Mole fraction of phase
P8123	20 °C[a]	Cementite	25	75.0	0.0	0.0	0.0	0.008
		M_2C	33.3	11.6	1.5	53.6	0.0	0.003
		M_6C	14.3	–	57.1	–	28.6	0.005
P8123	650 °C	NbC	49.9	0.2	7.6	42.3	0.0	0.004
		MoC	50	–	50	–	–	0.003
P8124	20 °C[a]	Nb_2C	33.3	0.0	0.0	66.7	0.0	0.003
		Fe_2(Nb,Mo)	–	–	10.9	22.5	66.7	0.009

[a]The equilibrium cannot be even approached at this low temperature after air cooling

temperature, as the MC-type carbides are the more stable phases. This correlates with the disappearance of cementite from the micrograph in Fig. 3.2b.

The equilibrium ferrite matrix composition at 20 °C calculated is Fe-0.37Si-0.88Mn. The composition of the Laves phase is given in Table 3.5. Silicon has an

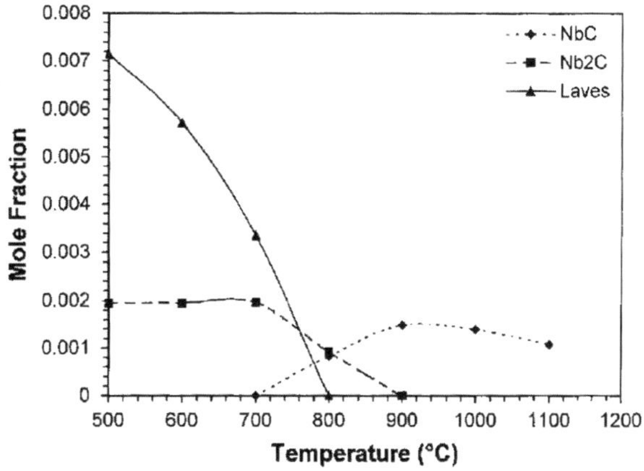

Fig. 3.6 Mole fractions of precipitate phases in P8240 steel as functions of temperature, as calculated thermodynamically. (From Sha et al. (2002), www.maney.co.uk/journals/mst and www.ingentaconnect.com/content/maney/mst.)

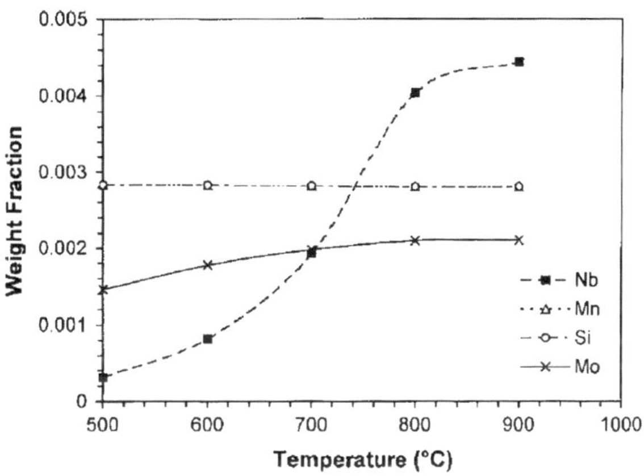

Fig. 3.7 Composition of ferrite phase in P8240 as function of temperature. Data points for Mn and Si overlap in the diagram. (From Sha et al. (2002), www.maney.co.uk/journals/mst and www.ingentaconnect.com/content/maney/mst.)

effect of lowering the solubility of niobium in iron and promotes the formation of the Laves phase. Silicon levels are around 15 % in Fe_2Nb precipitates for 1Si-2Nb steel. Silicon may decrease the lattice parameter of Fe_2Nb, reducing the required volume expansion of the matrix from 13.8 to 12.2 % and thus, facilitating nucleation. Silicon also results in a decrease in the niobium content of the Laves phase. These effects of silicon are not recognised in the thermodynamic database but are confirmed in experimental data (Sect. 3.3.4).

Equilibrium fractions of various phases as functions of temperature can be calculated thermodynamically, giving the variations of the mole fractions of precipitate phases (Fig. 3.6) and of the ferrite composition (Fig. 3.7) with increasing temperature. The Laves phase takes the form of $Fe_2(Nb,Mo)$, with the atomic fraction of Mo decreasing from 5 % at 500 °C to 2 % at 700 °C. At a temperature of 800 °C, two phases should co-exist, Nb_2C and NbC. Only NbC has been observed.

TiC with a slight deficiency in carbon at ~47 at.%, and Fe_2B phases exist in the equilibrium in P8241 steel up to 1,000 and 800 °C, respectively, but their quantities are very minute, owing to the low carbon and boron contents in the steel. In addition, Fe_2W is also present in calculated equilibrium, up to 500 °C, although it is not expected to form at such low temperatures owing to kinetic reasons. The steel is thus almost entirely solid solution strengthened.

The steels are unlikely to exhibit equilibrium microstructures, given the heat treatment duration and the cooling rate. However, thermodynamic calculations show the possible phases and their maximum amounts, and show the partition and therefore to some degree the effectiveness of the alloying elements.

3.3.3 Nippon Fire-Resistant Steels

3.3.3.1 Matrix Compositions and Cementite

There are several notable features. There is very little carbon in solid solution in the matrix. Secondly, the majority of molybdenum is in solid solution, although somewhat surprisingly, the molybdenum level in solution is lower in the Nb–Mo steel than in the Mo steel. The difference, with respect to bulk composition, in measured molybdenum content is related to the consumption of this element in grain boundary segregation and molybdenum rich precipitates. These will be discussed in more detail later in Sect. 3.3.3.2. Lastly, the silicon level in the heat treated Nb–Mo steel is significantly higher than the nominal level. This suggests that there is a low solubility of silicon in the cementite, which would increase the matrix level. The manganese level in the matrix of the Nb–Mo steel increases to almost the nominal level, after heat treatment. It is possible that the local segregation, which dissipated after heat treatment, in the as-rolled steel caused the difference, with respect to bulk composition, in measured manganese content.

Cementite compositions measured using atom probe are shown in Table 3.6. Cementite is not a very fine and densely distributed phase, compared to precipitates.

3.3.3.2 Grain Boundary Segregation and Precipitation

Field ion micrographs of a grain boundary in each type of steel are shown in Fig. 3.8. There are considerable levels of segregation of all interstitial and substitutional elements at grain boundaries in both steels.

Grain boundary segregation is a non-equilibrium phenomenon, so place to place variations in the extent of segregation are well expected. There are particularly high segregation levels for C and Mo, whereas Mn and Si have levels that are

Table 3.6 Cementite compositions (at.%) of the Nippon steels as-rolled

Steel	C	Fe	Mn	Si	Mo	Nb
Nb–Mo	26.9 ± 1.4	71.9 ± 1.4	0.3 ± 0.2	0.4 ± 0.2	0.4 ± 0.2	0.1 ± 0.1
Mo	23.1 ± 1.6	75.4 ± 1.6	0.6 ± 0.3	0.1 ± 0.1	0.9 ± 0.4	–

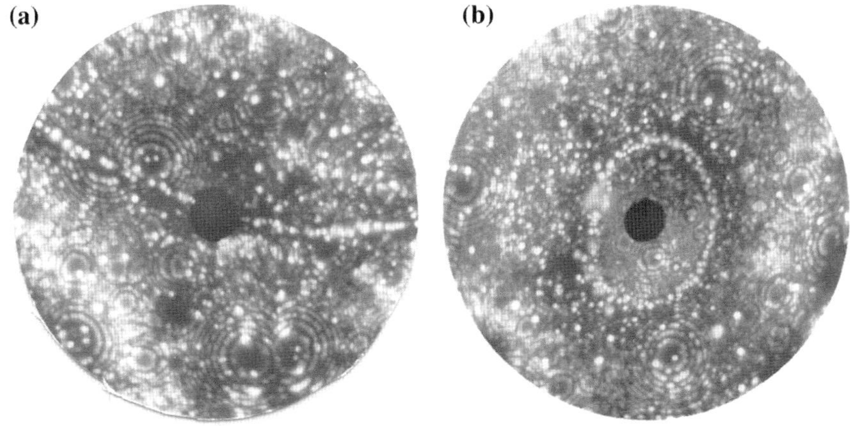

Fig. 3.8 Field ion micrographs of grain boundaries in two Nippon steels, both as-rolled. The distances across the images are 75 and 85 nm, respectively. **a** Nb–Mo steel; **b** Mo steel. (From Sha and Kelly (2004), www.maney.co.uk/journals/mst and www.ingentaconnect.com/content/maney/mst.)

more modest. The misfit energy of an atom provides the driving force for segregation. The misfit energy U of element is calculated from the following equation

$$U = 8\pi \mu r_0^3 \varepsilon^2 \tag{3.1}$$

where ε is the strain created by a foreign atom, μ is the shear modulus of the matrix and r_0 is the atomic radius of the matrix atom. The calculated values are shown in Table 3.7.

Carbon segregation at grain boundaries increases the yield strength by pinning grain boundary dislocations. This makes it more difficult for failure to propagate into the next grain. On the other hand, silicon has the opposite effect, acting in opposition to interstitial elements such as carbon and nitrogen by repelling them from the grain boundary, the so-called site competition effect. Low levels of silicon and high levels of carbon segregation would be beneficial to strength. In addition, the presence of slow diffusing elements such as molybdenum is beneficial to the hindrance of grain growth.

Increased rate of diffusion along grain boundaries is a well-known occurrence, boundary diffusion coefficients being approximately 105 times higher than matrix diffusion coefficients at a temperature of 0.6 T_m. Given this fact, it is almost certain that grain boundary precipitates will form as a secondary process to

Table 3.7 Misfit energy of elements

Steel	C	Mn	Si	Mo	Nb
U (kJ mol^{-1})	1420	26	149	182	280

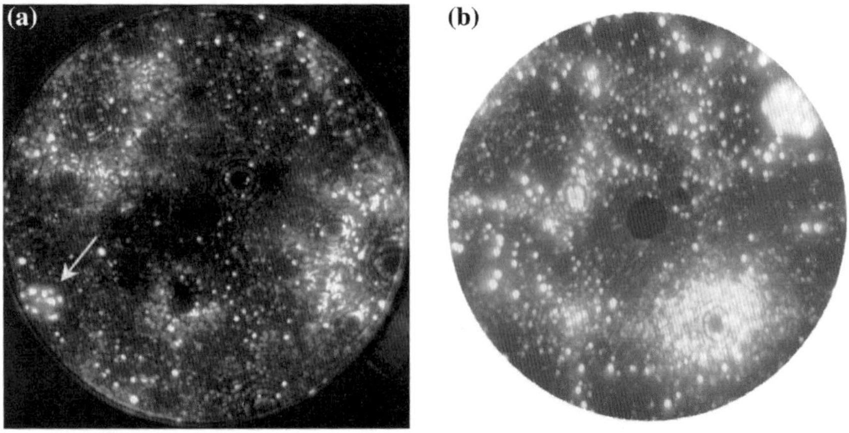

Fig. 3.9 Field ion micrographs of **a** an NbC precipitate in the Nb–Mo steel; **b** MoC precipitate in Mo steel (*top right* edge). (From Sha and Kelly (2004), www.maney.co.uk/journals/mst and www.ingentaconnect.com/content/maney/mst.)

segregation. There are similar segregation levels of molybdenum and carbon in a 0.2C–0.5Mo (wt%) steel.

Only MC-type precipitates are observed in the Nippon steels. Field ion micrographs of precipitates in steels are shown in Fig. 3.9. In addition to precipitates, clusters of molybdenum atoms are present in the steels before and, in larger numbers, after heat treatment.

3.3.4 Prototype Fire-Resistant Steels

The bulk compositions of the P8123 and P8124 steels are shown in Table 3.1. The compositions of the steels are also shown in Table 3.8 (at.%), along with the matrix (ferrite) compositions determined by thermodynamic calculations. At 20 °C, the equilibrium phases are ferrite (97.8 mol%), cementite (1.2 mol%), Fe_2Nb (0.5 mol%) and M_6C (Mo_4Fe_2C, 0.5 mol%). At 650 °C, the equilibrium changes to ferrite (99.3 mol%), NbC (0.4 mol%) and MoC (0.3 mol%). So, the

Table 3.8 Nominal compositions and calculated matrix compositions (20 °C) of prototype steels (at.%), with the balance being iron

Steel	Region	C	Si	Mn	Mo	Nb
P8123	Nominal	0.37	0.76	1.34	0.31	0.16
	Matrix	0.009	0.76	1.34	0.064	0.00
P8124	Nominal	0.09	0.72	0.88	0.09	0.38
	Matrix	0.00	0.72	0.88	0.08	0.05

Table 3.9 Precipitation compositions (at.%) determined by atom probe for P8123 and P8124 steels before and after heat treatmentheat treatment

Steel	Condition	Precipitate	C	Nb	Mo	Fe	Si	Mn
P8123	650 °C, 30 min	NbC	50 ± 3	35 ± 3	9 ± 2	6 ± 2	0	0
		MoC	46 ± 6	15 ± 4	33 ± 6	6 ± 3	0	0
P8124	As-rolled	Fe_2Nb	0	35.4 ± 1.5	1.5 ± 0.4	49.0 ± 1.6	13.0 ± 1.1	0.8 ± 0.3
	650 °C, 30 min	Fe_2Nb	0	33.8 ± 1.3	0.8 ± 0.3	50.6 ± 1.4	12.8 ± 0.9	1.5 ± 0.3

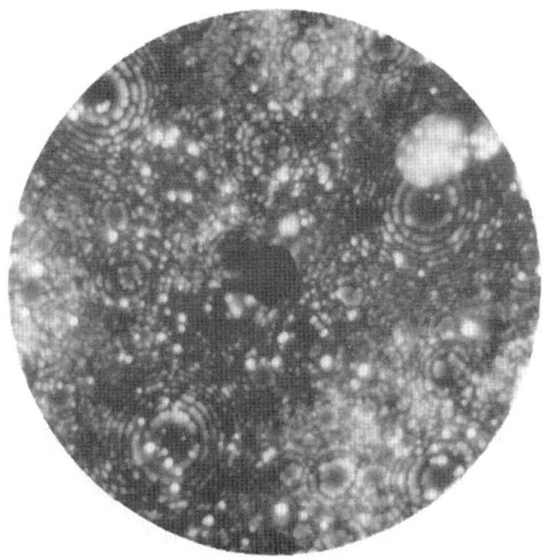

Fig. 3.10 Field ion micrograph of precipitate in P8123 steel after heat treatment. Distance across image is about 65 nm. (From Sha and Kelly (2004), www.maney.co.uk/journals/mst and www.ingentaconnect.com/content/maney/mst.)

carbon level in the as-rolled P8123 already bound up in cementite is released to form secondary carbides.

The average compositions of precipitates are shown in Table 3.9. After the heat treatment, NbC, MoC and Fe_2Nb are formed. A field ion micrograph of a typical precipitate in the heat treated P8123 steel is shown in Fig. 3.10. The total numbers of atoms analysed for different types of precipitates are reflected in the errors given in the table.

The apparent fine clusters in Fig. 3.10 are not all real clusters. Some are due to the strong field retention effects of elements such as molybdenum. However, it is expected that certain clusters do exist.

Due to the small size of carbide precipitates analysed by the atom probe, it is almost certain that there is some contribution from the matrix in the precipitate compositions, giving precipitate iron levels of about 6 %. In addition to the precipitates mentioned above, clusters of molybdenum atoms are also present in the P8123 steel before and after heat treatment. A typical cluster of molybdenum atoms is shown in Fig. 3.11a.

There are numerous extremely small precipitates present after heat treatment (Fig. 3.11b). Due to magnification effects of precipitates in field ion microscopy, it is notoriously difficult to estimate precipitate sizes from the micrographs. However, it is clear that there are many precipitates of around 1–2 nm in size.

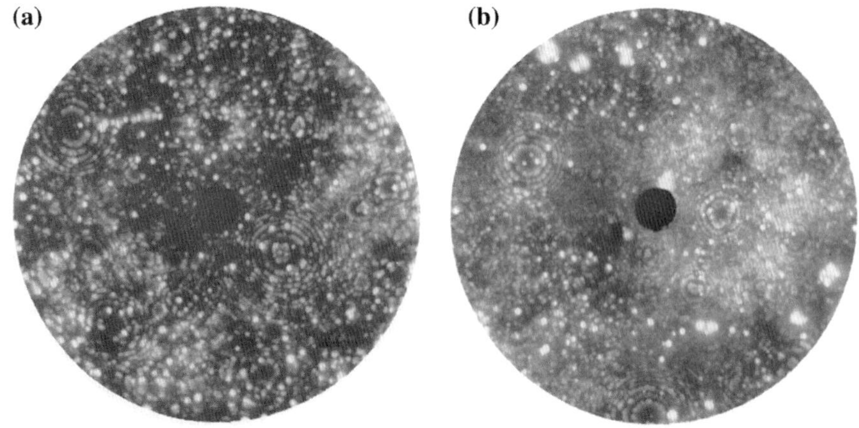

Fig. 3.11 **a** Field ion micrograph of molybdenum cluster in as-rolled P8123. Distance across image is about 75 nm. **b** Typically dense distribution of extremely fine MC precipitates after heat treatment. Distance across image is about 110 nm. (From Sha and Kelly (2004), www.maney.co.uk/journals/mst and www.ingentaconnect.com/content/maney/mst.)

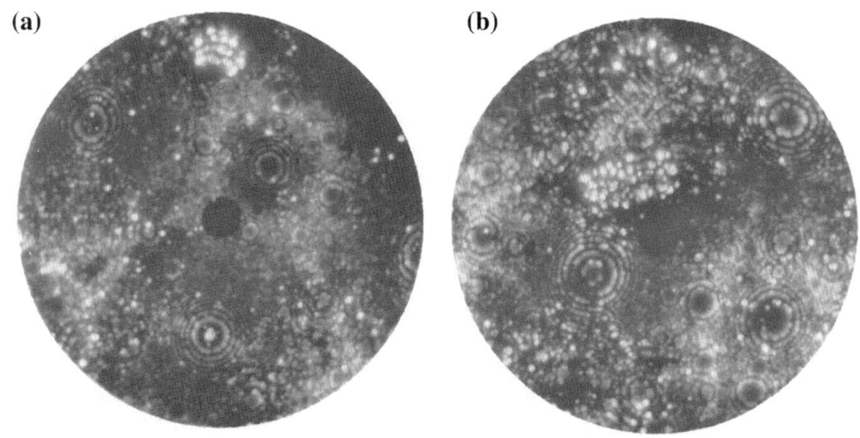

Fig. 3.12 **a** Field ion micrograph of Fe_2Nb precipitate in as-rolled P8124 steel. Distance across image is about 95 nm. **b** Field ion micrograph of Fe_2Nb precipitate in heat treated steel. Distance across image is about 65 nm. (From Sha and Kelly (2004), www.maney.co.uk/journals/mst and www.ingentaconnect.com/content/maney/mst.)

In the P8124 steel, only the ε-Laves phase, Fe_2Nb, is found, both in the as-rolled condition and after heat treatment at 650 °C for 30 min. The composition of these Fe_2Nb precipitates is included in Table 3.9. Field ion micrographs of typical Laves phase precipitates in the as-rolled and heat treated conditions are shown in Fig. 3.12. In addition, clusters of molybdenum atoms are present, although only in the heat treated condition.

The composition of the Laves phase by thermodynamic calculations (Sect. 3.3.2) is essentially stoichiometrically based on Fe_2Nb. For the P8124 steel, the calculated Laves phase composition is substantially different from the experimental composition. Experimental results show that the Laves phase has considerable levels of silicon and small amounts of manganese and molybdenum, none of which are present in the calculated Laves phase. The niobium content is also higher than that obtained by thermodynamic calculations.

3.4 Mechanical Properties

3.4.1 Strength

Yield strength (0.2 % proof strength) and tensile strength of the steels versus temperature are shown in Figs. 3.13 and 3.14, along with the values for the S275 steel for comparison. The steels were all in as-rolled state, with no heat treatments. The yield strength ratio of 60 % is a useful basis for comparison between the steels. Steels P8123 and P8124 reach 60 % strength at temperatures of 725 and 730 °C, respectively. This is a considerable improvement over the values of 625 and 645 °C exhibited respectively by the Nippon Nb–Mo and Mo steels and the value of 510 °C for the S275 steel. The steels exhibit work hardening up to 700 °C but such ceases to be apparent at 800 °C.

The low-alloying version of the Solution Steel is similar to P8124 but with lower alloying concentrations (including carbon). There is large stored energy in the P8240 steel because of rolling.

There are probably combinations of reasons for the strength loss of P8123 and P8124 steels above 700 °C. Grain boundary sliding may come into effect in this temperature range, as well as increased diffusion rates, leading to faster

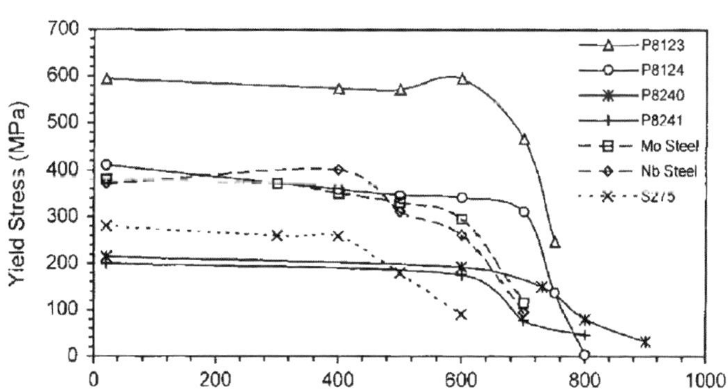

Fig. 3.13 Yield strength as function of temperature for fire-resistant steels and conventional constructional steel. (From Sha et al. (2002), www.maney.co.uk/journals/mst and www.ingentaconnect.com/content/maney/mst.)

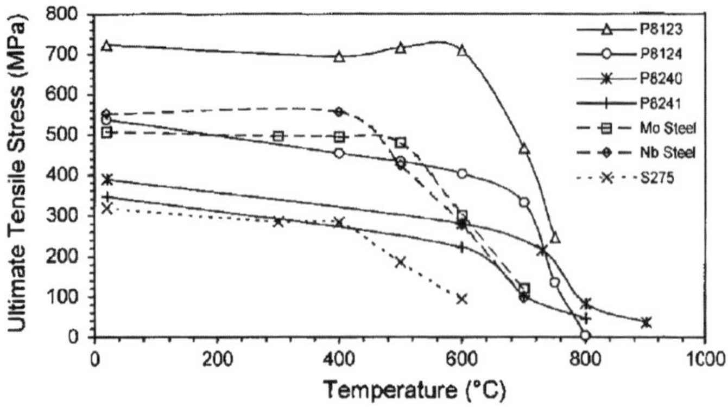

Fig. 3.14 Ultimate tensile strength as function of temperature for fire-resistant steels and conventional constructional steel for comparison. (From Sha et al. (2002), www.maney.co.uk/journals/mst and www.ingentaconnect.com/content/maney/mst.)

dislocation climb. For the P8123 steel, the loss of strength above 700 °C is also due in part to the progressive transformation of ferrite to austenite, which has been shown to start at around 704 °C from differential scanning calorimetry experiments. The transformation would have completed at 750 °C and precipitates coarsen rapidly and lose coherency at this temperature. There is a significant loss of strength associated with these phenomena.

For the P8124 steel, it is more difficult to assess the mechanism of strength loss. Thermodynamic calculations predicted that the Laves phase should go into complete solution at 756 °C. However, due to stabilisation of the Laves phase by silicon, an effect not considered in the database, this would not be expected to happen. TEM thin foil observation of tensile samples from 750 °C has confirmed that the Laves phase is present and its general morphology and distribution are unchanged. It is unlikely that grain boundary sliding contributes significantly to the loss of strength at 750 °C. There is an effect of inter-phase boundaries on grain boundary sliding. Grain boundary sliding, consisting of sliding and migration, is considerably reduced by the presence of inter-phase boundaries. Since the Laves phase grain boundary precipitates are present in the P8124 steel at 750 °C covering approximately 50 % of grain boundaries, the migration of these grain boundaries is highly unlikely. Therefore, the increased rate of diffusion and occurrence of vacancies and the resulting increase in dislocation climb are thought to be the main mechanisms of strength loss at 750 °C for the P8124 steel. Surprisingly, the Curie temperature, the change from the ferromagnetic to the paramagnetic conditions, may have a large impact on this mechanism.

Values of lattice friction stress (the resistance of the matrix to the motion of dislocations) for these steels at 600 °C are calculated to be 440 MPa for the P8123 steel and 192 MPa for the P8124 steel. These may be compared to 30, 72 and −92 MPa for the Nb–Mo, Mo and S275 steels respectively. The negative value of the S275 steel probably reflects the fact that grain boundary sliding occurs below this temperature or that the k_y value does not remain constant at these temperatures.

The high lattice friction stresses of the steels are due to the high volume fraction of precipitates, molybdenum clusters and the considerable amount of molybdenum and niobium in solid solution. In the P8123 steel, the particularly high friction stress is due to the distribution of extremely fine NbC precipitates. As for the Nippon steels, climb over these obstacles must occur for dislocations to move through the matrix. However, the process of climb is slower in the prototype steels.

3.4.2 Elongation and Fracture Surface

At room temperature, both P8123 and P8124 steels exhibit lower elongation than commercial steels. The loss of elongation for the steels at 600–700 °C is likely owing to carbide precipitation, which has been shown by DSC and thermodynamic calculations to occur at this temperature range. The rise in elongation after this temperature range suggests rapid overageing of these precipitates and/or enhanced climb at the higher temperatures. Low ductility at slow strain rates is a feature of age hardening alloys. The elongation of both P8123 and P8124 steels can be considered too low for a structural steel, being considerably lower than the commercial steels at almost every test temperature.

The poor elongation of the P8123 and P8124 steels was owing to the large amount of grain boundary precipitation of cementite (Sects. 3.3.2 and 3.3.4) and the Laves phase (Sect. 3.3). In addition, both the steels have high manganese contents, 0.87 % for the P8124 steel and 1.32 % for the P8123 steel. High Mn has a beneficial effect on the ductility of steels but this applies only to commercial steels that have a certain amount of sulphur. However, at high temperatures, manganese in solid solution decreases the ductility. Thus, this is a contributing factor to the poor high temperature elongation of the steels. It would therefore be advisable to reduce the manganese content in new steels, if they should be of high purity, containing virtually no sulphur. However, if the steel were to be produced commercially, manganese would certainly be required to control sulphide inclusions. A value of 0.2 % should be adequate.

The tensile fracture surfaces show that the poor ductility can be attributed to the presence of large Laves phase precipitates in the steels. Typical fracture surfaces for the P8123 and P8124 steels after conventional tensile testing at room temperature are formed by coalescence of dimples generally originated at large Laves phase particles, which are present on the fracture surfaces. The many small particles on the fracture surface for P8123 are carbides. The Laves phase particles are of the order of 5 μm in size.

3.4.3 Hardness and Creep

The hardness for the steels is shown in Fig. 3.15. The initial Vickers hardness values are 204 HV and 149 HV for the P8123 steel and P8124 steel respectively. The equivalent tensile strengths are 693 and 506 MPa, which are reasonably close to the experimental tensile values of 723 and 538 MPa for the P8123 and P8124 steel respectively. From 550–650 °C, P8123 steel shows a peak in hardness.

As this peak correlates well with microscopy observation of secondary precipitation and DSC precipitation peaks, it can be associated with MC precipitation and/or the formation of additional molybdenum clusters. The hardness values for the P8124 steel remain roughly constant over the whole temperature range, the small rise being within error limits but also perhaps indicative of a small amount of Laves phase or molybdenum cluster precipitation. This shows that the P8124 steel is indeed essentially non-ageing, in concordance with TEM and DSC observations.

A trend for all of the age-hardening steels is that the hardness value peaks during precipitation at around 650 °C, and then returns to a value that is slightly above the value before the precipitation peak, at say, 500 °C. The difference between the values after ageing at 500 and 700 °C could be explained by the presence of the greater number of precipitates which have precipitated between these temperatures. The Vickers hardness after heat-treatment at 650 °C for up to 3 h remains approximately constant.

Creep curves for P8240 and P8241 steels at 600 °C and 0.71 × σ_y, σ_y being room temperature yield stress, are similar to those from the other fire-resistant steels, namely P8123 and P8124, with the maximum creep strain reaching 0.18 % after three hours.

3.4.4 Connection Between Atom Probe Data and the Evolution of Mechanical Properties

Comprehensive mechanical properties of the steels at room and elevated temperatures, in the as-rolled state and after heat treatment, are given earlier. Of particular relevance to the atom probe data are the hardness measurements results of as-rolled steels and steels after ageing at 650 °C for 30 min, the conditions used for the steels in Sects. 3.3.3 and 3.3.4. The results for the two Nippon steels are given in Table 3.10.

The general observation is that the hardness of the two Nippon steels remains at about the same level as as-rolled hardness after heat treatment at 650 °C for 30 min. Note, however, that there is a significant decrease in hardness in the Nippon steels after treatment for the same length of time at all other temperatures (400–600 °C and 700 °C), and the hardness after the 650 °C treatment appears to

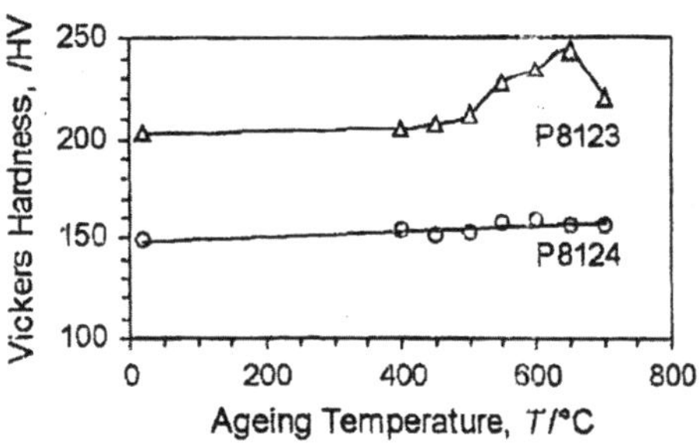

Fig. 3.15 Room temperature Vickers hardness values for the P8123 and P8124 steels after 30 min heat-treatment at various temperatures. (From Sha (2001).)

Table 3.10 Vickers hardness of Nippon fire-resistant steels before and after heat treatment

Steel	As-rolled	650 °C, 30 min
Nb–Mo	166	163
Mo	148	144

be a peak. The decrease in hardness is thought to be due to the recovery process of the deformed as-rolled steels. Therefore, ageing at 650 °C has caused significant hardening, although the hardness values appear to be similar to as-rolled hardness. Such hardening is caused by the additional precipitation and grain boundary segregation observed through atom probe field ion microscopy.

3.4.5 High Temperature Transient Tensile Properties

Section 3.4.6 is concerned with the properties of fire-resistant steels at high temperatures using transient tensile testing. Each country requires a similar minimum standard performance from structure during a fire over a specified period. For example when designing a structure for use by the public, in some specific situations that structure must be able to withstand a fire for two hours without major structural failure occurring. This means that structural steel members must usually be able to withstand air temperatures in excess of 1,000 °C over the required period.

The importance of metals as constructional materials is almost invariably related to their load bearing capacity in either tension or compression and their ability to withstand deformation without fracture. It is usual to assess these properties by tensile tests in which the modulus of elasticity, the yield or proof stress, the tensile strength and the percentage elongation are determined. A stress–strain curve is plotted using stress obtained by dividing the load by the original cross-sectional area of the specimen, and strain obtained as the extension divided by the original heated length. This curve is known as an engineering stress–strain curve and rises to a maximum stress level and then falls off with increasing strain until it terminates as the specimen breaks. The maximum stress level is known as the tensile strength of the specimen. Stress–strain curve shows a linear relationship from the origin up to the proportional limit point, also called elastic limit. This initial linear deformation is the elastic deformation of the material. This deformation is recovered once the load is removed from the specimen. This behaviour is called the elastic behaviour. The point at which the line begins to deviate from this linear characteristic is known as the elastic limit. After this point, any deformation that occurs will be plastic and so it is not recoverable.

In many structures, it is common for a loaded steel component to be in fire situation subjected to a change in temperature and it is important to know how the resulting deformation of the material will develop. For this reason, tensile tests under transient heating conditions have been devised. In such tests, the load on a steel specimen is maintained constant while its temperature is increased at a given rate and the changes in gauge length are constantly recorded. Typical tensile curves are shown up to 2 % strain derived from transient tests at a heating rate of 10 °C per minute. The principal features comprise a small initial elastic extension, when the load is applied at room temperature, followed by a very gradual increase in length as the temperature is increased. For the purpose of fire engineering design, strength reduction factors can be derived from the transient tensile curves. These factors can be found by dividing the stresses at 2, 1.5 and 0.5 % strains by the yield strength of the material at room temperature (approximately 20 °C). Then strength reduction factor versus temperature curves can be plotted. The relationship between the elongation at fracture and temperature also can be derived from the transient tensile data.

3.4.6 High Temperature Transient Tensile Test

The building standards in the UK, Western Europe and North America are different from the standard in Japan, for fire resistance of steel structures. The Western codes are based on the strength reduction factors of steels, which are defined as the ratios of tensile stresses of steels at specific strain levels (e.g. 2 % for beams), at high temperatures, to room temperature yield stress. Normally, the tensile stresses at different strain levels are obtained easily with tensile tests, when a sample is held at a constant temperature while the stress–strain curve is obtained. However, in European building design codes, this test method is not acceptable for obtaining such stresses. Instead, a transient test must be used.

In the transient test, a tensile sample is held at a constant stress, or load, and the temperature is increased constantly, at a specified rate of 10 °C per minute. The strain is measured at short intervals. So, this is similar to a creep test, except that the temperature in a creep test is constant. The transient test will result in a strain-temperature curve. By conducting many such tests for a given material at many different stress levels, stress–strain curves for various temperatures can be built. British and European codes require the strength reduction factors to be obtained from stress–strain curves built in this way.

Essentially, the amounts of experimentation are similar for the two types of tests. The normal constant temperature tests (called steady-state test in Eurocodes) require many tests at various fixed temperatures with constantly increasing strain in each test, while the transient tests require many tests at various fixed stress levels with constantly increasing temperature in each test. However, the stress–strain curves built with transient tests usually give lower strength because in transient tests the samples are heated for a considerable length of time before the strain is taken, and this usually causes softening. For example, in a stress–strain curve at 620 °C, the steel will have been heated from 20 to 620 °C for an hour before the data are taken. However, the fire-resistant steels may well behave differently.

The author has been conducting research on fire-resistant steels since 1995. From the long contacts with British and European building industry, the largest resistance of using the Nippon fire-resistant steels is because of a lack of knowledge and confidence. Although the Nippon steels are widely publicised, there are no transient test data, and therefore, there are no strength reduction factors data. The building designers cannot use Nippon steels when no appropriate materials strength data are available. The steel companies can dismiss Nippon steels for this simple reason. It is widely regarded in the West that the steels are only designed for Japanese market. The transient test results will demonstrate clearly the performance of Nippon steels, using test methods accepted in Western Europe.

3.5 Summary

Japanese Industrial Standards (JIS), "Rolled steels for welded structures" specifies chemical composition and mechanical properties at room temperature and does not specify mechanical properties at high temperature. Fire-resistant steel satisfies specification on JIS. The three fire-resistant steels, Nb-Mo steel, Mo steel and Mo steel (2) were made by Nippon Steel. The compositions of the steels are given in Table 3.1. The yield strength, tensile strength and elongation are also shown in Table 3.1, compared with the normal requirement for structural steel of yield strength between 300 and 400 MPa and an elongation of 20 %. A brief description of each alloying element used in the steels is shown in Table 3.11. Nb–Mo steel and Mo steel (2) are G3106 SM490A with tensile strength at room temperature higher and equal to 490 MPa because of the amount of manganese in their chemical composition. Mo steel is SM400A with tensile strength at room temperature higher and equal to 400 MPa. Structural steel should satisfy the weldability, and weld test like maximum hardness test in welded heat-affected zone (HAZ). The hardness of the heat-affected zone is lower for the fire-resistant steel than the conventional steel.

Given that carbon-based steels are characterised by ferrite to austenite transformation temperatures of 700–750 °C, and the relative instability of carbide precipitates above 650 °C, their use at high temperature is questionable. The greater resistance to ageing of the intermetallic Laves phase, Fe_2Nb, indicates that the use of intermetallic precipitates may be of great potential benefit for fire-resistant steels.

The better high temperature tensile strength and creep properties of the Nippon steels than conventional structural steels are due to the very fine distribution of MC precipitates, Mo clusters and molybdenum in solid solution. Grain boundary segregation also plays a role in improvement of high temperature mechanical properties. In addition, there is a strong secondary wave of precipitation at approximately 650 °C. These, probably coherent, precipitates have the effect of producing a very fine and stable dispersion.

For the prototype P8123 steel, 0.08C-0.38Si-1.32Mn-0.54Mo-0.26Nb, the high temperature strength is due to the distribution and secondary precipitation of ultra fine MC precipitates, as well as the molybdenum and niobium in solid solution. The mixed composition of these precipitates has the effect of producing a very fine dispersion of stable, coherent precipitates, which coarsen very slowly. For the

Table 3.11 Brief description of each alloying element used in the fire-resistant steels

Element	Description
C, carbon	An essential ingredient in steel. Higher carbon content increases the yield point and hardness and reduces ductility and weldability
Si, silicon	Higher silicon content increases the strength of the steel at high temperatures by raising the ferrite–austenite transition temperature
Mn, manganese	Manganese is similar in many ways to iron and is widely used in steel as deoxidant. Manganese can contribute to temperature embrittlement. It lowers the ferrite–austenite transition temperature and can therefore be used to stabilise silicon content in steel
Mo, molybdenum	The principal use of molybdenum worldwide is in alloy steel. It can help to reduce temper brittleness in steel. Molybdenum raises the high temperature strength and creep resistance of steel, but reduces the yield strength at room temperature. It does this by reducing the austenite-pearlite transformation temperature
Nb, niobium	Niobium improves the steel strength at room temperature and at high temperature without affecting weldability. It also improves creep strength

P8124 steel, 0.02C-0.36Si-0.87Mn-0.16Mo-0.63Nb, the high strength is due to precipitation of the Laves phase, which is stable and resists coarsening. The loss of strength at high temperatures is proposed to stem mainly from dislocation climb, and, in the case of the P8123 steel, the ferrite to austenite phase transformation, which reduces the strength dramatically.

The experimental steels developed, with nearly completely ferritic microstructures, have demonstrated high elevated temperature strength compared to the Nippon steels. The conclusion to be drawn from this is that good high temperature strength is not dependent on the presence of acicular transformation products such as acicular ferrite or bainite. The Nippon steels have such products that have been claimed to be responsible for good high temperature strength. The P8123 experimental steel performed extremely well, but is clearly unsuitable as a structural steel for building construction purposes due to its very high room temperature yield strength, 594 MPa. However, the P8124 experimental steel represents a change in direction for fire-resistant steels. It has low carbon content, is thermally stable and has a totally ferritic microstructure, so it does not have the overageing problems of ferrite/pearlite steels. The upper temperature limit for phase stability and good strength imposed by the ferrite to austenite phase transformation is very high, at around 920 °C. It is difficult to separate the effect of its large grain size from its high friction stress and the volume fraction of Laves phase particles. Transmission electron microscopy (TEM) has shown that they are present in this steel even after heat treatment at 750 °C. Therefore, it is thought that the Laves phase particles are responsible for the good high temperature strength of the P8124 steel, as well as the large grain size and the high friction stress of the elements in solid solution.

It has also been shown that the elongation of the P8123 and P8124 steels is not adequate for structural steels and that grain boundary precipitation is the cause of the poor elongation. Therefore, this property needs to be improved in future steels. It is of scientific interest to design a type of steel based almost entirely on solid solution effects. Solid solution steel has never been used for high temperature use in the past to the author's knowledge.

The experimental steels almost certainly provide 30-min fire resistance for an unprotected I-beam, achieving the target fire-resistance set out before their design.

References

Sha W (2001) Mechanical properties of structural steels with fire resistance. In: Hanada S, Zhong Z, Nam SW, Wright RN (eds) The fourth Pacific Rim international conference on advanced materials and processing. The Japan Institute of Metals, Sendai, pp 2707–2710

Sha W, Kelly FS (2004) Atom probe field ion microscopy study of commercial and experimental structural steels with fire resistant microstructures. Mater Sci Technol 20:449–457. doi:10.1179/026708304225012305

Sha W, Kelly FS, Browne P, Blackmore SPO, Long AE (2002) Development of structural steels with fire resistant microstructures. Mater Sci Technol 18:319–325. doi:10.1179/026708301225000789

Chapter 4
Heat-Resistant Steel

Abstract The microstructure evolution of 10Cr ferritic/martensitic heat-resistant steel during creep at 600 °C is a main topic in this chapter. The 10Cr steel has higher creep strength than conventional ASME-P92 steel. The martensitic laths coarsen with time and eventually develop into subgrains during creep. Laves phase grows and clusters along the prior austenite grain boundaries during creep and causes the fluctuation of solution and precipitation strengthening effects. The microstructure evolution could be accelerated by stress. Laves phase is one of the most significant precipitates in ferritic/martensitic heat-resistant steels. Cobalt in the steel could accelerate the growth of Laves phase and coalescence of the large Laves phase would lead to the brittle intergranular fracture. Another topic of this chapter is the microstructural evolution during short-term thermal exposure of 9/12Cr heat-resistant steels, as well as mechanical properties after exposure. The tempered martensitic lath structure, as well as the precipitation of carbide and MX-type carbonitrides in the steel matrix is stable after 3,000 h exposure at 600 °C. During short-term thermal exposure process, the change of mechanical properties is mainly caused by the formation and growth of Laves phase precipitates.

4.1 Microstructure Before Creep and Creep Rupture Strength

After heat treatment, both the 10Cr steel and the P92 steel (Table 4.1) have tempered-martensite microstructures, with similar prior austenite sizes of about 20 μm. Prior austenite grain boundaries and lath boundaries are decorated by precipitates, $M_{23}C_6$ or MX, after tempering.

Figure 4.1 shows the relationship between the creep rupture strength and the rupture life at 600 °C in logarithmic plots for the two steels. The creep rupture strength generally decreases with the rupture life. However, the creep rupture strength of the 10Cr steel is higher than that of the P92 steel in the entire testing range. Under the creep rupture strength of 210 MPa, the rupture life of the 10Cr steel can reach about 8,354 h, which is much longer than that of the P92 steel, about 431 h. Therefore, the creep rupture strength of ASME-P92 could be improved by revising the chemical composition, like proper increase of the contents of chromium and tungsten as well as addition of cobalt.

W. Sha, *Steels*, DOI: 10.1007/978-1-4471-4872-2_4,

Table 4.1 Chemical compositions of heat-resistant steels (wt%)

Steels	C	Si	Mn	Cr	Mo	W	Co	Cu	Ni	V	Nb	N
P92	0.11	0.37	0.46	8.77	0.42	1.73	–	0.15	0.41	0.17	0.057	0.048
9Cr	0.09	0.31	0.50	8.58	0.40	1.65	1.64	–	0.39	0.18	0.060	0.040
10Cr	0.088	0.31	0.50	10.42	0.40	2.55	2.19	–	0.33	0.18	0.056	0.058

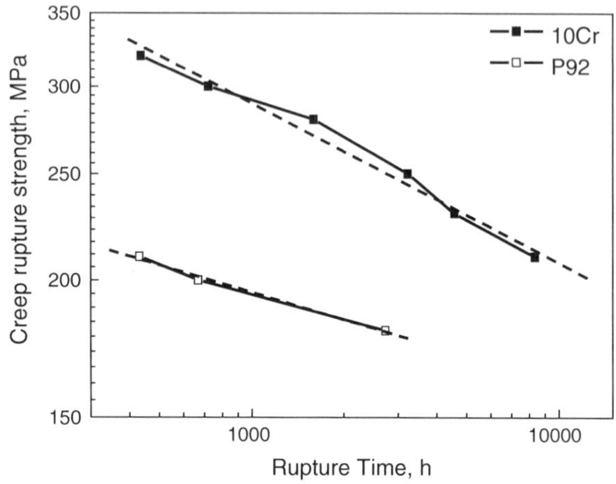

Fig. 4.1 Creep rupture strength versus rupture time of the steels at 600 °C. Reprinted from Hu et al. (2011) with permission from Elsevier

Moreover, two slope changes are seen from the curve of the creep rupture strength versus rupture time of the 10Cr steel. The slope of the curve first decreases after about 723 h and then increases after about 1,599 h. Since strength is dependent on microstructure, the two slope changes should be related to the microstructure evolution during creep.

4.2 Effect of Microstructure Evolution on Creep Rupture Strength

4.2.1 Martensitic Lath Structure

After normalising and tempering, the 10Cr steel acquires a tempered martensitic lath structure strengthened by $M_{23}C_6$-type carbides and MX-type carbonitrides. Most $M_{23}C_6$-type carbides are rod-shaped particles, 30–60 nm in width and 60–300 nm in length, mainly precipitated along lath boundaries. Most MX-type carbonitrides are spherical particles, 20 nm in size, and usually formed at dislocation sites inside the lath matrix. The high density of these fine precipitates could prevent the dislocation movement and the migration of the lath boundaries, resulting in a decrease in the creep rate. Therefore, they are desirable in the initial microstructure in order to improve the microstructure stability.

The mean width of the martensite laths increases with increase of the rupture life. The widened laths are further expanded and they finally evolve into subgrains after 8,354 h. Both lath widening and formation of subgrains during creep could be attributed to the continuous movement and annihilation of dislocations in the matrix and at the lath boundaries. The behaviour of dislocations follows a thermodynamic spontaneous process, since they are driven by the free energy difference between the martensite phase and the thermodynamically stable ferrite phase. Accompanied with lath widening and formation of subgrains usually is the decrease of both lath boundary area and dislocations density inside laths, which should be one of the main reasons for the slight decrease of creep rupture strength in Fig. 4.1.

4.2.2 Laves Phase

Creep deformation at elevated temperature should result from the migration of dislocations and subgrain boundaries (Maruyama et al. 2001; Hald 2008). Homogeneous distribution of fine and thermally stable precipitates can effectively pin the movement of dislocations and retard the coarsening of subgrains, resulting in the reduced creep rate. For the 9–12 wt% chromium ferritic/martensitic heat-resistant steels, there are mainly three types of precipitates. Both MX particles, where M could be Nb, V and Ti, and X could be C or N and $M_{23}C_6$, where M could be Cr, will precipitate after tempering treatment while $(Fe,Cr)_2(W,Mo)$, namely Laves phase, with a hexagonal crystal structure, can form during the long time creep. Research on these precipitates has shown that MX is the most stable and $M_{23}C_6$ is the second most stable precipitates. Both of them are small and very useful to the long-term stability of microstructure. However, Laves phase was observed to have the largest size and the highest coarsening rate. Therefore, the Laves phase is receiving more and more attention as a crucial component of the microstructure. The consumption of tungsten and molybdenum atoms in forming Laves phase will bring about counteraction between the solid solution strengthening and precipitation strengthening. Much work has been conducted to study the effect of Laves phase in this case. Lee et al. (2006) suggested that the precipitation strengthening due to the Laves phase could compensate the loss of solid solution strengthening in the early stage but the coarse Laves phase failed to compensate that loss in the later stage. Abe et al. (2007) seemed to find the similar result that the precipitation of Laves phase could reduce the creep rate but large coarsening of Laves phase promoted the creep rate at the acceleration creep stage after reaching a minimum creep rate in 9Cr-WVTa steels at 650 °C. However, they also found that large coarsening of Laves phase did not take place within their testing range at 600 °C so the precipitation of Laves phase could reduce the minimum creep rate and the steels had a longer rupture life, which was different from the opinion held by Lee et al. (2006). The work of Hald (2008) has revealed that size of the Laves phase remained steady, nearly 0.1 μm, after about 10,000 h creep exposure; therefore, he suggested that precipitation strengthening of the Laves phase had more benefit to the microstructure stability and creep strength than the solid solution strengthening by tungsten and molybdenum atoms in P92 steel crept at 600 °C. This may explain the above beneficial effect of Laves phase on minimum creep at 600 °C found by Abe et al. (2007). All the research shows that the size of Laves phase is of great significance to the creep properties. Therefore, it is very important to carefully investigate the growth and coarsening behaviour of Laves phase in heat resisting steel.

In steel with high cobalt contents, the mean size of Laves phase was found to grow to 0.2 μm in a relatively short period of time of 1,598 h at 600 °C. Although Laves phase with the average size beyond 0.13 μm was discovered to trigger a fracture mode transition from ductile transgranular to brittle intergranular in the work of Lee et al. (2006) on P92 steel, brittle intergranular fracture did not always appear, in other steels. Therefore, the precipitation behaviour of Laves phase as well as its effect on the creep behaviour could be complex and will be investigated in Sects. 4.3–4.5 by considering the effect of cobalt on the nucleation and growth of Laves phase.

Microstructural examination of the 10Cr steel in the deformed section of the creep specimens suggests that Laves phase does not form after tempering (Hu et al. 2011). The evolution of Laves phase in the steel during creep shows that both the volume fraction and the mean size of Laves phase increase with the rupture life.

Tungsten, molybdenum, chromium and carbon all have homogeneous distribution in the 10Cr steel after tempering, whereas, after creep for 8,354 h, accumulation of tungsten, molybdenum, and chromium takes place in some regions. These tungsten-rich or molybdenum-rich regions are related to the formation of Laves phase. The chromium-rich or carbon-rich regions seem to be near or surrounding the tungsten-rich or molybdenum-rich regions. The granular or film-like Laves phase is formed adhering to the massive $M_{23}C_6$ along grain boundaries. Therefore, these chromium-rich or carbon-rich regions should be related to the $M_{23}C_6$-type carbides.

The Laves phase both inside the grains and along the prior austenite grain boundaries has larger size than the $M_{23}C_6$-type carbide, after creep for 8,354 h. Since Laves phase forms long after the $M_{23}C_6$-type carbide in the steel, it is implied that Laves phase must grow faster than $M_{23}C_6$-type carbide. After creep for 8,354 h, the mean size of Laves phase reaches about 0.3 μm.

Apart from the volume fraction and mean size, Laves phase also tends to cluster along the prior austenite grain boundaries. This clustering behaviour of Laves phase would greatly reduce the effect of Laves phase preventing the migration of the lath boundaries, and hence accelerate the lath widening and its evolution to subgrains.

In addition to the precipitation strengthening by $M_{23}C_6$-type carbides and MX-type carbonitrides, the solution strengthening from both tungsten and molybdenum atoms is an important strengthening mechanism for 9–12 %Cr ferritic/martensitic heat-resistant steels. However, formation of Laves phase is inevitable for this type of steels (Hasegawa et al. 2001). Since tungsten and molybdenum are the main elements of forming Laves phase, Fe_2W and Fe_2Mo, the formation of Laves phase would inevitably consume the dissolved tungsten and molybdenum atoms in the matrix and weaken solution strengthening.

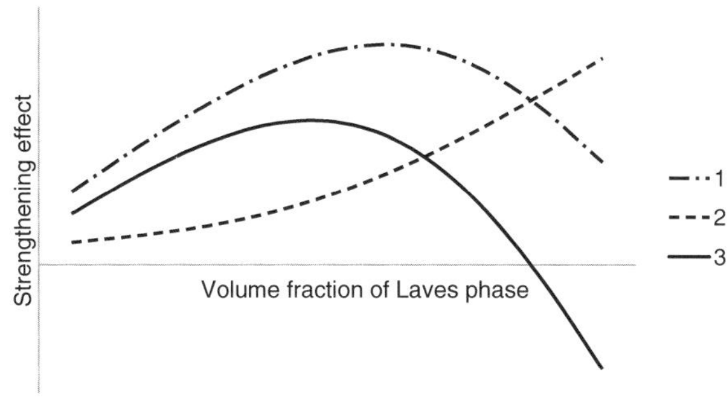

Fig. 4.2 Sketch diagram demonstrating the relationship between creep rupture strength and volume fraction of Laves phase. Curve 1: increase in creep rupture strength for precipitation strengthening effect caused by formation of Laves phase; curve 2: decrease in creep rupture strength for loss in solution strengthening caused by reduction in amount of solute atoms tungsten and molybdenum; curve 3: combined effect of curves 1 and 2

The two slope changes occurring on the creep rupture strength versus rupture time curve of the 10Cr steel as observed in Fig. 4.1 can also be well explained from the view of the precipitation behaviour of Laves phase. A sketch diagram qualitatively illustrating the relationship between the volume fractions of Laves phase and strengthening effect is presented in Fig. 4.2.

As mentioned above, on the one hand, formation of Laves phase consumes the solute atoms, tungsten and molybdenum, in the steel matrix, so the loss in solution strengthening from tungsten and molybdenum should be increased along with the increase of the volume fraction of Laves phase, as depicted by curve 2.

However, on the other hand, precipitation strengthening from Laves phase is not a monotonic increasing function of the volume fraction of Laves phase. It is affected by both the volume fraction and the mean size of Laves phase. The relationship between precipitation strengthening and the volume fraction of Laves phase could include two phases, as shown by curve 1 in Fig. 4.2. At the first stage, when the mean size of Laves phase is fine enough and can provide effective strengthening, before the maximum point of curve 1, precipitation strengthening from Laves phase is definitely increasing with the volume fraction. At the second stage, when Laves phase grows into large size, precipitation strengthening effect from Laves phase begins to decline.

The interaction between the solution strengthening and the precipitation strengthening can be demonstrated by combining curves 1 and 2 into curve 3, which illustrates the entire strengthening effect change along with the volume fraction of Laves phase. At the beginning, the creep rupture strength can be enhanced by the formation of fine Laves phase particles, arrived at the peak of curve 3, and then it starts to drop due to the fast growth of Laves phase into large size. Thus, the two slope changes on the creep rupture strength versus rupture time curve are qualitatively interpreted.

4.2.3 Effect of Stress on Microstructure Evolution

Figure 4.3 shows the Vickers hardness versus creep rupture strength of the 10Cr steel and the three sites selected for Vickers hardness measurement. Site A is located near the fracture surface. Site B is within the homogenous deformation part in the gauge section and about 10 mm away from the fracture surface. Site C is at the head portion. The gauge section and head portion are two parts with different states. The former was under the creep state while the latter could be considered to be under a simple ageing state.

The hardness at sites A and B decreases with the rupture life. However, site C shows a much smaller and slower hardness decrease, compared with site A or B, in the time range from 437 to 8,354 h. It can be deduced that the applied stress in gauge section accelerates the decline of hardness.

The local brightness of a back-scattered electron (BSE) image in SEM is in direct proportion to the average atomic number of the atoms in that region. The atomic number of tungsten, molybdenum and chromium are 74, 42 and 24, respectively. As a result of containing a large amount of tungsten and molybdenum atoms, Laves phase should have higher brightness in back-scattered electron images, which makes it easy to distinguish Laves phase from the $M_{23}C_6$-type carbides and MX-type carbonitrides.

In addition to back-scattered electron image, TEM diffraction pattern and energy dispersive X-ray spectrometry (EDS) analysis can distinguish Laves phase from $M_{23}C_6$-type carbide. However, opposite to the BSE image, the local

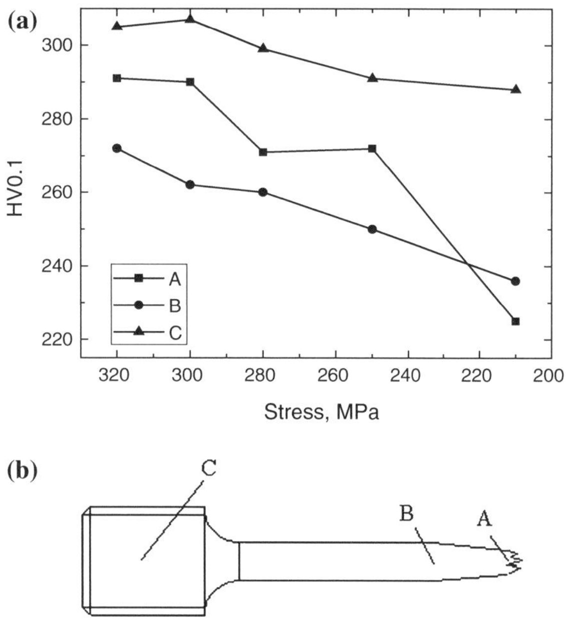

Fig. 4.3 Vickers hardness versus creep rupture strength of the 10Cr steel (**a**) and the sites for Vickers hardness measurement on the creep specimen (**b**). Reprinted from Hu et al. (2011), with permission from Elsevier

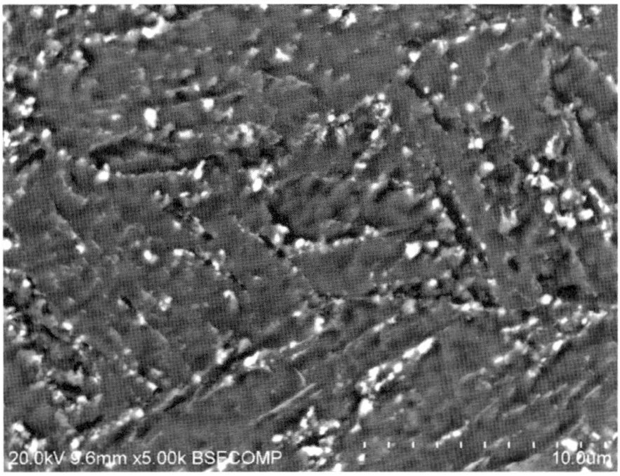

Fig. 4.4 SEM back-scattered electron micrograph of the 10Cr steel aged for 8,354 h. Reprinted from Hu et al. (2011) with permission from Elsevier

brightness of a TEM bright-field image is in inverse proportion to the atomic number of the atoms in that region, which gives Laves phase a darker appearance than $M_{23}C_6$-type carbide.

After creep for 8,354 h, the microstructure evolution at the head portion of specimen is slower than that at the gauge section (Figs. 4.4 and 4.5). Smaller amount of Laves phase forms in the ageing part than in the crept part. Moreover, the martensitic lath structure is retained and no transformation of lath into subgrains is found in the ageing part. Therefore, smaller hardness degradation happens at the head portion due to the slower microstructure evolution.

Sawada et al. (2001) found that, compared with the head portion, loading stress could accelerate the growth rate of MX carbonitrides in the gauge section of the specimen. They thought that the mobile dislocations could act as fast diffusion

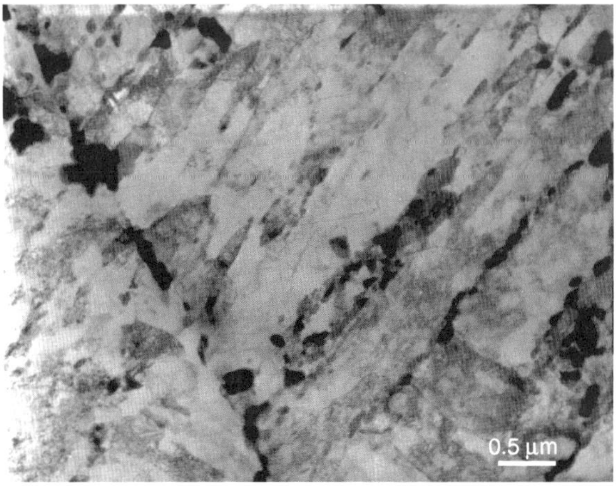

Fig. 4.5 TEM micrograph of the 10Cr steel aged for 8,354 h. Reprinted from Hu et al. (2011) with permission from Elsevier

pipes once they came across the MX carbonitrides. Fast diffusion of atoms led to the fast growth rate of these MX carbonitride precipitates. The applied stress can both increase the number of mobile dislocations and promote their movement, which will lead to two results:

(1) Due to the same pipe diffusion effect, Laves phase in the gauge section has large sizes and great amount;
(2) The laths would be much easier to transform into subgrains by dislocation migration.

In summary of Sects. 4.1 and 4.2, microstructure evolution takes place in 10Cr steel during creep, mainly exhibiting two aspects: (1) martensitic laths widening and evolving into subgrains, and (2) Laves phase growth and clustering along the prior austenite grain boundaries. The two slope changes on the creep rupture strength-life curve of the 10Cr steel are related to the precipitation behaviour of Laves phase. Stress accelerates the microstructure evolution.

4.3 Thermodynamic Calculations

The equilibrium state calculations show that no Laves phase should be found at 1,050 °C, the temperature for normalising. However, Laves phase should be present as an equilibrium phase at 760 °C, the temperature for tempering, and 600 °C, the temperature for creep. The equilibrium mass of Laves phase at 600 °C is around 1.6 %, larger than 0.6 % at 760 °C. It will take time for mass of Laves phase to reach the equilibrium. The volume percentage of Laves phase as a function of time at 760 °C and 600 °C is shown in Fig. 4.6. The Laves phase starts to form at 760 °C after a short incubation period of only 2,000 s, but its volume percentage can barely reach over 0.05 % even after tempering for 90 min. Therefore, it can be suggested that in the as tempered steel the dominant precipitates in the microstructure are MX and $M_{23}C_6$, and most of Laves phase in the steel should form during creep exposure at 600 °C. For the tempered steel under creep at 600 °C, Laves phase starts to precipitate after a long incubation period of almost 139 h, and then its volume percentage increases rapidly and reaches a plateau of about 1.6 % after nearly 8,334 h.

It is also illustrated in Fig. 4.6 that the formation of Laves phase can be clearly separated into two stages. One is the nucleation stage, which is characterised by the incubation period; the other is the growth stage, during which the volume percentage increases rapidly with time. When the growth of Laves phase is over, the coarsening stage begins. However, the volume percentage of Laves phase will keep constant in the coarsening period.

It is hinted by this calculation that Laves phase at the creep rupture are still at the growth stage since the longest rupture life is 8,354 h. Therefore, in the following, Laves phase will be discussed at the growth stage.

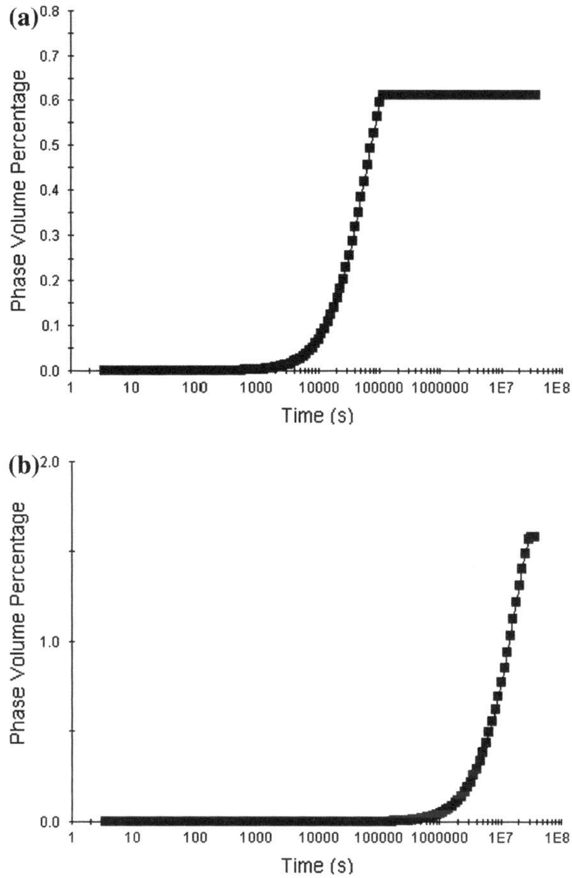

Fig. 4.6 Diagrams showing change of Laves phase volume percentage with time at (**a**) 760 °C and (**b**) 600 °C. With kind permission from Springer Science+Business Media: Frontiers of Materials Science in China, Hu et al. (2009)

4.4 Initial Microstructure and Growth of Laves Phase During Creep

Figure 4.7 is an optical micrograph of the initial microstructure of the tempered martensite. The prior austenite grain boundaries and lath boundaries are outlined by particles of $M_{23}C_6$ and MX precipitated during tempering. The δ-ferrite phase which often forms in the high Cr martensitic steels and is adverse to the creep rupture strength is scarcely found in this steel. The prior austenite grain size of the steel is about 15 μm.

Figure 4.8 shows the relationship between the applied stress and the rupture life in double logarithmic plot. In such a plot, the creep rupture strength linearly decreases as the rupture life extended. Experimentally, it was found that the amount of Laves phase increases with the endurance life from 430 to 3,230 h (Hu et al. 2009), consistent with the above prediction shown in Fig. 4.6b by thermodynamic calculations.

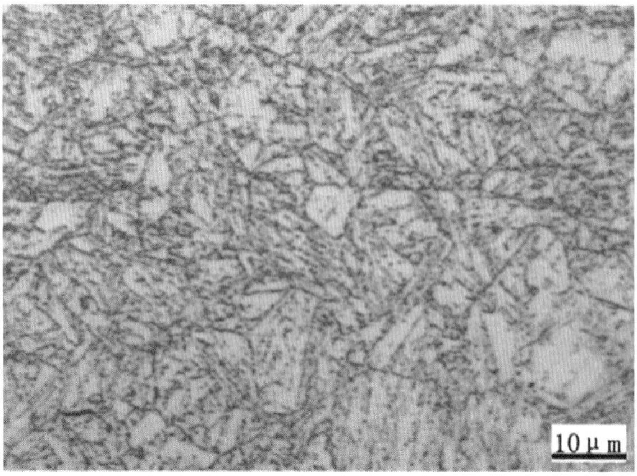

Fig. 4.7 Optical micrograph of the initial microstructure of the steel. With kind permission from Springer Science+Business Media: Frontiers of Materials Science in China, Hu et al. (2009)

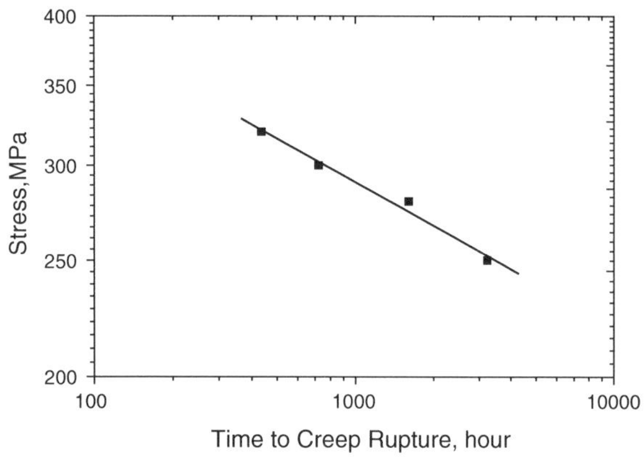

Fig. 4.8 The endurance curve of the steel in double logarithmic coordinates. With kind permission from Springer Science+Business Media: Frontiers of Materials Science in China, Hu et al. (2009)

Li (2006) pointed out that Laves phase would nucleate at martensite lath boundaries at first, keeping coherent with a lath, and then grow into the adjacent lath without a rational orientation. This morphology characteristic is helpful for identifying the Laves phase. In terms of morphology, there are three types of Laves phase: fine plate-shape and coarse rectangular shape Laves phase at lath boundaries, as well as coarse block-shape Laves phase in the matrix. According to Li (2006), the former two types of Laves phase should have a coherent orientation relationship with the adjacent lath. According to the growth process of Laves phase in the steel through the TEM observation with rupture life from 436 to 3,230 h (Hu et al. 2009), at the early stage of creep exposure, Laves phase prefers to distribute along lath boundaries. At the later stage of creep exposure, an increasing number of fully grown Laves phase distributes randomly inside the laths or subgrains. The mean size of Laves phase is more than 0.2 μm after 3,230 h which was much larger than that of $M_{23}C_6$, about 50 nm, under the same condition, which indicates that Laves phase is much easier to grow into large size than $M_{23}C_6$.

4.5 Effects of Size on Creep Behaviour and Cobalt

As part of their evolution process, the mean width of the martensitic laths increases with the endurance life during creep and the lath is evolved into the subgrain structure in some places after 3,230 h of creep exposure. Lath widening is caused by the migration of lath boundaries by absorbing free dislocations. On the one hand, reduction of the dissolved tungsten and molybdenum atoms due to the formation of Laves phase during creep would decrease the pinning effect and accelerate the migration of dislocations. On the other hand, the coherent Laves phases at laths boundaries could inhabit the migration of lath boundaries and retard the coarsening of laths. However, the fine coherent Laves phases would grow into incoherent ones, which would greatly decrease the effect of pinning dislocations and lath boundaries. Therefore, at the early stage of creep exposure, the loss of solid solution strengthening by reduction of tungsten and molybdenum in matrix could be compensated by the precipitation strengthening of the fine coherent Laves phase, whereas, at the later stage, large incoherent Laves phase would fail in this compensation. This change in the effect of Laves phase on creep strength of the steel agrees with the results on P92 obtained by Lee et al. (2006).

It was also found by Lee et al. (2006) that the Laves phase with average size beyond 0.13 μm could trigger the ductile-to-brittle fracture mode transition, the major cause of breakdown of creep rupture strength. However, in this steel discussed in this section, the mean size of Laves phase exceeds this threshold after only 1,598 h creep exposure (Hu et al. 2009). In addition, the steel ruptured at 3,230 h under stress of 250 MPa exhibits ductile fracture characterised by dimples, as shown in Fig. 4.9. The dimple morphology on the creep rupture surfaces suggests ductile transgranular fracture. Therefore, there must be reasons other than the large Laves phase that should be responsible for the fracture mode transition from ductile transgranular to brittle intergranular. The coarsened Laves phases are the favoured nucleation sites for creep cavities. On the short-term creep exposure,

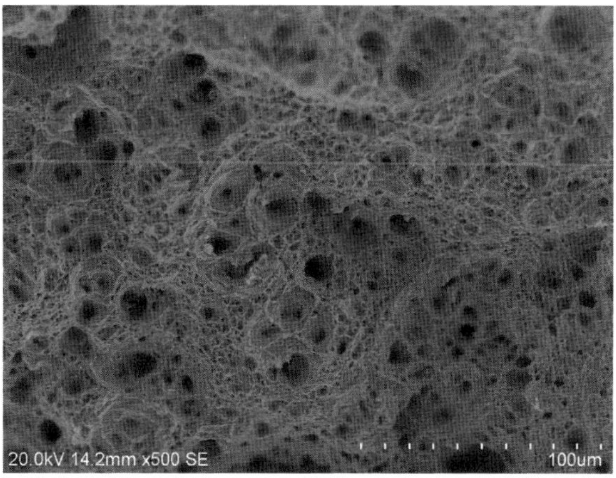

Fig. 4.9 Fracture surface of the steel ruptured at 3,230 h under 250 MPa. With kind permission from Springer Science+Business Media: Frontiers of Materials Science in China, Hu et al. (2009)

there are numbers of large Laves phase particles within grains. These large Laves phase precipitates within grains first generate a great number of creep cavities inside grains, which are then linked together through propagation of transgranular cracks before rupture, resulting in the ductile transgranular fracture. However, Laves phase would aggregate along the prior austenite gain boundaries during creep exposure, as shown in Fig. 4.10. If the mass of Laves phase in the microstructure is assumed to be constant after long-term creep exposure, the density of the coarsened Laves phase would increase along grain boundaries but decrease inside grains. The high density of the coarsened Laves phase along the prior austenite grain boundaries produced dense creep cavities along grain boundaries, which should be associated with the brittle intergranular fracture. In fact, Lee et al. (2006) observed that creep cavities were usually attached to large Laves phase precipitates on grain boundaries when brittle intergranular fracture took place, but they did not give an explanation why coarsened Laves phase particles on grain boundaries became the preferential nucleation sites for creep cavities.

The growth rate is controlled by diffusion. It is believed that cobalt can retard the diffusion of metal atoms in steels, because the diffusion coefficient is dependent on the Curie-temperature that is raised by cobalt addition. Various precipitation phenomena are retarded by cobalt. Therefore, the addition of cobalt is supposed to inhibit the growth of Laves phase. Hald (2008) found that the size of Laves phase reached nearly 0.1 μm after about 10,000 h creep exposure and almost stayed constant at 600 °C in P92 steel. Lee et al. (2006) also found that the size of Laves phase increased to nearly 0.13 μm after about 25,000 h creep-exposure at 600 °C in P92 steel. However, the average size of Laves phase has arrived at about 0.2 μm after only 1,598 h in the present new steel. The fact that Laves phase shows faster growth rate in this steel with cobalt addition than that in P92 steel without cobalt conflicts with those results that cobalt had an effect of retarding precipitation.

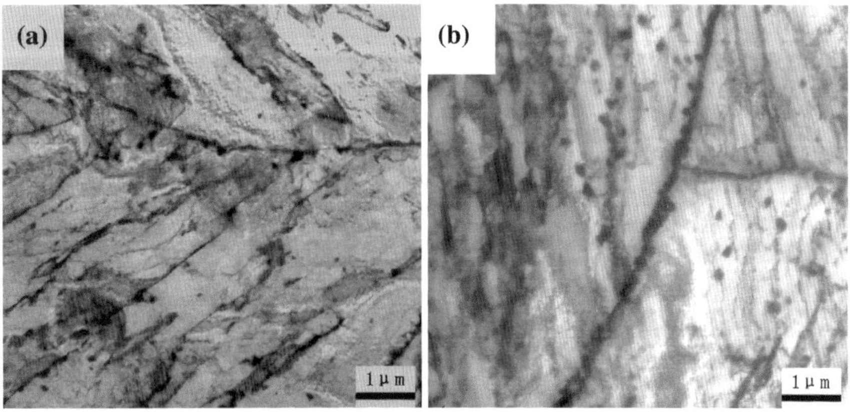

Fig. 4.10 TEM images showing the aggregation of Laves phase along prior austenite grain boundaries. **a** 436 h, 320 MPa; **b** 1,598 h, 280 MPa. With kind permission from Springer Science+Business Media: Frontiers of Materials Science in China, Hu et al. (2009)

Cobalt is also an important element added in maraging steels and there were also arguments on the influence of cobalt on their precipitate behaviour. Experiment on maraging steel suggests that cobalt could lower the solubility of molybdenum in matrix and promote the precipitation of Laves phase (Fe_2Mo). In the work on Fe-10 %Cr-6 %W alloys conducted by Cui et al. (2001), cobalt was found responsible for enhancement of growth of the Laves phase. Comparing the growth rate of Laves phase between P92 steel and the present new steel, it can be inferred that cobalt in this steel could accelerate rather than inhabit the growth of Laves phase, which may be explained by considering that cobalt can prevent the formation of δ-ferrite phase so that more tungsten and molybdenum can be kept in the matrix after normalisation, which could provide both more atoms needed for the growth of Laves phase and the driving force.

In summary of Sects. 4.3–4.5, in 10 % Cr heat-resistant steel containing cobalt, the Laves phase grows to a large size in a relatively short exposure time. With increase of the creep exposure time, Laves phase not only grows larger but also loses its coherent relationship with the adjacent lath, which could cause a loss of precipitation strengthening. In addition, the increasing number of Laves phase precipitates could decrease the solid solution strengthening effect due to the consumption of dissolved tungsten and molybdenum atoms. Decreases in both precipitation strengthening and solution strengthening effects should be the two reasons for degradation of creep rupture strength. The serious aggregation of large Laves phase precipitates along the prior austenite grain boundaries rather than the large size Laves phase precipitates should be considered responsible for the transition of ductile transgranular fracture to the brittle intergranular fracture. Addition of cobalt is believed to accelerate the growth rate of Laves phase.

4.6 Microstructural and Mechanical Properties of Short-Term Thermally Exposed Steels

The long-term creep rupture strength and microstructure stability of the 9/12Cr heat-resistant steels have been improved by an alloying and optimised heat-treatment procedure (Gustafson and Ågren 2001). The most recent improvements have been achieved by cobalt or tungsten alloying (Helis et al. 2009; Yamada et al. 2003). For example, it has been confirmed that cobalt is one of the important alloying elements to suppress the δ-ferrite formation during the high temperature normalising process in 9/12Cr heat-resistant steels. In contrast, it is also believed that addition of 2–3 % cobalt could drastically improve the short-term creep strength. Similarly, a lower creep rate and higher creep rupture strength in tungsten-containing heat-resistant steels can be attributed to the effects of both precipitation of Laves phase (Fe_2W) or μ phase (Fe_7W_6), as well as solid solution strengthening resulting from the addition of tungsten, which produces larger lattice misfit than molybdenum.

However, during thermal exposure, the steels containing cobalt or tungsten will undergo a microstructural change, leading to the degradation of their mechanical properties. It was found that the $M_{23}C_6$ and MX carbides coarsened, and Laves phase precipitated, during long-term exposure of a 12Cr steel containing tungsten (Kadoya et al. 2002). Most heat-resistant steels containing a high content of tungsten or cobalt show ductile-to-brittle transition with an increase of the rupture life, and it has been found that creep cavities are easily nucleated at coarse precipitates of Laves phase along grain boundaries. These findings suggest that, on the one hand, tungsten or cobalt can promote the precipitation during short-term thermal exposure, but, on the other hand, it can accelerate the coarsening of the precipitates during the long-term exposure process. When the coarsened Laves phase precipitates or carbides are greater than a critical size, the cavity formation is triggered and the consequent brittle intergranular fracture occurs (Lee et al. 2006; Abe 2004; Sawada et al. 2001). During the exposure process, the precipitated particles may crack or become detached from the matrix, and cavities may form in the microstructure. These might eventually affect the fracture mode of the ferritic steels during subsequent deformation, at ambient temperature.

Previous research has often focused on the mechanical properties and microstructure evolution of 9/12Cr heat-resistant steels after long-term exposure or under deformation creep. Few studies have discussed the change of microstructure under short-term thermal exposure. The study on microstructure evolution under short-term thermal exposure will play a complementary role in comprehensively understanding the effect of temperature and stress on the structural stability of 9/12Cr heat-resistant steels. In Sects. 4.7–4.10, based on the chemical composition of ASME-P92, through tungsten and cobalt additions, two ferritic/martensitic steels were designed and then fabricated at the laboratory scale. Room-temperature mechanical properties and microstructure evolution of the 9/12Cr heat-resistant steel during thermal exposure will be analysed. The aim of these sections is to identify the reason for the low impact toughness of the short-term thermally exposed 9/12Cr heat-resistant steels. The effect of tungsten and cobalt additions on microstructure evolution and fracture characteristics of 9/12Cr heat-resistant steels is also discussed.

4.7 Effect of Thermal Exposure on Mechanical Properties

The room-temperature mechanical properties of the thermally exposed (at 600 °C) steels (Table 4.1) are summarised in Fig. 4.11. Short-term thermal exposure at 600 °C has small influence on tensile properties, including yield stress (YS), ultimate tensile strength (UTS), elongation to fracture (A) and reduction in area (Z). The UTS of 10Cr steel increases after exposure at 600 °C for 500 h, and it tends to be stable with a subsequent increasing in exposure time. The UTS of P92 and 9Cr steels is lower than that of 10Cr steel before and after exposure. It is assumed that the difference in UTS for different steels should not be related to the

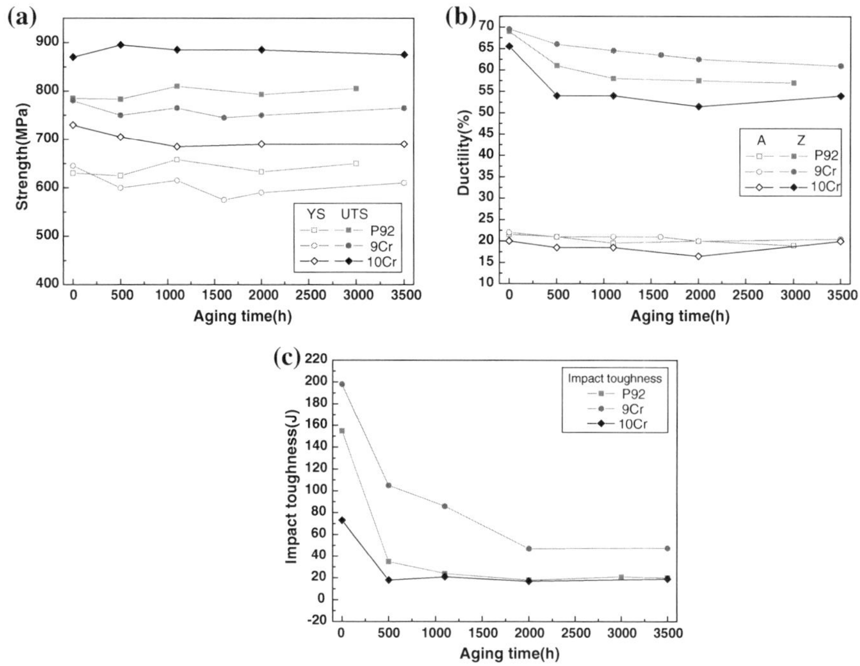

Fig. 4.11 Changes of (**a**, **b**) tensile and (**c**) impact properties with exposure time. Standard prismatic specimens ($10 \times 10 \times 55$ mm^3) with central V-shaped notch (2 mm in depth) were used for Charpy impact test. With kind permission from Springer Science+Business Media: Wang et al. (2012)

cobalt content but tungsten content. The reason for it can be analysed as follows. It is generally believed that 9/12Cr heat-resistant steels can be solution strengthened by additions of tungsten and cobalt, but for the current P92 and 9Cr steels, the addition of cobalt in 9Cr steel does not lead to increase of UTS. The 9Cr steel has 1.64 % cobalt, compared with the zero cobalt in the P92 (Table 4.1), but the UTS of the 9Cr steel is lower (Fig. 4.11a). If cobalt had a strengthening effect, then the 9Cr steel should have higher strength. Only when the content of tungsten is increased to 2.55 %, in 10Cr steel, is an increase approximately 100 MPa in UTS achieved.

Therefore, the effect of thermal exposure on tensile properties of the three steels is not significant. However, the influence of thermal ageing on the impact toughness is remarkable, as shown in Fig. 4.11c. The impact toughness of all three steels decreases dramatically when the exposure time is 500 h, and then the toughness tends to be stable with prolonged exposure time. Among the three steels, the 9Cr steel shows higher impact toughness than the other two steels. Thus, the impact data support the well known observation that most materials within a given microstructural class exhibit higher toughness with a decrease in strength (Chap. 7; Kim et al. 2008).

4.8 Effect of Thermal Exposure on Fracture Characteristics

Overall views on the impact fracture surfaces of the P92 steel after thermal exposure at 600 °C for different times are shown in Fig. 4.12. The fracture mode changes from ductile-to-brittle with increasing exposure time. The fracture mode of the tempered steel is characterised by transgranular ductile dimple tearing resulting from the coalescence of microvoids, but when the exposure time is 500 h, 100 % cleavage fracture is triggered. Especially, when the exposure time is extended to 3,000 h, an intergranular fracture characteristic is found near the cracking initiation site, as shown in Fig. 4.12c.

Figure 4.13 gives general views on the impact fracture surfaces of the 9Cr steel before and after thermal exposure. The 9Cr steel before thermal exposure shows the same fracture mechanism as the P92 steel. However, after thermally exposed for 500 h, the impact fracture surface is composed of both ductile zone and brittle zone. The area of the brittle zone increases with increasing exposure time. When the exposure time is 3,000 h, the entire fracture surface is brittle quasi-cleavage.

The fracture morphology of the 10Cr steel is different from the other two steels. Before thermal exposure, the fracture mode of 10Cr steel shows mixed rupture characteristics of brittle quasi-cleavage and dimples. After thermal exposure, it changes to completely brittle quasi-cleavage, as shown in Fig. 4.14.

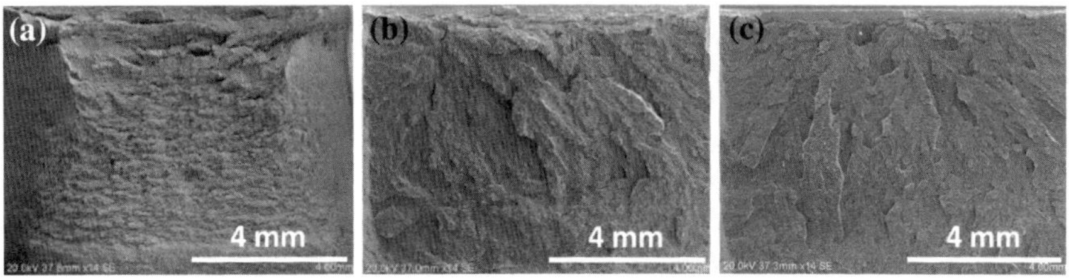

Fig. 4.12 SEM fractographs of impact specimens of the P92 steel after exposure for different periods of time. **a** 0 h; **b** 500 h; **c** 3,000 h. With kind permission from Springer Science+Business Media: Wang et al. (2012)

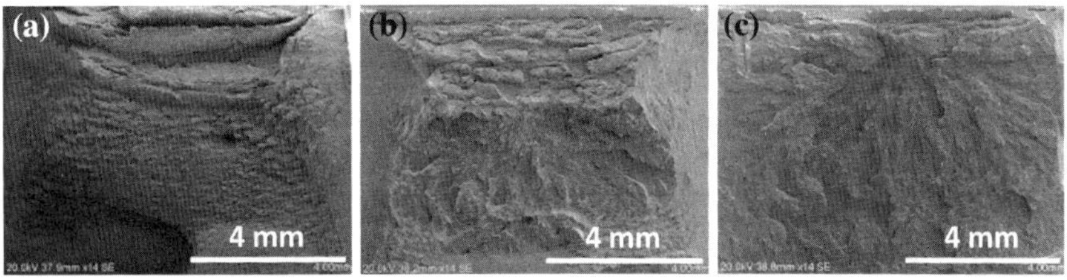

Fig. 4.13 SEM fractographs of impact specimens of the 9Cr steel after exposure for different periods of time. **a** 0 h; **b** 500 h; **c** 3,000 h. With kind permission from Springer Science+Business Media: Wang et al. (2012)

Fig. 4.14 SEM fractographs of impact specimens of the 10Cr steel after exposure for different periods of time. **a** 0 h; **b** 500 h; **c** 3,000 h. With kind permission from Springer Science+Business Media: Wang et al. (2012)

4.9 Effect of Thermal Exposure on Microstructure

During the thermal exposure, in contrast to tensile properties of the three steels, the impact toughness and the fracture mode exhibit considerable variations. Microstructural evolution under thermal exposure should be the main reason.

Figures 4.15a–c and 4.15d–f show SEM images of the three steels after thermal exposure for 500 and 3,000 h, respectively. Large amounts of fine precipitates distribute at the grain boundaries of prior austenite, martensite laths, and interior of the laths. The precipitates are carbide enriched with iron, chromium and tungsten. Also, nanosize MX-type carbonitrides form inside the martensite lath matrix. Helis et al. (2009), Kadoya et al. (2002), Lee et al. (2006), Abe (2004), Sawada et al. (2001) and Blach et al. (2009) found that additions of tungsten and cobalt could accelerate the growth of both carbide and MX carbonitride during thermal exposure, but these results came from the conditions of long-term exposure process or creep process. Here, the size of the carbide and MX carbonitride precipitates does not increase significantly during short-term thermal exposure process.

Ghassemi-Armaki et al. (2009) found that the coarsening rate of martensite lath was low during the exposure process. The three steels here also keep the structure of martensite as tempered after thermal exposure for 3,000 h, as shown in Fig. 4.16. These micrographs show little change in the martensitic structure of the tempered steels. The chemical compositions of the precipitates observed at the martensite interfaces are discussed subsequently in this section. Regarding the crystal structure of the precipitates, we will show a diffraction pattern of Laves-phase precipitates in Sect. 4.10. Their role on strengthening will also be discussed in Sect. 4.10. Without the effect of stress during the thermal exposure process, the carbide and MX could effectively retard the movement of dislocations at lath boundary or lath interior. Even after 3,000 h of thermal exposure, no obvious recovery could be triggered.

If no coarsening of carbide and MX happens and the recovery mechanism also is excluded, then what is the main reason for the difference of impact toughness after thermal exposure? In addition to carbide and MX, a change of another precipitate, Laves phase, should not be neglected (Kadoya et al. 2002;

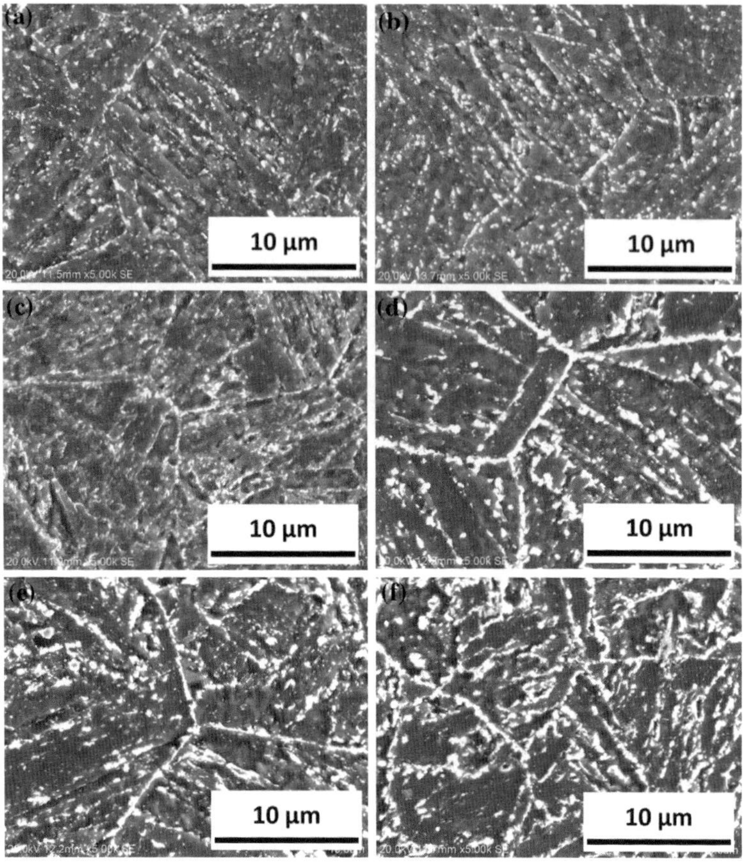

Fig. 4.15 SEM images of the microstructures after thermal exposure for 500 h and 3,000 h. **a** 500 h, P92 steel; **b** 500 h, 9Cr steel; **c** 500 h, 10Cr steel; **d** 3,000 h, P92 steel; **e** 3,000 h, 9Cr steel; **f** 3,000 h, 10Cr steel. With kind permission from Springer Science+Business Media: Wang et al. (2012)

Lee et al. 2006; Cui et al. 2001; Thomas Paul et al. 2008). Figure 4.17 shows the SEM BSE images of the three steels after thermal exposure for 500 and 3,000 h. As the brightness of SEM BSE image of a precipitate is determined by the average atomic weight of the compound, $M_{23}C_6$ type carbide should not be bright particles because its average atomic weight is lower than that of the matrix. In contrast, although ultrafine particles such as (V,Nb)(N,C) can be resolved in the SEM BSE images, the size of the precipitates does not increase significantly. As a result, Laves phase (Fe_2W) with bright contrast in Fig. 4.17 can be distinguished easily. The bigger particles with bright contrast in Fig. 4.17 contain tungsten element, which should be the Laves phase in the steels. With increasing exposure time, the number, volume fraction and size of Laves-phase precipitates increase.

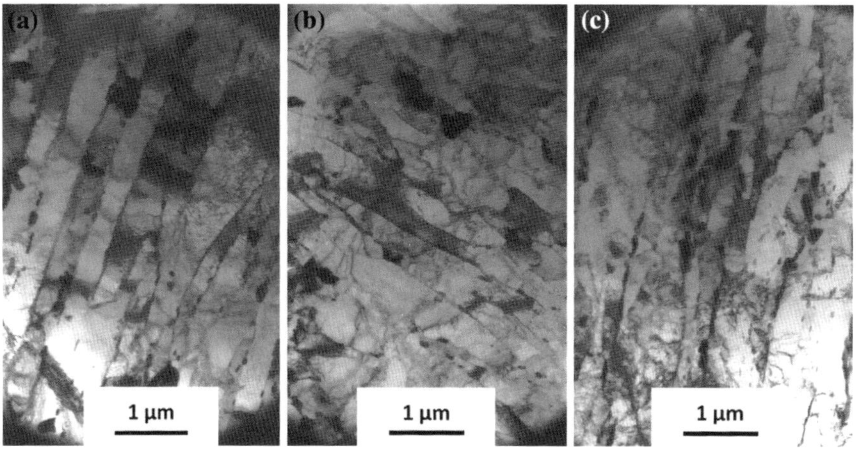

Fig. 4.16 TEM micrographs of martensitic lath structures after thermal exposure for 3,000 h, showing little change in the martensitic structure of the tempered steels. **a** P92 steel; **b** 9Cr steel; **c** 10Cr steel. With kind permission from Springer Science+Business Media: Wang et al. (2012)

Meanwhile, the EDS analysis in Fig. 4.18 show the different compositions of Laves-phase precipitates in P92 steel, 9Cr steel and 10Cr steel after 3,000 h thermal exposure. Microstructure characterisation in the form of the EDS spectra of the precipitates indicates the enrichment of tungsten in them. It should be noted that, in the EDS analysis in connection with an SEM, the spatial resolution is of the order of 1 μm. As these precipitates are much smaller than this size, the EDS data are not the actual precipitate compositions but are compositions of the precipitate and the surrounding matrix, with a total volume of the order of 1 μm^3. The precipitate under analysis will only be a small fraction of this volume, and the magnitude of this small fraction varies from one precipitate to another, depending on the exact size of the precipitate being measured. Therefore, the variation of the measurement results of different precipitates is dominantly controlled by this factor, and not the statistical factors. For this reason, the statistical errors would be much smaller than the errors caused by the inherent spatial resolution of SEM–EDS, when measuring composition of particles much small than this resolution. The contents of tungsten and likely molybdenum in Laves-phase precipitates in the three steels are underestimated, because of the contribution from the matrix surrounding the small precipitates. An increase of tungsten or cobalt content in steel can promote the precipitation of Laves phase, and accelerate the incorporation of tungsten in Laves phase. Other than the newly formed Laves phase the number and size of the original precipitates, i.e. carbide and MX in the three steels almost remain the same as in the tempered state, even after exposure for 3,000 h. Therefore, the remarkable decrease of impact toughness should be attributed to the precipitation and growth of Laves phase in the steels.

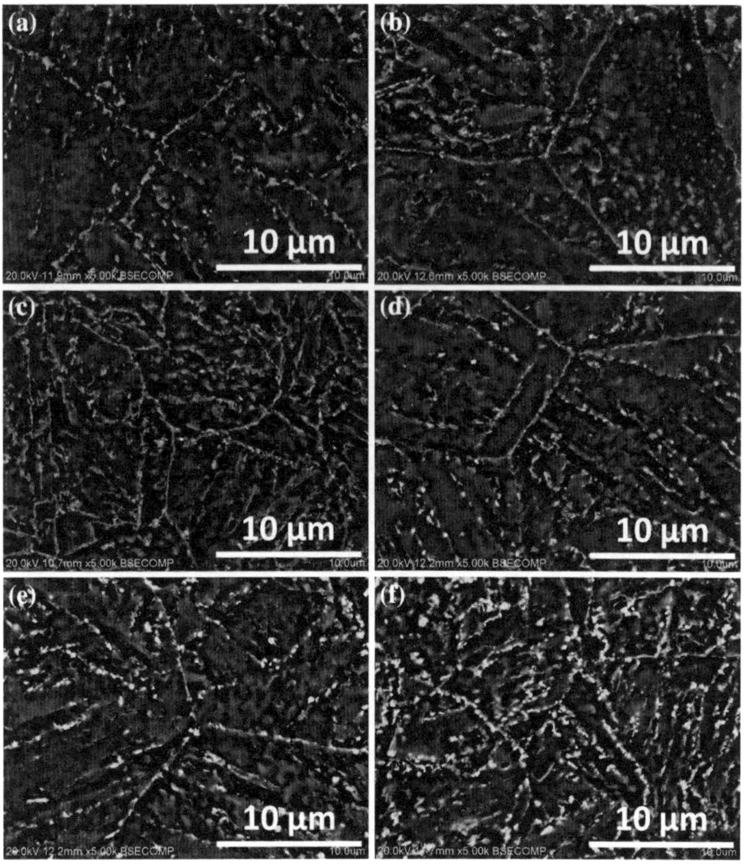

Fig. 4.17 SEM BSE images of the steels after thermal exposure for 500 h and 3000 h. The bright particles in the figures are Laves phase. **a** 500 h, P92 steel; **b** 500 h, 9Cr steel; **c** 500 h, 10Cr steel; **d** 3,000 h, P92 steel; **e** 3,000 h, 9Cr steel; **f** 3,000 h, 10Cr steel. With kind permission from Springer Science+Business Media: Wang et al. (2012)

4.10 Effect of Microstructure Evolution on Mechanical Properties

During short-term thermal exposure process, the effect of cobalt and tungsten on the size and distribution of carbide and MX could be neglected, and the difference of mechanical properties could be mainly caused by the precipitation of Laves phase in the steel. However, the evolution process of Laves phase precipitation is affected by cobalt and tungsten in the steels.

There were disagreements on the role of cobalt and tungsten in the microstructural evolution of 9/12Cr steels. With the addition of tungsten, the diffusion coefficient of the steel matrix would be decreased, the dislocations movement

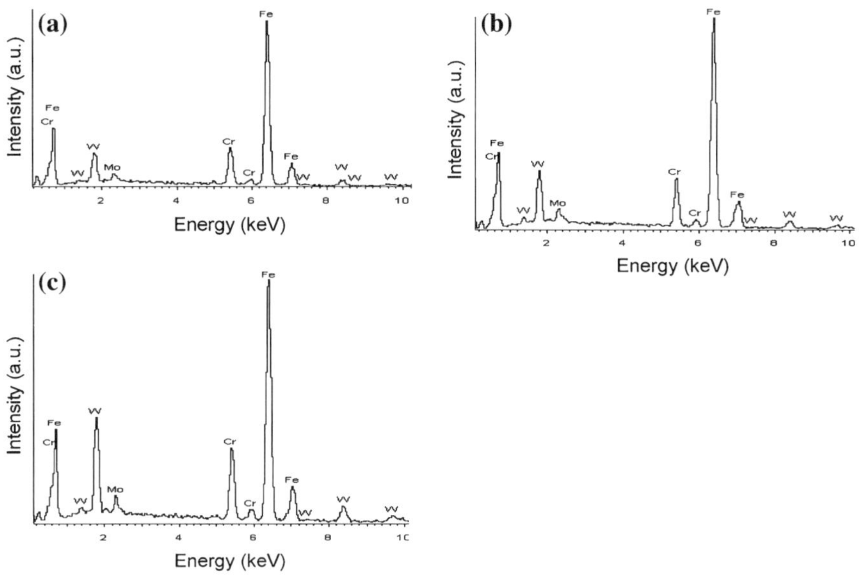

Fig. 4.18 SEM-EDS analysis of Laves phase in the microstructures after ageing for 3,000 h. **a** P92; **b** 9Cr; **c** 10Cr. With kind permission from Springer Science+Business Media: Wang et al. (2012)

and coarsening of carbide and recovery of martensite could be retarded, and eventually, the creep rupture strengths of the steels were increased (Abe 2004; Hasegawa et al. 2001). However, after long-term thermal exposure, the existence of tungsten would promote the formation and growth of Laves-phase precipitates in the steels, which would damage the impact property of the steels and the solution-strengthening effect of tungsten (Fernández et al. 2002). Especially when cobalt and tungsten are together added in 9/12Cr steels, the synergistic effect of the two elements on the microstructure and mechanical properties can be more remarkable. On the one hand, cobalt can retard the diffusion of metal atoms in steels because the diffusion coefficient of metal atoms is dependent on the Curie-temperature that is raised by cobalt addition. Therefore, the addition of cobalt can inhibit the coarsening of Laves-phase precipitates in the steels. On the other hand, the addition of cobalt can promote the precipitation of tungsten-containing compound from the steel matrix. It is well known that cobalt is an important element added in maraging steels because cobalt can decrease the solubility of molybdenum in the martensite matrix (He et al. 2002). Then, there should be more molybdenum taking part in the ageing reaction to strengthen the maraging steel, which is called the synergistic effect of cobalt and molybdenum in such type of steel. Regarding 9/12Cr steels, a synergistic effect of cobalt and tungsten is possible, as tungsten is similar to molybdenum in some ways.

As to the 10Cr steel with higher tungsten and cobalt contents, the driving force for the nucleation of Laves phase should be increased, and a large amount of Laves phase could precipitate from the steel matrix. As shown in Fig. 4.17, the amount of Laves-phase precipitates in the 10Cr steel is much larger than that in 9Cr and P92 steels, for both times of exposure, 500 and 3,000 h. As a result, 10Cr steel has a higher tensile strength. However, why does 10Cr steel show lower impact toughness compared with the other two steels? It is believed that the incoherent Laves phase would trigger the fracture mode transition from ductile-to-brittle, as shown in Figs. 4.12, 4.13, 4.14. Lee et al. (2006) found that when the average size of Laves-phase precipitates exceeded 0.13 μm, the fine coherent Laves-phase precipitates would grow into the incoherent ones, which would greatly decrease their pinning effect on dislocations movement and lath boundaries. As shown in Fig. 4.17, with increasing exposure time and contents of tungsten and cobalt, more and more Laves-phase precipitates clustered along grain boundaries. Figure 4.19 shows the morphology of Laves phase precipitates in the 10Cr steel after exposure for 3,000 h, with size exceeding 0.1 μm. The large size and high density of Laves-phase precipitates along the grain boundaries would produce cavities at first and then lead to the brittle intergranular fracture.

It should be noted that Sects. 4.6–4.10 are not about creep strength, i.e. strength at the elevated temperature, but rather, they investigate the room-temperature strength and impact toughness after exposure at the elevated temperature. None of the discussions involve loading at the elevated temperature. Good ductility but poor impact toughness is an interesting phenomenon after the short-ageing process of these steels.

In summarising, Sects. 4.6–4.10 present the evolution of microstructure and mechanical properties of different 9/12Cr heat-resistant steels after short-term

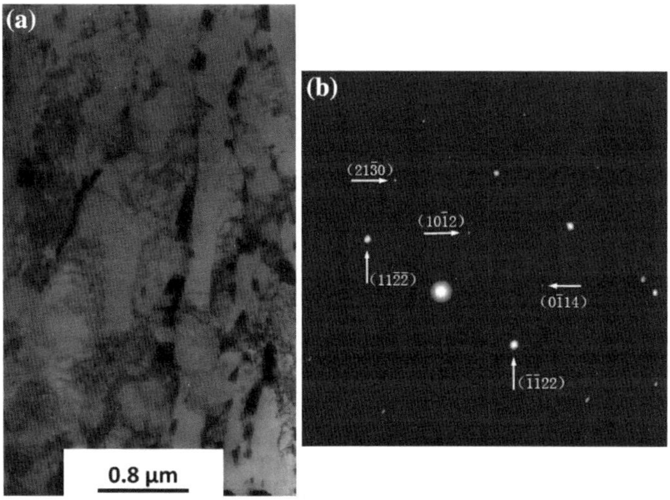

Fig. 4.19 Laves phase in the 10Cr steel thermally exposed for 3,000 h. **a** TEM bright-field image of coarse cubic Laves phase precipitates distributed at the lath boundaries, as well as coarse block Laves phase in the steel matrix; **b** diffraction pattern. With kind permission from Springer Science+Business Media: Wang et al. (2012)

thermal exposure. The tensile strength of 10Cr steel with higher cobalt and tungsten contents is higher than that of 9Cr with cobalt addition and commercial P92 steels, which have lower contents of cobalt and tungsten, but its impact toughness is lower than the other two steels. During short-term thermal exposure process up to 3,000 h, the effect of cobalt and tungsten on the size and distribution of carbide and MX can be neglected. The change of mechanical properties is mainly caused by the precipitation of Laves phase in the steels. On short-term thermal exposure, with increase of cobalt and tungsten contents, cobalt could promote the segregation of tungsten along the martensite lath to form Laves-phase precipitates. A large size and high density of Laves-phase precipitates along the grain boundaries could lead to the brittle intergranular fracture of the steels.

References

Abe F (2004) Coarsening behavior of lath and its effect on creep rates in tempered martensitic 9Cr–W steels. Mater Sci Eng A 387–389:565–569. doi:10.1016/j.msea.2004.01.057

Abe F, Semba H, Sakuraya T (2007) Effect of boron on microstructure and creep deformation behavior of tempered martensitic 9Cr steel. Mater Sci Forum 539–543:2982–2987. doi:10.4028/www.scientific.net/MSF.539-543.2982

Blach J, Falat L, Ševc P (2009) Fracture characteristics of thermally exposed 9Cr-1Mo steel after tensile and impact testing at room temperature. Eng Fail Anal 16:1397–1403. doi:10.1016/j.engfailanal.2008.09.003

Cui J, Kim IS, Kang CY, Miyahara K (2001) Creep stress effect on the precipitation behavior of Laves phase in Fe-10 % Cr-6 % W alloys. ISIJ Int 41:368–371. doi:10.2355/isijinternational.41.368

Fernández P, Hernández-Mayoral M, Lapeña J, Lancha AM, De Diego G (2002) Correlation between microstructure and mechanical properties of reduced activation modified F-82H ferritic martensitic steel. Mater Sci Technol 18:1353–1362. doi:10.1179/026708302225007411

Ghassemi-Armaki H, Chen RP, Maruyama K, Yoshizawa M, Igarashi M (2009) Static recovery of tempered lath martensite microstructures during long-term aging in 9–12 % Cr heat resistant steels. Mater Lett 63:2423–2425. doi:10.1016/j.matlet.2009.08.024

Gustafson Å, Ågren J (2001) Possible effect of Co on coarsening of $M_{23}C_6$ carbide and Orowan stress in a 9 % Cr steel. ISIJ Int 41:356–360. doi:10.2355/isijinternational.41.356

Hald J (2008) Microstructure and long-term creep properties of 9–12 % Cr steels. Int J Pres Ves Pip 85:30–37. doi:10.1016/j.ijpvp.2007.06.010

Hasegawa T, Abe YR, Tomita Y, Maruyama N, Sugiyama M (2001) Microstructural evolution during creep test in 9Cr–2W–V–Ta steels and 9Cr–1Mo–V–Nb steels. ISIJ Int 41:922–929. doi:10.2355/isijinternational.41.922

He Y, Yang K, Qu W, Kong F, Su G (2002) Strengthening and toughing of a 2800-MPa grade maraging steel. Mater Lett 56:763–769. doi:10.1016/S0167-577X(02)00610-9

Helis L, Toda Y, Hara T, Miyazaki H, Abe F (2009) Effect of cobalt on the microstructure of tempered martensitic 9Cr steel for ultra-supercritical power plants. Mater Sci Eng A 510–511:88–94. doi:10.1016/j.msea.2008.04.131

Hu P, Yan W, Sha W, Wang W, Guo Z, Shan Y, Yang K (2009) Study on laves phase in an advanced heat-resistant steel. Front Mater Sci Chin 3:434–441. doi:10.1007/s11706-009-0063-7

Hu P, Yan W, Sha W, Wang W, Shan Y, Yang K (2011) Microstructure evolution of a 10Cr heat-resistant steel during high temperature creep. J Mater Sci Technol 27:344–351. doi:10.1016/S1005-0302(11)60072-8

Kadoya Y, Dyson BF, McLean M (2002) Microstructural stability during creep of Mo- or W-bearing 12Cr steels. Metall Mater Trans A 33A:2549–2557. doi:10.1007/s11661-002-0375-z

Kim BC, Park SW, Lee DG (2008) Fracture toughness of the nano-particle reinforced epoxy composite. Compos Struct 86:69–77. doi:10.1016/j.compstruct.2008.03.005

Lee JS, Armaki HG, Maruyama K, Maruki T, Asahi H (2006) Causes of breakdown of creep strength in 9Cr-1.8W-0.5Mo-VNb steel. Mater Sci Eng A 428:270–275. doi:10.1016/j.msea.2006.05.010

Li Q (2006) Precipitation of Fe_2W laves phase and modeling of its direct influence on the strength of a 12Cr-2W steel. Metall Mater Trans A 37A:89–97. doi:10.1007/s11661-006-0155-2

Maruyama K, Sawada K, Koike J (2001) Strengthening mechanisms of creep resistant tempered martensitic steel. ISIJ Int 41:641–653. doi:10.2355/isijinternational.41.641

Sawada K, Kubo K, Abe F (2001) Creep behavior and stability of MX precipitates at high temperature in 9Cr-0.5Mo-1.8W-VNb steel. Mater Sci Eng A 319–321:784–787. doi:10.1016/S0921-5093(01)00973-X

Thomas Paul V, Saroja S, Vijayalakshmi M (2008) Microstructural stability of modified 9Cr-1Mo steel during long term exposures at elevated temperatures. J Nucl Mater 378:273–281. doi:10.1016/j.jnucmat.2008.06.033

Yamada K, Igarashi M, Muneki S, Abe F (2003) Effect of Co addition on microstructure in high Cr ferritic steels. ISIJ Int 43:1438–1443. doi:10.2355/isijinternational.43.1438

Wang W, Yan W, Sha W, Shan Y, Yang K (2012) Microstructural evolution and mechanical properties of short-term thermally exposed 9/12Cr heat-resistant steels. Metall Mater Trans A 43 A, pp 4113–4122. doi:10.1007/s11661-012-1240-3

Chapter 5
Nitride-Strengthened Ferritic/Martensitic Steel

Abstract Nitride-strengthened reduced activation ferritic/martensitic (RAFM) steel is anticipated to have higher creep strength because of the remarkable thermal stability of nitrides. Such steels with different manganese contents are designed based on the chemical composition of Eurofer 97 steel but the carbon content is reduced to an extremely low level. The larger amount of vanadium-rich nitrides and more dissolved chromium in the matrix could be responsible for the strength being similar to Eurofer 97 steel. The steels have the microstructure of full martensite with fine nitrides dispersed homogeneously in the matrix and display extremely high strength but poor toughness. Compared with the steel with low carbon content (0.005 % in wt%), the steel with high carbon content (0.012 % in wt%) has not only the higher strength but also the higher impact toughness and grain-coarsening temperature. The complicated Al_2O_3 inclusions are responsible for the initiated cleavage fracture by acting as the critical cracks. The final part of the chapter is concerned with conventional nitride-strengthened martensitic heat resistant steel. It was surprising to reveal that the half-size impact energy greatly increases from several Joules to nearly 100 J, when the tempering temperature is increased from 650 to 750 °C.

5.1 Microstructure, Nitride Precipitation, Hardness and Effect of Tempering Temperature

The structural materials in future fusion reactor need to withstand high temperature under long-term loading and irradiation (Li et al. 2010). Currently, reduced activation ferritic/martensitic (RAFM) steels are the leading candidate structural materials for fusion reactors due to the good thermal properties and superior swelling resistance compared with austenitic stainless steel.

The typical microstructure of the RAFM steels is the tempered martensite with a large number of precipitates (Tanigawa et al. 2011). There are two main types of precipitates in RAFM steels, $M_{23}C_6$ (M is for Cr, Fe, W, etc.) and MX (M is for V, Ta, etc. and X is for C, N). However, it has been proved that the coarsening rates of these two kinds of precipitates are different during creep and the coarsening rate of $M_{23}C_6$ carbide is much higher than that of the MX-type nitride (Sawada et al. 2001). It has been recognised that one of the effective methods of enhancing creep

W. Sha, *Steels*, DOI: 10.1007/978-1-4471-4872-2_5,

strength is to achieve microstructure with good thermal stability by using thermally stable precipitates (Sklenička et al. 2003). Taneike et al. (2004) reported that when the carbon content was decreased to a very low level, the time before rupture was significantly increased in the heat-resistant steel for power plant. It could be attributed to the elimination of thermally unstable $M_{23}C_6$ carbides. Therefore, strengthening the RAFM steels with only nitrides which have excellent thermal stability can extend the creep life.

In the microstructure and composition design of nitride-strengthened RAFM steels, instead of cobalt or tungsten, manganese was employed to suppress δ-ferrite more prone to form due to the carbon elimination, taking into account of the reduced activation property and the formation of Laves phase Fe_2W. Fully martensitic microstructure could be obtained by 3 % manganese addition. However, the high manganese content leads to such severe problems as the low A_{c1} temperature, manganese segregation along prior grain boundaries and many MnS inclusions. These effects are so detrimental that the nitride-strengthened RAFM steels of high manganese content show poor performance (Hu et al. 2010). Fortunately, after decreasing manganese content to 1.4 %, the nitride-strengthened RAFM steels now can be tempered at temperature as high as 750 °C and thus show excellent toughness. This chapter presents the microstructure and mechanical properties of the optimised nitride-strengthened RAFM steel. A more thorough review of the much work published on ferritic/martensitic steels, including the low-activation variety, is given in Sect. 1.4.1. To make a ferritic/martensitic steel 'low activation', both molybdenum and niobium are removed and replaced by tungsten and tantalum. Cobalt should be avoided, too. The new point in the present chapter is the new steel compositions designed. Such steel compositions have never been systematically examined before.

A full martensite microstructure cannot be produced in NS1 steel (Table 5.1) by either air-cooling or water-cooling (Hu et al. 2010). A large amount of δ-ferrite is inevitable in the microstructure of NS1 steel. The grain size of NS1 steel heat treated at 1200 °C is larger than that at 1050 °C. It is known that the Vickers hardness of martensite is higher than that of δ-ferrite. Therefore, Vickers hardness can be used to distinguish microstructures with different hardness. For NS1 steel heat-treated at 1200 °C, two values of Vickers hardness, about HV170 and HV300, are obtained as shown in Fig. 5.1, suggesting that two different phases, δ-ferrite and martensite, coexisted in the microstructure.

A large number of fine nitrides with average size of 50 nm are homogeneously distributed in the martensite (Hu et al. 2010). Most nitrides are cubic in shape,

Table 5.1 Compositions of low activation heat resistant steels (wt%)

Steel	C	Si	Mn	Cr	W	V	Ta	N
NS1	0.006	0.16	0.47	9.11	1.47	0.21	0.11	0.043
NS2	0.005	0.05	3.73	9.06	1.49	0.15	0.12	0.039
NS3	0.012	0.06	4.00	8.92	1.49	0.15	0.13	0.042
NS4	0.003	–	1.40	8.72	1.55	0.15	0.09	0.039

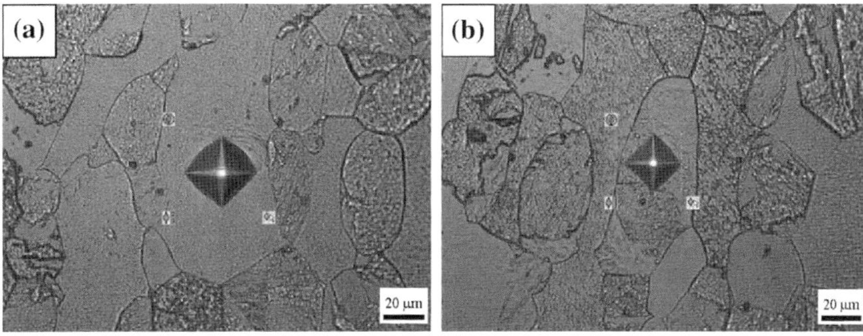

Fig. 5.1 The indents on the different phases. **a** A δ-ferrite grain with HV174 and **b** a martensite grain with HV291. Reprinted from Hu et al. (2010) with permission from Elsevier

although some are acicular. However, the nitrides in δ-ferrite are clustered and all acicular. The δ-ferrites with two different shapes are in the microstructure, blocky δ-ferrite and elongated δ-ferrite. The area of each δ-ferrite grain seems to be divided into two zones, the central zone and the edge zone. The nitride precipitates are concentrated in the central zone while fewer nitrides are in the edge zone. The width of this edge zone is between 0.5 and 1 μm.

Heterogeneous precipitation behaviours of $M_{23}C_6$, MX and Laves phase in δ-ferrite were also found by Yoshizawa and Igarashi (2007), Yamada et al. (2003) and Hu et al. (2009). Although the mechanism of the heterogeneous precipitation is still not clear, it is certain that δ-ferrite facilitates the heterogeneous precipitation. In NS1 steel, the martensite transformed from high-temperature austenite is rich in austenite forming elements and δ-ferrite retained from high temperature to room temperature is rich in ferrite forming elements such as vanadium. Therefore, the nitrides in δ-ferrite should be rich in vanadium. Yamada et al. (2003) found that localised precipitation of fine vanadium-enriched MX with high density took place within δ-ferrite. It is also revealed that the vanadium nitrides always precipitate on other pre-existed precipitates to make a vanadium-wing like shape. This should be the reason for the acicular shape of the nitrides in δ-ferrite. However, in martensite the nitrides will be mostly TaN together with VN. Therefore, the shape of most nitrides in martensite is cubic, although some are acicular.

As a high-temperature phase, δ-ferrite has not only the bcc structure with big interstitial space but also the larger crystal lattice, which benefits atom diffusion greatly. Regarding precipitation hardening, the edge zone with few nitrides would be the weak part during creep. Therefore, δ-ferrite should be eliminated from the microstructure in order to ensure high creep rupture strength.

Full martensitic microstructure is achieved in NS2, NS3 and NS4 steels (Table 5.1) after normalising at 980 °C, with a dispersion of fine MX type precipitates in the martensite (Yan et al. 2012). The lath width is about 0.6 μm, in the case of NS4 steel, as an example. The martensitic matrix has high-density dislocations, with few large $M_{23}C_6$ carbides along the grain boundaries or lath boundaries. In NS4 steel, the dislocations are dense when the steel is tempered at 650 and 700 °C, while the amount of them reduces when tempered at 750 °C. Also, there is no recrystallisation. Most precipitates are of the cubic shape, which identifies them as the MX type precipitates. In NS4 steel, the precipitate amount increases with the tempering temperature in the range of 650–750 °C. There are rare precipitates in the matrix when the steel is tempered at 650 °C. However, when the tempering temperature is increased to 700 °C, the number of precipitates greatly increases. Subsequently, when the tempering temperature is increased to 750 °C, the nanosize nitrides are observed to keep on increasing in quantity. Besides, the shape of the nitrides is sharper and clearer, which should indicate that the nitrides are becoming incoherent with the matrix.

Energy dispersive X-ray spectrometry (EDS) can be used in the transmission electron microscope, to analyse the composition of precipitates. The precipitates in the lower carbon NS2 steel, however, are finer. It is difficult to obtain selected-area electron diffraction patterns of precipitates, because of the strong magnetic nature of the steels, causing electron beam movement.

The nitrides interact with dislocations. Some dislocations are pinned strongly at two ends by the fine nitrides in the matrix, showing a bow-like shape. The end points of the dislocation bows are where the nitrides are located. The fine nitrides have a strong pinning effect on the dislocations.

Dislocation strengthening contributes a lot to the high strength (see Sect. 5.3). However, we observe small nitrides interacting with dislocations, as shown in Fig. 5.2. The nitrides in the steels are small, approximately ten nanometres, and they pin the dislocation movement effectively. Even in high temperature, the dislocations cannot move. Therefore, the strength at 600 °C is still high (see Sect. 5.3).

When the NS4 steel is tempered at 700 °C, the impact toughness at −20 °C is as poor as 3 J while when the tempering temperature is increased to 750 °C, the impact toughness at −20 °C is nearly 100 J. The nitride precipitation is far from complete at the tempering temperature of 700 °C, for if the nitride were completely precipitated when tempered at 700 °C the impact toughness should be as good as when tempered at 750 °C.

The precipitate particles can be identified by morphology. As to the chemical composition, measurement is very difficult due to the small size of nitrides.

The composition was designed taking consideration of avoiding Z phase. Therefore, the nitrogen content was strictly controlled to be under 0.05 %. In addition, the 9 % chromium content would not facilitate the formation of the Z phase before 10000 h creep. Hald and other researchers have reported extensively on this topic (Sawada et al. 2007; Danielsen and Hald 2004, 2006, 2007; Cipolla et al. 2010; Hald 2008). Z phase is not serious in 9 % chromium steels, but should be taken into account in 12 % chromium steels.

The precipitation strengthening depends on the distance between precipitate particles. Based on the measurement of the distance between precipitate particles, the precipitation strengthening due to nitrides can be discussed and then the precipitation strengthening with the solid solution strengthening due to solute nitrogen can be compared.

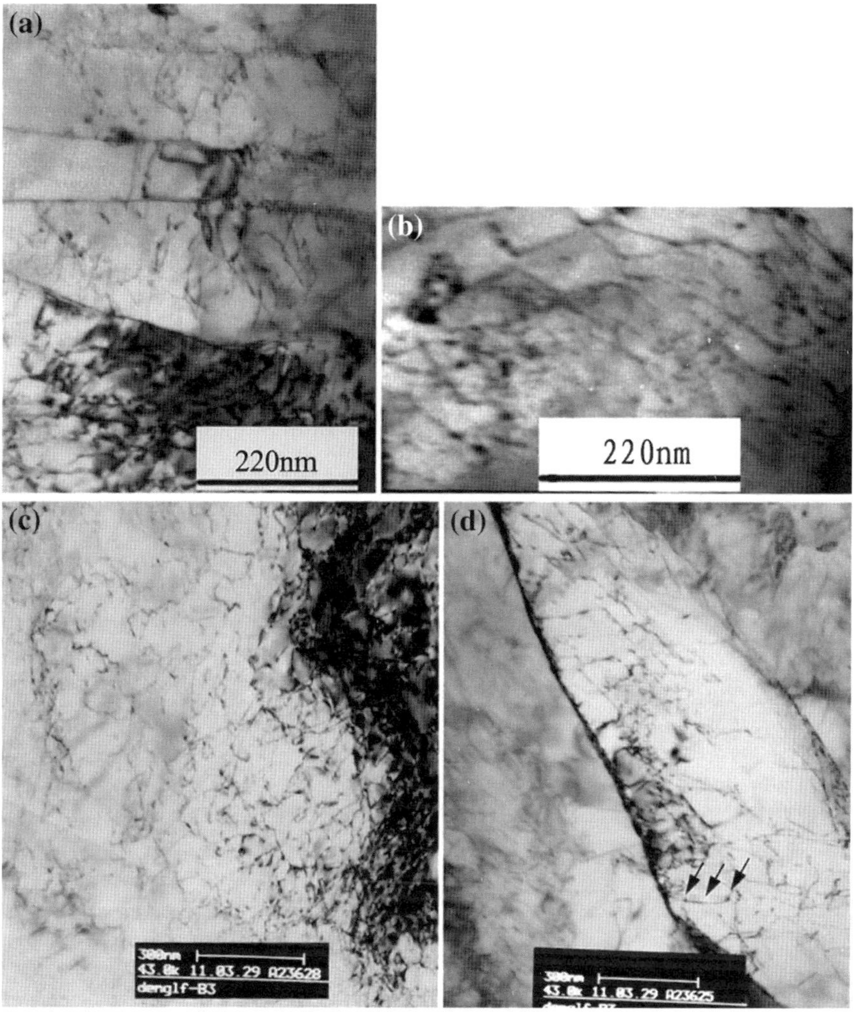

Fig. 5.2 Small nitrides interacting with dislocations in the NS3 steel, normalised at 980 °C and tempered at 650 °C. Examples of precipitates pinning dislocations, and the resulting dislocation bowing, can be found. One example is in **d**, indicated by a set of three *parallel arrows*. The two *outside arrows* point to the two precipitates pinning the dislocation in between, pointed to by the *middle arrow*. The dislocation is bowed. With kind permission from Springer Science+Business Media (Yan et al. 2012, Fig. 2)

Interparticle spacing cannot be simply taken from measurements on SEM and TEM micrographs, because SEM is a surface or sectioning technique, while TEM specimens have finite thickness. In simple terms, the average distance between particles on SEM or TEM micrographs is not straightforwardly the average particle spacing.

The dependence of hardness on tempering temperature is shown in Table 5.2. The hardness decreases little with tempering temperature. The peak temperature of nitride precipitation is nearly 750 °C. The tempering temperatures are not high enough to make the nitrides precipitate from the matrix. That is also the reason of the poor toughness of the matrix.

5.2 Impact Toughness, its Dependence on Tempering, and Phase Transformation

The ductile to brittle transition temperature (DBTT) is a very important property for RAFM steels. As the radiation exposure will greatly increase the DBTT, low DBTT is often regarded as a basic requirement for RAFM steels, which has drawn much attention in research in this field. The DBTT of the traditional Eurofer 97 steel, CLAM steel and F82H steel is about −90 °C (Reith et al. 2003), −91 °C (Baluc et al. 2007) and −60 °C (Jitsukawa et al. 2002), respectively. These steels possess higher impact toughness and exhibit ductile fracture at room temperature. However, NS1 steel has much lower impact toughness at room temperature than these three steels. It displayed an impact toughness of only 12 J on average and brittle cleavage fracture at room temperature.

It has been sufficiently proved that large amount of δ-ferrite would lead to the reduction of impact toughness for various factors such as (1) the low toughness nature of δ-ferrite compared to tempered martensite (Ryu et al. 2006), (2) the δ-ferrite/martensite interface as a crack propagation path (Hu et al. 2009; Ryu et al. 2006) and (3) precipitation-caused embrittlement. Hu et al. (2009) have proved that when the volume fraction of δ-ferrite exceeded 4 %, the room temperature impact toughness of 10 wt% Cr martensitic heat-resistant steel would decrease to lower than 15 J. Therefore, the existence of large amounts of δ-ferrite in the NS1 steel should be one of the critical reasons for the low impact toughness.

In addition to the δ-ferrite, another disadvantageous factor for the impact toughness was found by fractography on the impact fracture surfaces. The brittle cleavage fracture is initiated from the inclusions in NS1 steel. The energy dispersive spectrum (EDS) clearly shows that the inclusion is rich in tantalum (Fig. 5.3). A number of tantalum-rich inclusions randomly distribute in NS1 steel, as shown in Fig. 5.4. Therefore, it could be inferred that the impact toughness of NS1 steel would be further decreased by the tantalum-rich inclusions that serve as the initiation site of the cleavage fracture during the impact process. However, the reason for the existence of tantalum-rich inclusions in NS1 steel cannot be ascertained yet. Elimination of δ-ferrite and tantalum-rich inclusions from the microstructure of nitride strengthened steel is expected to be the effective way to improve the impact toughness.

The room temperature impact energy of two steels is given in Table 5.2. Both steels displayed poor impact toughness. NS3 steel displayed an impact energy value of approximately 12 J and NS2 steel 2 J.

Table 5.2 Room temperature mechanical properties of the steels tempered at 600 or 650 °C

Steel	Normalising (30 min) + tempering (90 min) temperatures (°C)	HRC	Yield strength (MPa)	Tensile strength (MPa)	Elongation (%)	Reduction of area (%)	RT Charpy V energy (J, half size)	−20 °C Charpy V energy (J, half size)
NS2	980 + 600	30.7	865.8	946.7	17.7	66.7	1.5	1.3 ± 0.4
	980 + 650	28.1	828	915	18	70	1.5 ± 0.7	1.5 ± 0.7
NS3	980 + 600	31.8	915.5	1003.2	15.3	64.4	8 ± 6	2
	980 + 650	30.2	855	952	16	66	12	1.5 ± 0.7

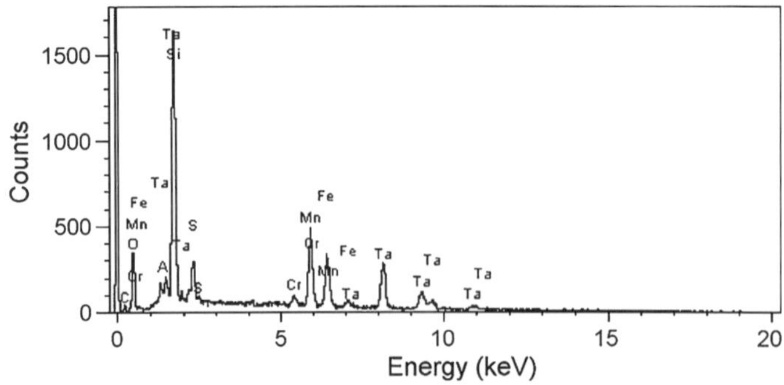

Fig. 5.3 EDS of the inclusion. Reprinted from Hu et al. (2010) with permission from Elsevier

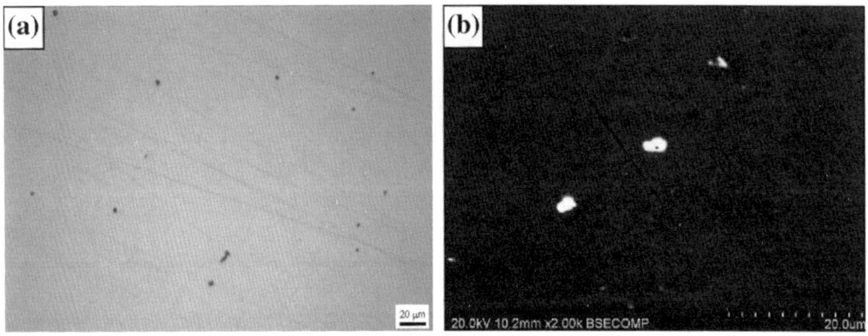

Fig. 5.4 Inclusions in NS1 steel. **a** Optical micrograph and **b** back-scattered SEM. Reprinted from Hu et al. (2010) with permission from Elsevier

NS3 steel has not only the higher strength but also the higher impact energy. As we know, grain refinement is one way to improve both strength and toughness. Therefore, it is believed that NS3 steel has higher tensile strength and room-temperature impact toughness than NS2 steel because of the finer grains. On one hand, carbon reduction leads to tantalum-rich inclusions in NS2; on the other hand, the grain grows larger when normalised at a high temperature caused by the absence of tantalum carbonitrides, which would decrease impact toughness.

The A_{c1} temperature of NS2 and NS3 steels is as low as 670 and 682 °C, respectively, because manganese reduces A_{c1} remarkably. A tempering temperature higher than A_{c1} should be avoided. That is the reason why only 600 and 650 °C are employed as the tempering temperatures. Because the A_{c1} temperature is low, the tempering temperatures are not too low for the steels. The manganese content does not need to be as high as thermodynamic calculations indicate, more than 3 %. Just 1 % manganese will be enough to suppress the δ-ferrite and gain full martensite. It can be observed from Table 5.2 that the toughness does not improve after increasing the tempering temperature from 600 to 650 °C in these two steels.

The austenitising starting temperature (A_{c1}) of NS4 steel (Table 5.1) is as high as 770 °C. Therefore, the steel could be tempered at high temperatures. Even under

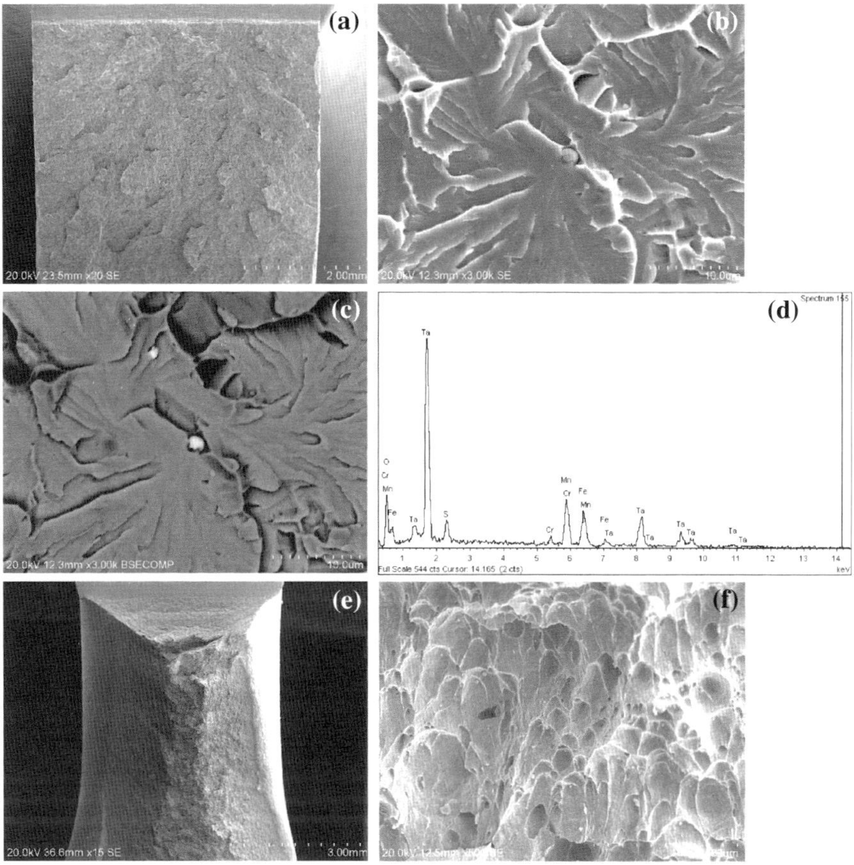

Fig. 5.5 Fractography of NS4 steel at room temperature tempered at (**a–d**) 650 °C and (**e**, **f**) 750 °C. With kind permission from Springer Science+Business Media (Zhou et al. 2012, Fig. 6)

the slow cooling rate of 0.05 °C/s, martensitic transformation can occur in the steel, which means that it is very easy for the steel to acquire a martensitic microstructure.

After tempering at 650 °C for 90 min, the room temperature impact energy of NS4 steel is as low as only 2 J. However, the room temperature impact energy abruptly increases to around 100 J with the increase of tempering temperature from 650 °C to 700 °C, and 750 °C. The –20 °C impact toughness of the steel tempered at 700 °C remains as poor as when tempered at 650 °C. When the tempering-temperature further increases to 750 °C, however, the impact-toughness at −20 °C is greatly improved again. At −40 °C, even the steel tempered at 750 °C gives the poor impact energy of several Joules.

Brittle cleavage fracture is observed on the impact fracture surface of the steel tempered at 650 °C and tantalum-rich inclusions are found to initiate the cleavage fracture, as shown in Figs. 5.5a–d. With the increasing of tempering temperature, the toughness is remarkably increased. It displays excellent room temperature impact energy of 107 J when tempered at 750 °C and ductile fracture characteristic shown in Figs. 5.5e and 5.5f. When the tempering temperature is increased from 650 to 750 °C, the ductile–brittle transition temperature (DBTT) is significantly decreased from above room temperature to −30 °C.

The impact toughness depends a lot on the tempering temperature. The microstructure change described in Sect. 5.1 can also explain the dependence of room-temperature impact toughness on the tempering temperature. With the increase of tempering temperature, the dislocation density is decreased either by their own thermal interaction or by acting as the nucleation sites of nitrides. Hence, the matrix becomes tough and can accommodate larger deformation and absorb more energy, which means high impact toughness (Kimura et al. 2010).

Low DBTT is one of the critical issues for RAFM steels. The DBTT is decreased from above room temperature to −30 °C when the tempering temperature increases from 650 to 750 °C. Generally, it can be assumed that the cleavage fracture stress slightly changes with temperature at fracture, and the tempering temperature for the steel. Thus, it is easy to understand in a phenomenological way that the DBTT will decrease with the increase of the tempering temperature, because of the decrease of yield strength (Sawada et al. 2003). As discussed above, the reduction of dislocation density and the nitride precipitation are the reason for the decrease of yield strength. So, it actually can be interpreted that the DBTT decrease can be really associated with the two changes in microstructure.

5.3 Tensile Properties and Effects of Chemical Composition and Tempering Temperature

Table 5.3 shows the tensile properties for NS1 steel, Eurofer 97 steel (Reith et al. 2003) and CLAM steel (Li et al. 2006) at room temperature and 600 °C. It should be noted that these three steels experienced different heat treatments. At room temperature, NS1 has almost the same strength as Eurofer 97 steel and CLAM steel. At 600 °C, NS1 steel has higher yield strength and tensile strength than Eurofer 97 steel and CLAM steel while its elongation and reduction of area are lower than Eurofer 97 steel and CLAM steel.

In theory, strength is usually related to both chemical composition and microstructure of the steel. Dislocation hardening, precipitation hardening and solution hardening are three dominant strengthening mechanisms for high chromium ferritic/martensitic heat resistant steels used as structural materials for ultra super critical power generation (Maruyama et al. 2001). High chromium RAFM steels are also strengthened by these three mechanisms. According to the three traditional strengthening mechanisms, NS1 steel should have lower tensile strength than Eurofer 97 steel and CLAM steel. First, owing to failing to obtain a microstructure of full martensite, the strength increase produced by dislocation hardening for NS1 steel should be lower than that of Eurofer 97 steel or CLAM steel of full martensite. Second, the carbon content is reduced to extremely low in NS1 steel, so the strength increase produced by solution hardening should also be lower than that of Eurofer 97 steel or CLAM steel. Third, much tantalum is consumed in tantalum-rich inclusions instead of TaN nitrides, which would decrease the strength increase produced by precipitation hardening for NS1 steel compared with that of Eurofer 97 steel or CLAM steel. However, the NS1 steel has not only almost the same strength as Eurofer 97 steel or CLAM steel at room temperature, but also higher strength at 600 °C. Other factor(s) must be responsible for the strength.

As it is known, steel strength is also related to heat treatment procedures. As listed in Table 5.3, with the same tempering treatment, NS1 steel was subjected to a higher solution temperature and longer holding time than Eurofer 97 steel or CLAM steel. The solubility of vanadium and tantalum is predicted to increase with increasing normalising temperature (between 900 and 1070 °C for vanadium, and between 980 and 1150 °C for tantalum) in ferritic steels (Shen et al. 2009). Therefore, it can be inferred that more vanadium and tantalum should be dissolved in the NS1 steel, compared with Eurofer 97 steel or CLAM steel. Hence, more vanadium/tantalum-rich precipitates should form after tempering in NS1 steel. Since the carbon content is controlled to under an extremely low level in NS1 steel, these vanadium/tantalum-rich precipitates should be mainly nitrides. Taneike et al. (2004) have proved that the amount of dissolved chromium in the matrix could increase by about 1 wt% because precipitates were almost nitrides after tempering when the carbon content was decreased from 0.12 wt% to close to zero. However, taking the tantalum-rich inclusions into account, higher solution temperature and longer holding time should produce more vanadium-rich nitrides rather than tantalum-rich nitrides in NS1 steel, compared to Eurofer 97 steel or CLAM steel. Therefore, strength increase due to more vanadium-rich nitrides and more dissolved chromium atoms in the matrix might at least compensate for the strength loss caused by the three factors mentioned above. Lu et al. (2009) also found that the hardness (proportional to the tensile strength) of a Eurofer 97 steel increased gradually with increasing normalising temperature from 980 to 1150 °C.

δ-ferrite should be eliminated from the microstructure due to its detrimental effects on the creep rupture strength (Yoshizawa and Igarashi 2007; Yamada et al. 2003) and impact toughness (Hu et al. 2009; Ryu et al. 2006) of ferritic/martensitic steels. In order to enable a full martensite microstructure in the steel, the effect of chemical composition on the microstructure is considered. It is well-known that addition of some austenite forming elements can effectively prevent the formation of δ-ferrite during solution treatment. Nickel, copper, cobalt and manganese are all austenite forming elements. However, nickel, copper and cobalt should be eliminated or minimised in RAFM steels because they might produce long-lived radioactive isotopes during irradiation. Therefore, only manganese can be selected as an austenite forming element in RAFM steels. Moreover, addition of manganese can make a right shift of the continuous cooling transformation (CCT) curve of the steel to decrease the critical cooling rate for martensite transformation during continuous cooling. Therefore, NS2 steel containing 3.73 wt% manganese was designed to use the effect of manganese

Table 5.3 Tensile properties of three types of steel

Steel	Heat treatment	Temperature	Yield strength (MPa)	Tensile strength (MPa)	Elongation (%)	Reduction of area (%)
NS1	Normalised at 1050 °C for 50 min and tempered at 760 °C for 90 min	Room	542	642	22	77
		600 °C	337	375	20	80
Eurofer 97 (Reith et al. 2003)	Solution treated at 980 °C for 27 min, air-cooled and tempered at 760 °C for 90 min	Room	537	652	21	80
		600 °C	277	292	29	94
CLAM (Li et al. 2006)	Solution treated at 980 °C for 30 min, water quenched and tempered at 760 °C for 90 min	Room	514	668	25	77
		600 °C	293	334	29	87

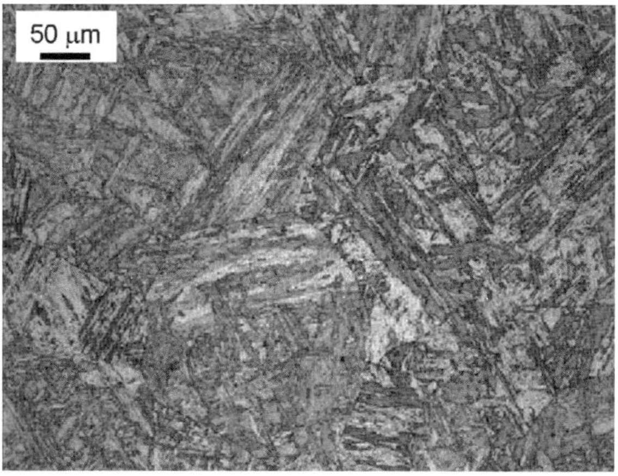

Fig. 5.6 Optical micrograph showing the microstructure of NS2 steel. Reprinted from Hu et al. (2010) with permission from Elsevier

on the formation of δ-ferrite, compared with NS1 steel. NS2 steel (Fig. 5.6) has succeeded in obtaining a full martensite microstructure with Vickers hardness of about HV280, after normalising at 1200 °C for 50 min.

Summarising the Sects. 5.1–5.3 so far, the steel with low manganese content could not have full martensite microstructure and avoid the δ-ferrite formation after solution treatment. However, if the manganese content in the steel is increased to 3.73 wt%, a microstructure of full martensite could be obtained. Besides the δ-ferrite, tantalum-rich inclusions serving as the initiation site for cleavage fracture during impact process should be an important cause for the extremely low impact toughness of the steel at room temperature. More vanadium-rich nitrides might be formed in the steel for higher solution temperature and longer holding time. More chromium should be remained in the matrix due to the extremely low level carbon content. These two factors should be responsible for the similar room temperature strength and the higher strength at 600 °C for the steel compared with Eurofer 97 steel or CLAM steel.

The mechanical properties of two steels are given in Tables 5.2 and 5.4. The two steels have high strength at both room temperature and 600 °C; this observation of high strength may be explained in several ways. The high content of

Table 5.4 High temperature tensile properties at 600 °C of the steels tempered at 600 or 650 °C

Steel	Normalising (30 min) + tempering (90 min) temperatures (°C)	Yield strength (MPa)	Tensile strength (MPa)	Elongation (%)
NS2	980 + 600	493	572	17
	980 + 650	485	553	18
NS3	980 + 600	520	606	15
	980 + 650	515	580	18

manganese should be one of the most significant reasons. It is usually thought that the substitution solution strengthening effect of manganese is not strong or effective. The dissolved manganese, while giving solution strengthening, will not increase DBTT (Yong 2006). This is the reason why high-strength, low-alloy steels have 1–2 % manganese to increase the hardenability (Jun et al. 2006). However, in these two steels, the manganese is added to such a high level of the order of 4 %, so the solution strengthening contribution should not be neglected. Thus, the addition of high level manganese should be the main reason of high strength. The second important reason lies in the nitrogen solution-strengthening contribution. It has been reported that only half of vanadium in this type of steels can form nitrides (Abe et al. 2007; Taneike et al. 2004; Sawada et al. 2004). As mentioned, approximately 75–90 % tantalum does not form carbonitrides but just exists in solution state or inclusions. Therefore, the steel might receive a significant strength contribution from the free nitrogen solution strengthening.

Although the strength decreases significantly when the temperature is increased from room temperature to 600 °C, the ductility is not increased as expected. The strength reduction should result mainly from the decreased dislocation density at high temperature. It could be assumed that even at 600 °C the dislocation movement is still prevented strongly by the fine nitrides so that the ductility could hardly increase. This is perhaps why there is virtually no difference in elongation at room temperature and 600 °C (Tables 5.2 and 5.4). In addition, the high density of residual dislocations may have caused high flow stress levels.

The high temperature strength, at 600 °C, after different tempering temperatures is shown in Table 5.4. In the tempering temperature range of 600–650 °C, the strength does not change much with it. It seems that higher tempering temperature resulted in slightly lower yield strength and ultimate tensile strength. The explanation could be the same as the hardness change; see Sect. 5.1.

The microstructure of materials deformed at 600 °C characterised under an optical microscope is shown in Fig. 5.7. The steels deformed at the high temperature show no difference in microstructure deformation from the room-temperature tensile samples.

The tensile properties of NS4 steel are shown in Table 5.5. The yield strength at room temperature is decreased from 799 to 734 MPa, and then to 565 MPa when the tempering temperature is increased from 650 to 700 °C, and then to 750 °C. It seems that the yield strength declines more significantly when the tempering temperature is increased from 700 to 750 °C. The reduction is 169 MPa. However, the similar trend cannot be seen in the yield strength at 600 °C. At 600 °C, the reduction of yield strength is 62 and 73 MPa, when the tempering temperature is increased from 650 °C to 700 °C and from 700 °C to 750 °C, respectively. The difference is small and can be neglected.

Eurofer 97 has a very high A_{c1} temperature. Therefore, it can be tempered at 760 °C. However, our steel has a lower A_{c1} temperature, 770 °C. The tempering temperature should not be higher than the A_{c1} temperature and it is not proper to make a comparison with Eurofer 97 under exactly the same heat treatment condition. The important point is that they both can be used at 600 °C. Further, if the manganese content is reduced to 1 %, the A_{c1} is increased to around 820 °C.

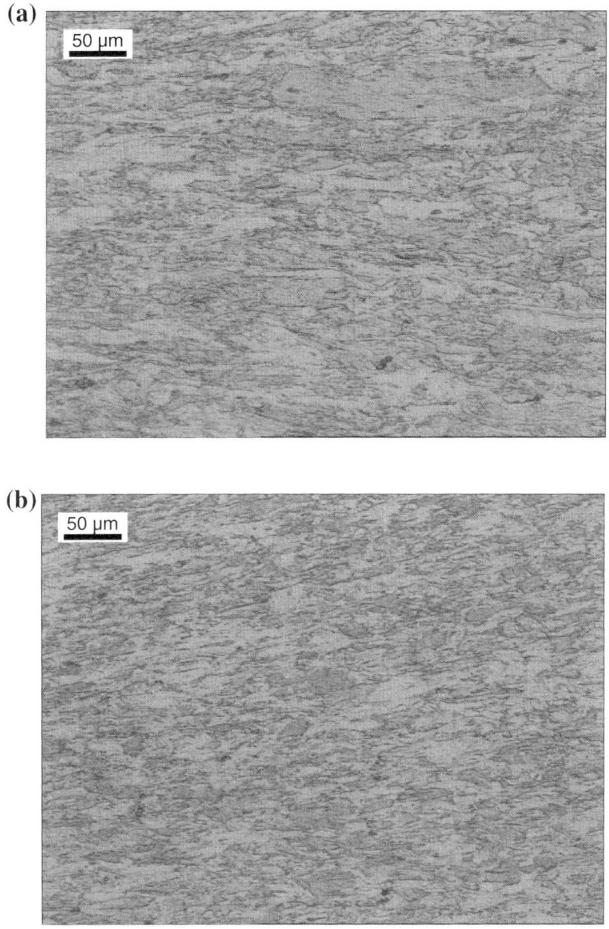

Fig. 5.7 Optical micrographs of microstructure after deformation at 600 °C. **a** NS2, **b** NS3. With kind permission from Springer Science+Business Media (Yan et al. 2012, Fig. 7)

The strength decreases with the increase of tempering temperature. Strength decrease could be attributed to several reasons such as recrystallisation (Mungole et al. 2008), decreased dislocation density (Pešička et al. 2003), and reduction of the amount of alloying elements in solid solution (Sawada et al. 2004). The strengthening effect of free nitrogen atoms in the matrix is much higher than the precipitation hardening of nitrides. Thus, the yield strength quickly drops when tempered at 750 °C. Recrystallisation does not play a role at all in the strength decrease, because it does not happen in the tempered microstructure. As to the dislocation density, a given tempering temperature in the range of 650–750 °C should lead to a corresponding dislocation density, because the tempering temperatures of 650 °C and higher is high enough for dislocations to move and annihilate, and then reach a stable density. Hence, if the decrease of dislocation density has the largest effect, the strength should decrease gradually with the increase of the tempering temperature. However, as illustrated in Table 5.5, the yield strength

Table 5.5 Tensile properties of NS4 steel

Testing temperature (°C)	Tempering temperature (°C)	Yield strength (MPa)	Ultimate tensile strength (MPa)	Elongation (%)
RT	650	799	884	19
	700	734	809	19
	750	565	725	21
600	650	484	537	18
	700	422	468	21
	750	349	389	19

decreases by 169 MPa when the tempering temperature is increased from 700 to 750 °C, much larger than 65 MPa when the tempering temperature is increased from 650 to 700 °C. Therefore, apart from the dislocation density, it is reasonable to look for a further cause for the large decrease of strength.

Nitrogen is as effective as carbon in solid–solution strengthening. Nitrogen solid–solution strengthening increases both the yield strength and ultimate strength strongly, which should be one of the important reasons for the high room-temperature strength of steels tempered at 650 °C. When the tempering temperature is increased to 750 °C, the important change in the microstructure is the precipitation of nitrides. A large number of nitrides form along the boundaries and in the matrix. On one hand, the nitride precipitation consumes dislocations by nucleating on them and cause strength loss (Ghassemi-Armaki et al. 2009). On the other hand, the formation of nitrides consumes a lot of free nitrogen which has provided solid-solution strengthening to a great degree, and thus brought about a large strength decrease. Although the nitride precipitation strengthening may contribute to the strength, it cannot compensate for the loss of the solid–solution strengthening of nitrogen atoms. Therefore, the strength decreases much quicker when the steel is tempered at 750 °C, as demonstrated in Table 5.5.

However, the high temperature strength and room temperature ultimate tensile strength seem to linearly decrease with the increase of the tempering temperature. It seems to indicate that the solid-solution strengthening does not affect the high temperature strength and the ultimate tensile strength as much as the yield strength at room temperature.

5.4 Inclusions

The inclusions in two steels are shown in Figs. 5.8 and 5.9. The inclusions in NS3 steel are composed mainly of large Al_2O_3 and MnS inclusions. Electron diffraction is difficult because the steels are magnetic and the inclusions are orientated randomly. We could, however, identify the inclusions by their morphology and the chemical composition by EDS. The large Al_2O_3 particles play the role of the cores and they are coated completely with MnS inclusion, as shown in Fig. 5.9. The inclusion sizes range from 2 to 5 μ. Because the MnS inclusion is soft and deformable (Poulachon et al. 2002), it is

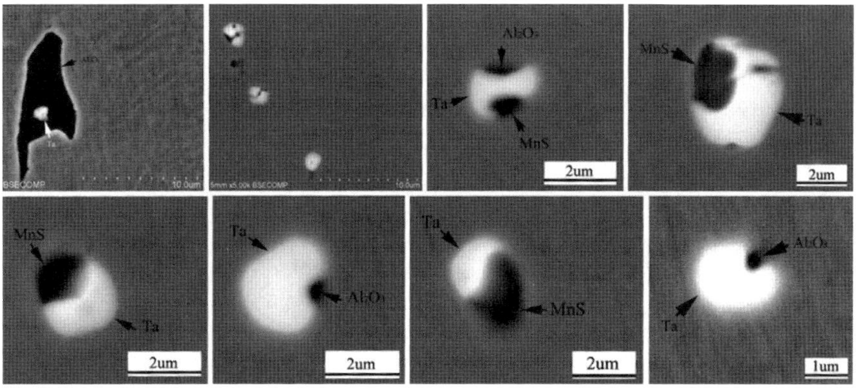

Fig. 5.8 Morphologies of inclusions in NS2 steel. With kind permission from Springer Science+Business Media (Yan et al. 2012, Fig. 4)

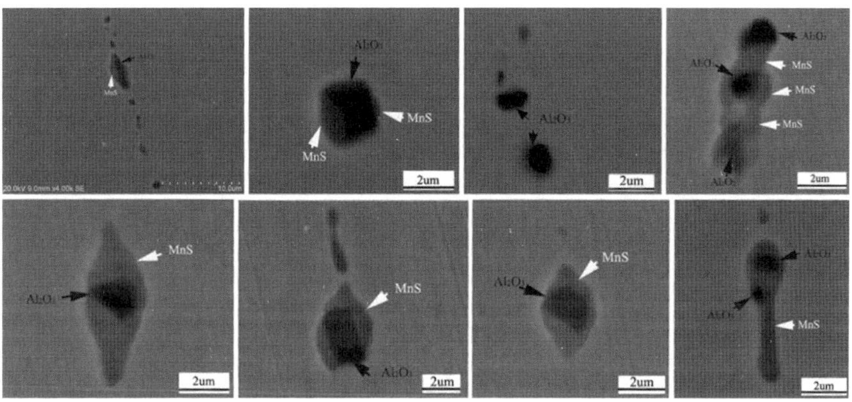

Fig. 5.9 Morphologies of inclusions in NS3 steel. With kind permission from Springer Science+Business Media (Yan et al. 2012, Fig. 3)

possible that the large Al_2O_3 inclusion will not be detrimental to the toughness after being wrapped by MnS. Besides the inclusions with this characteristic, some Al_2O_3 inclusions are attached by little MnS, as shown in one of the micrographs in Fig. 5.9. Because the SEM observation is conducted on only one polished surface, such inclusion might also be wrapped by MnS inclusion, which could be revealed if the observation were carried out from another direction. Even if that is the case, in at least one dimension, the inclusion appears as a large Al_2O_3 inclusion attached by little or no MnS inclusion. This observation is critical to understanding the effect of these inclusions on the toughness of NS3 steel, to be discussed subsequently.

The interaction between the Al_2O_3 particles and MnS inclusion is likely as follows. The Al_2O_3 inclusion could exist even in the molten steel (Zhang and Thomas 2003). The formation temperature of MnS inclusion in NS3 steel is also high because the manganese content is increased to such a high level. Therefore, the

MnS inclusion would take advantage of the pre-existing Al_2O_3 particles by nucleating on them. Thus, the Al_2O_3 inclusions are coated with MnS inclusion, which leads to the previous characteristic of the inclusions in NS3 steel.

The Al_2O_3 particles are not coated with MnS inclusions in NS2 steel. Small-size Al_2O_3 particles are covered by large amount of tantalum and small size undissolved tantalum particles, and MnS particles are attached to large-size Al_2O_3 inclusions. The tantalum-rich inclusions, 2–7 μm in diameter, are in great numbers, appearing as white particles in the backscattered electron images. The small-size Al_2O_3 particles and MnS inclusions are the core of the tantalum rich inclusions.

The pre-existing fine Al_2O_3 particles and small MnS inclusions would facilitate the formation of the tantalum-rich inclusions by acting as nucleation sites. These fine Al_2O_3 particles and small-size MnS inclusions are then isolated from each other by tantalum, and they cannot produce the morphology in NS3 steel. These tantalum-rich inclusions show another characteristic in that they are always spherical. Tantalum seems to change the irregular shape Al_2O_3 particles and small MnS inclusions into spherical inclusions after wrapping them. The tantalum-rich inclusions thus are not likely to trigger cleavage fracture because of the spherical shape. Although the tantalum-rich inclusions contained dangerous Al_2O_3 particles, their sizes are too small to act as initiators. Those big size Al_2O_3 particles are too large to be covered completely by tantalum. Therefore, tantalum or MnS inclusions seem to be attached to the large Al_2O_3 particles at some local area, and so the large Al_2O_3 particles can serve as harmful initiators.

Despite the preceding proposal, the exact mechanism of tantalum-rich inclusion formation is not clear. Some researchers have found that, for an unknown reason, 75–90 % tantalum in steel is in solid solution state, without forming carbonitrides as expected (Klueh 2005). Therefore, as mentioned, the dissolved tantalum may segregate with such beneficial conditions as pre-existing Al_2O_3 particles or MnS inclusions. Furthermore, it is difficult to tell whether the tantalum-rich inclusions are segregated or undissolved. There is no such tantalum-rich inclusion in NS3 steel, which indicates that the difference of carbon content may be a clue to the formation of tantalum-rich inclusion because both steels were prepared by the same furnace and process. However, another investigation is needed to confirm this.

When tempered at 650 °C, the steel has very high strength but poor toughness. Usually, higher strength is related to smaller critical crack size (Blach et al. 2009). It means that when tempered at 650 °C, although the steel has high strength, it can only tolerate very small size inclusions. Such inclusions larger than the critical crack size will be harmful to the impact toughness. As shown in Figs. 5.5a–d, the inclusions identified to be Ta-rich by EDS serve as pre-existing cracks and initiated the cleavage fracture, which do harm the impact toughness. Thus, the impact toughness could be further improved by purification.

However, when tempered at 750 °C, the strength of the steel is decreased and the matrix becomes tough. The critical crack size of the steel is increased. Therefore, the matrix can accommodate large size inclusions. Hence, the Ta-rich inclusions are observed at the bottom of some dimples, as shown in Fig. 5.5f.

The tantalum-rich inclusions are purely undissolved tantalum, and not Fe_2Ta or FeTa. These particles are observed in all the conditions of the steel. They exist even when melting. However, the pure undissolved tantalum is rather hard to accept. Tantalum should have enough time and temperature to get into the matrix. Investigating the maximum solubility of tantalum in pure iron would help us to know whether tantalum is too much for the steel to dissolve. If it turns out to be the case, the tantalum-rich inclusions could be eliminated by reducing tantalum content.

5.5 Dependence of the Prior Austenite Grain Size on Normalising Temperature

The prior austenite grain sizes of two steels at different normalising temperatures are presented in Figs. 5.10 and 5.11, respectively. At 980 °C, NS3 steel has a small grain size of approximately 15 μm, whereas NS2 steel has a grain size of 25 μm.

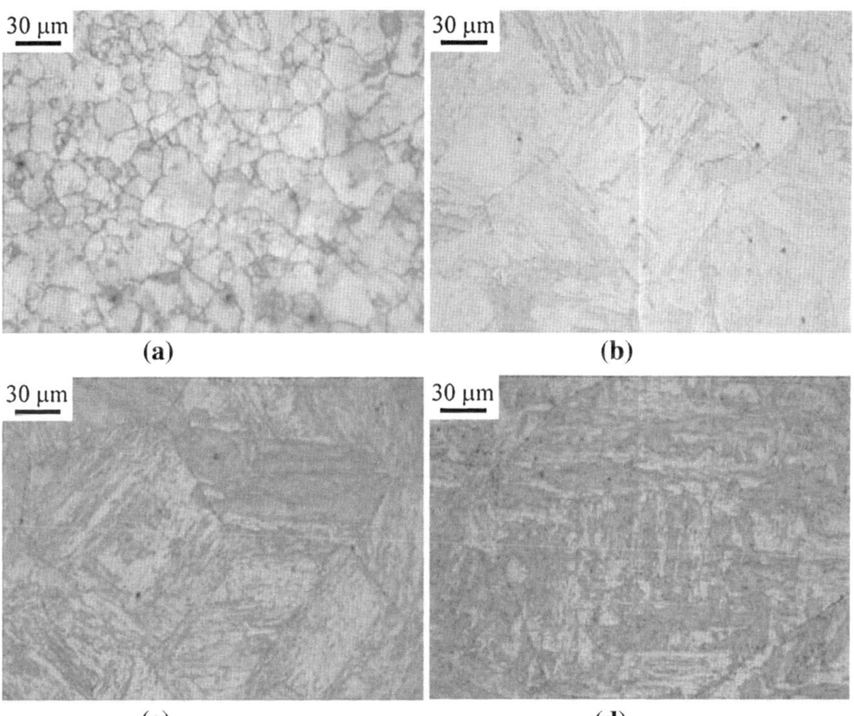

Fig. 5.10 Dependence of the prior austenite grain size on normalising temperature in NS2 steel. **a** 980 °C, **b** 1050 °C, **c** 1100 °C, **d** 1200 °C. With kind permission from Springer Science+Business Media (Yan et al. 2012, Fig. 6)

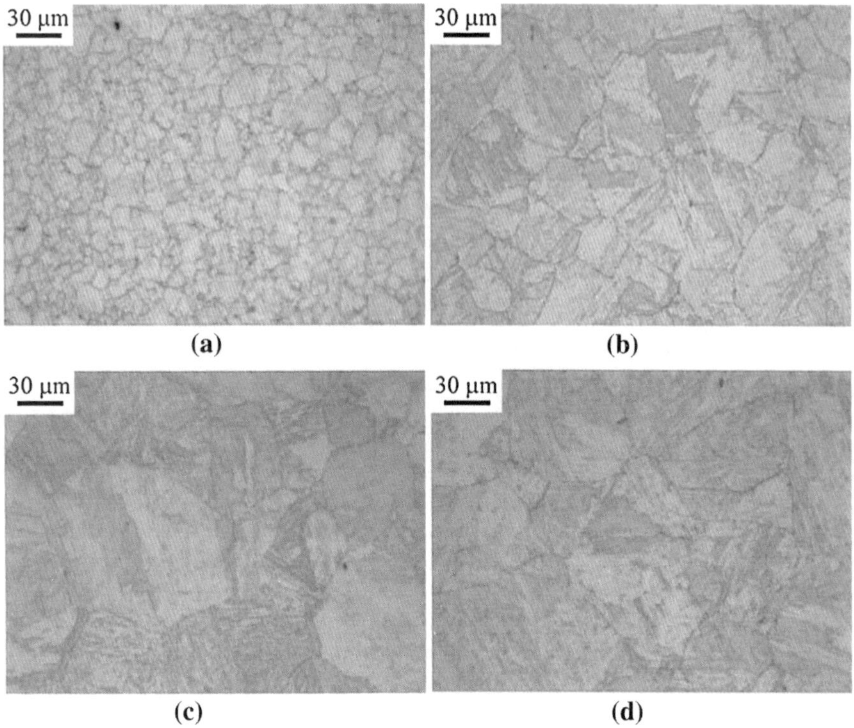

Fig. 5.11 Dependence of the prior austenite grain size on normalising temperature in NS3 steel. **a** 980 °C, **b** 1050 °C, **c** 1100 °C, **d** 1200 °C. With kind permission from Springer Science+Business Media (Yan et al. 2012, Fig. 5)

When the normalising temperature is increased to 1050 °C, the grain size of NS3 steel is increased to approximately 50 μm, whereas the grain size of NS2 steel is increased to 80 μm. When the normalising temperature is further increased to 1200 °C, the grain size of NS3 steel is increased slightly, to approximately 60 μm, whereas the grain size of NS2 steel is increased greatly to approximately 200 μm.

It is obvious that, at the same normalising temperature, the prior austenite grain size of NS3 steel is smaller than that of NS2 steel. Also, the austenite grain of NS3 steel grows less with increasing normalising temperature than that of NS2 steel.

Grain size is dependent on such factors as the solution temperature and the precipitate pinning effect. In NS2 steel, many tantalum-rich inclusions nearly exhausted the tantalum content. Therefore, on the one hand, only a limited amount of TaN particles could be formed to provide the pinning effect on austenite grain size, which is not as strong as that in the NS3 steel. On the other hand, NS3 steel has not only the higher carbon content but also no tantalum-rich inclusions. Plenty of carbonitrides of tantalum in NS3 steel could prevent the grain boundary migration. Thus, NS3 steel had smaller austenite grain than NS2 steel at the same normalising temperature.

5.6 Inclusion Initiating Cleavage Fracture

Generally, once yield strength is pronouncedly promoted, the critical crack size will be diminished and steels become sensitive to small flaws such as large size carbonitrides and inclusions. Inclusions of several microns will be large enough to initiate the brittle cleavage fracture on the rigid zone ahead of the notch, where the stress is concentrated (Yan et al. 2007). The fractography analysis on the fracture surface of the broken impact specimens is conducted by using scanning electron microscopy. Energy dispersive spectrometry reveals that in both steels the coarse Al_2O_3 inclusions initiate the brittle cleavage fracture at the head top area of the Charpy impact specimens and reduce the toughness, as shown in Figs. 5.12, 5.13, 5.14.

The initiators of the impact fracture of NS3 steel are composed of Al_2O_3 inclusion and MnS inclusion, as shown in Fig. 5.12e, which was consistent with the morphology of inclusions in Fig. 5.9. The SEM images of the initiator in Figs. 5.14d, e show that the big (approximately 4 μm) Al_2O_3 inclusion is attached by the small tantalum particles, which agrees with the big Al_2O_3 inclusion morphology in Fig. 5.8.

In NS2 steel, the tantalum-rich inclusions do not play a role in triggering the cleavage fracture. As shown in Figs. 5.13b, c, the black Al_2O_3 inclusion marked with A is right at the origination whereas the tantalum-rich inclusion marked with B is right below the origination. This finding illustrates that not the tantalum-rich inclusion but the Al_2O_3 inclusion truly initiated the cleavage fracture. This can also be proved from the other tantalum-rich inclusions dispersed near the origination without triggering the cleavage fracture. For example, in Figs. 5.13b, c, not far from the tantalum-rich inclusion marked with B, another tantalum-rich inclusion marked with C seems dormant, having no responsibility for the initiation. As we can observe, most tantalum-rich inclusions are spherical and with no sharp angle. In addition, the tantalum-rich inclusion might be soft and deformable. All these characteristics should make the tantalum-rich inclusion rather harmless. However, the Al_2O_3 inclusion is refractory, hard and not deformable (Hesabi et al. 2006). Additionally, the Al_2O_3 inclusion has a much different thermal expansion coefficient from the steel (Yllmaz et al. 2005) matrix, which would bring about a concentrated stress at the interface. Therefore, it stands a good chance that the large size Al_2O_3 inclusion would have a harmful effect on toughness (Zhang et al. 2002). By comparing Figs. 5.13b, d, it could be observed that the tantalum-rich inclusion marked with B is broken into two parts. It could be imagined that the cleavage crack propagated through the tantalum-rich inclusion marked with B from the matrix after it was initiated by the Al_2O_3 inclusion. The curved interfaces between the tantalum-rich inclusion and the matrix in Figs. 5.13b, d could have accommodated some deformation energy.

The high oxygen content should be responsible for the large amount of Al_2O_3 inclusions (Zhang and Thomas 2003). However, sometimes even when the gases are controlled strictly to low levels, inclusions still cannot be avoided completely because of the local composition fluctuation, especially in large ingots. Therefore,

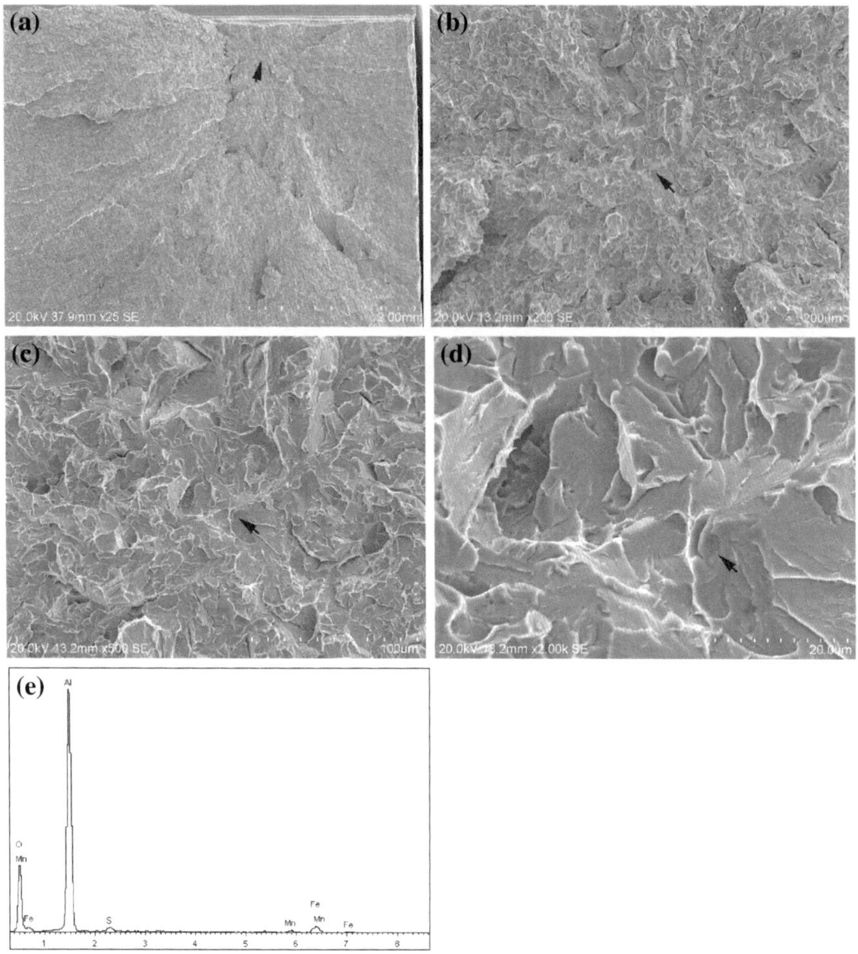

Fig. 5.12 Impact fractography of NS3 steel. **a** SEM image showing the origin of cleavage fracture, **b** magnification of the origin area in **a**, **c** initiator in the cracking origin, **d** magnification of the initiator in the cracking origin in **c**, **e** EDS of the inclusion. With kind permission from Springer Science+Business Media (Yan et al. 2012, Fig. 8)

improving the toughness of nitride-strengthened steels should be the main topic in future research, for example optimisation of alloying addition of not only manganese but also tantalum, as well as decreasing sulphur and oxygen to below 10 ppm.

In summary of Sects. 5.1–5.6, concerning the microstructure and mechanical properties of the nitride-strengthened RAFM steels, these steels can achieve full martensitic microstructure after normalising and tempering, by reduction of carbon and addition of manganese (1.4 %). The steels, in the main, show high strength at both room temperature and 600 °C, but their impact toughness is poor. The strength decreases with the increase of tempering temperature. With the increase of

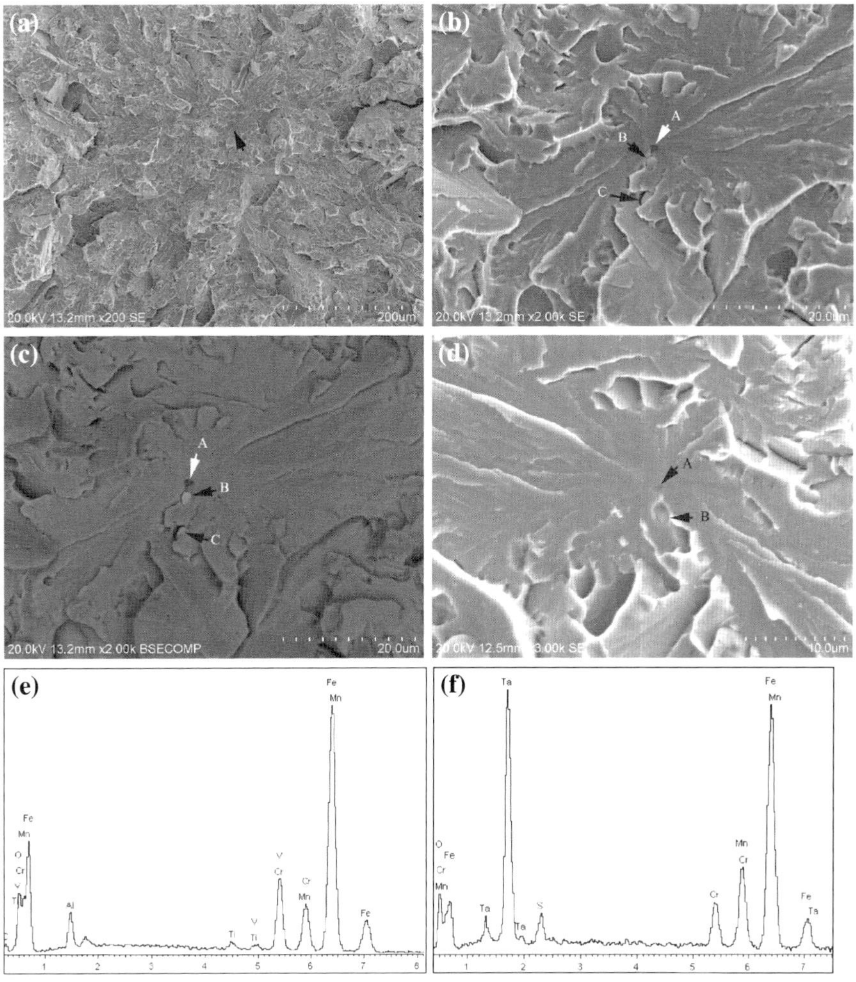

Fig. 5.13 Inclusions at the origin of the impact fracture surface of NS2 steel, initiating cleavage fracture. **a** SEM image showing the origin of cleavage fracture, **b** initiator in the cracking origin, **c** backscattered image of **b**, **d** initiator on the matching broken impact sample, **e** EDS of the inclusion A; **f** EDS of the inclusion B. With kind permission from Springer Science+Business Media (Yan et al. 2012, Fig. 9)

tempering temperature, the room-temperature impact toughness increases dramatically. In addition, the DBTT is decreased, for instance from above room temperature to −30 °C when the tempering temperature is increased from 650 to 750 °C, which might be interpreted, in a phenomenological way, as due to the decrease of yield strength. NS3 steel with higher carbon content has smaller grain size under the same normalising temperature, which might be related to the pinning effect of the larger amount of precipitates because of the higher carbon content and no

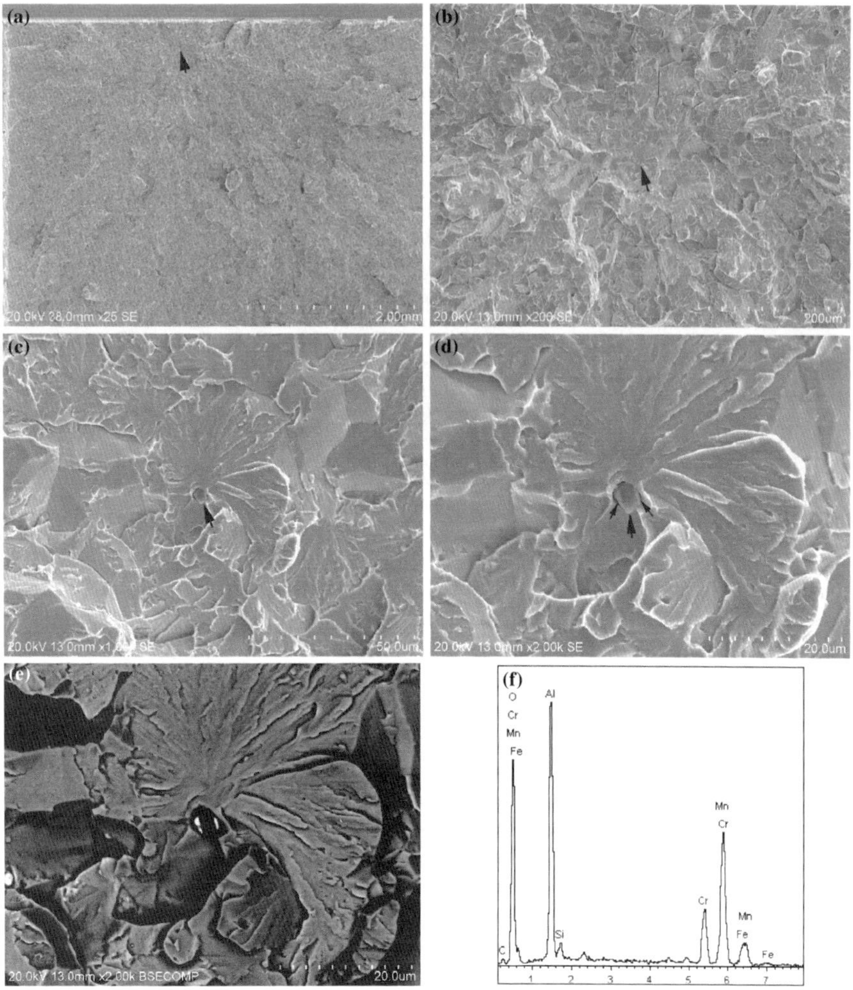

Fig. 5.14 Coarse inclusions at the cleavage impact fracture origin in NS2 steel. **a** SEM image showing the origin, **b** magnified **a**, **c** inclusion at the origin, **d** magnified **c**, **e** backscattered image of **d**, where the bright particles are rich in tantalum, **f** EDS result showing Al_2O_3 inclusion. With kind permission from Springer Science+Business Media (Yan et al. 2012, Fig. 10)

tantalum-rich inclusions. Tantalum-rich inclusions do not initiate the cleavage fracture in NS2 steel whereas Al_2O_3 inclusions do. Greatly increased strength could make the steels sensitive to inclusions, which initiate cleavage fracture and decrease toughness. In another steel, however, the tantalum-rich inclusions serve as initiators in the origin of cleavage fracture on the brittle broken impact fracture surface while they also are found at the bottom of some of the dimples on the surface of the ductile broken impact fracture. The detrimental effect depends on whether their size is larger than the critical crack size of the steel or not.

5.7 Conventional Nitride-Strengthened Heat Resistant Steel

Different from the previous sections, this section is concerned with conventional nitride-strengthened martensitic heat resistant steel, i.e., without the low activation properties.

5.7.1 Microstructure and Nitride Precipitation

The steel (Table 5.6) normalised at 980 °C for 30 min has full martensitic microstructure. After normalising, the steel is tempered at 650, 700 and 750 °C for 90 min. Almost no precipitates are formed when the steel is tempered at 650 °C, as illustrated in Fig. 5.15a. The tempering temperature of 650 °C is not high enough for the nitrides to precipitate. However, when the tempering temperature is increased to 700 °C, the precipitates are noticed in the matrix, as shown

Table 5.6 Composition of conventional nitride-strengthened martensitic steel (wt%)

C	Cr	Mn	W	V	Co	Nb	N	Al
0.005	8.63	1.06	1.53	0.19	1.47	0.062	0.033	<0.01

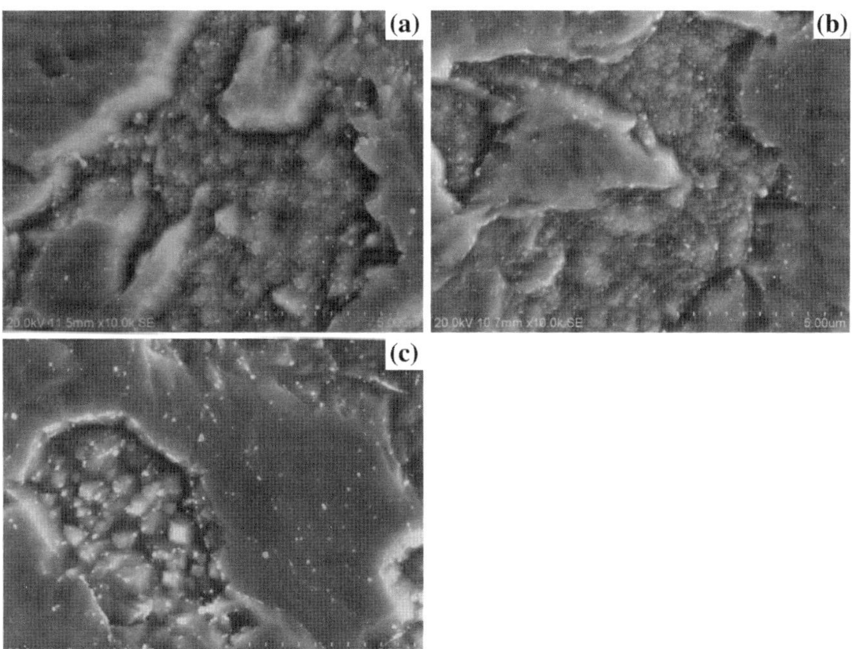

Fig. 5.15 The tempered microstructure of the steel at **a** 650 °C, **b** 700 °C, **c** 750 °C. With kind permission from Springer Science+Business Media (Zhang et al. 2012, Fig. 1)

in Fig. 5.15b. Finally, when the tempering temperature is increased to 750 °C, Fig. 5.15c, the quantity of the precipitates increases promptly and the nano-sized precipitates are in a sharper and clearer shape than those tempered at 700 °C. 750 °C is widely accepted as the peak precipitation temperature of nitrides.

Since the carbon content in the steel is decreased to such a low level of 0.005 % in wt% (Table 5.6), it is very difficult to form carbides in the steel. This is proved by the microstructure shown in Fig. 5.15 that no such big size carbides as $Cr_{23}C_6$ are observed under SEM. However, the nitrogen is at a high level in the steel. Therefore, it is reasonable to believe that the precipitates formed in the steel are nitrides of niobium and vanadium, which are very fine and in the cubic shape shown in Fig. 5.16.

So, the precipitates formed in the matrix during tempering are MX type nitrides. The tempering temperature is critical to the nitride precipitation. Martensitic strengthened by only thermally stable nitrides is a desirable microstructure. Therefore, it is logical to speculate that the nitride-strengthened martensitic heat resistant steel should have good long term creep strength due to the microstructure stability (Sawada et al. 2004).

5.7.2 Mechanical Properties, DBTT, and Fractography

The precipitation behaviour of nitrides is certain to affect the mechanical properties. The strength changing with the tempering temperature is shown in Fig. 5.17. The strength rapidly decreases with increasing tempering temperature, especially

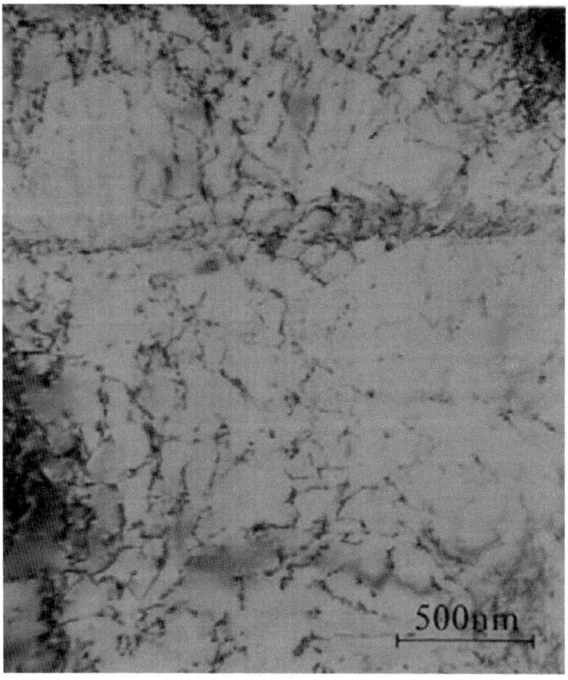

Fig. 5.16 TEM image of the steel tempered at 750 °C for 90 min, showing the MX type nitrides. With kind permission from Springer Science+Business Media (Zhang et al. 2012), Fig. 2

when tempered at 750 °C. The room temperature yield strength nearly decreases by 100 MPa when the tempering temperature is raised from 650 to 700 °C and by 150 MPa from 700 to 750 °C. It is obvious that the room temperature yield strength decreases more quickly when the steel is tempered at 750 °C. The commercial P92 steel was reported in literature to have the yield strength of 345 MPa and the tensile strength of 390 MPa when tempered at 750 °C (Mungole et al. 2008). Compared with this P92, the experimental steel tempered at 750 °C still has relatively higher yield strength of 515 MPa and tensile strength of 625 MPa. It is proved that the experimental steel can have comparable room temperature strength with the P92 steel.

The high temperature yield strength of 600 °C also decreases when the steel is tempered at 750 °C, as shown in Fig. 5.17. The yield strength of 600 °C is reduced by 69 MPa when the tempering temperature is increased from 650 to 700 °C and by 86 MPa from 700 to 750 °C. The experimental steel tempered at 750 °C displays high temperature yield strength of 307 MPa and tensile strength of 342 MPa, which are also comparable to those of the commercial P92.

Figure 5.18 demonstrates the toughness and the DBTT dependence on the tempering temperature. The half size CVN impact specimens tempered at 650 °C can only absorb 18 J energy at room temperature and 2 J at −20 °C, which indicates that the steel tempered at 650 °C has a high DBTT of above room temperature. When the steel is tempered at 700 °C, the CVN specimen can take in energy up to 86 J at room temperature but still decreases to 3.5 J at −20 °C, which indicate that the steel tempered at 700 °C presents a DBTT of about 0 °C. However, when the tempering temperature is increased to 750 °C, the steel can display not only good toughness of 96 J at room temperature, but also high toughness of 110 J at −20 °C, 89 J at −40 °C but 11 J at −60 °C, which indicate that the steel tempered at 750 °C has a low DBTT of around −50 °C. These results clearly demonstrate that the DBTT of the steel would be greatly decreased by increasing the tempering temperature.

The fractography analysis under SEM shows that the steel broken at low impact energy exhibits cleavage fracture characteristic, while the steel broken at high impact energy presents dimple fracture characteristic, as shown in Fig. 5.19. The

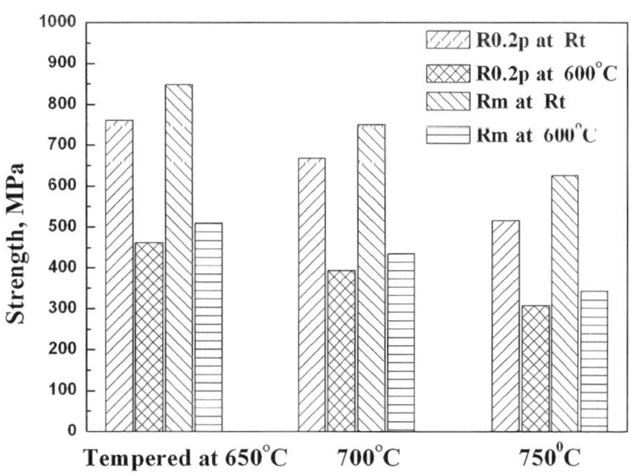

Fig. 5.17 Strength of the steel tempered at different temperatures. With kind permission from Springer Science+Business Media (Zhang et al. 2012, Fig. 3)

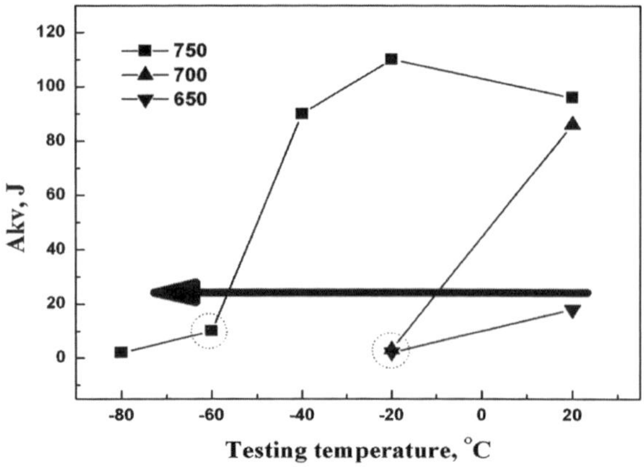

Fig. 5.18 Half size Charpy value of the steel at different temperatures. With kind permission from Springer Science+Business Media (Zhang et al. 2012, Fig. 4)

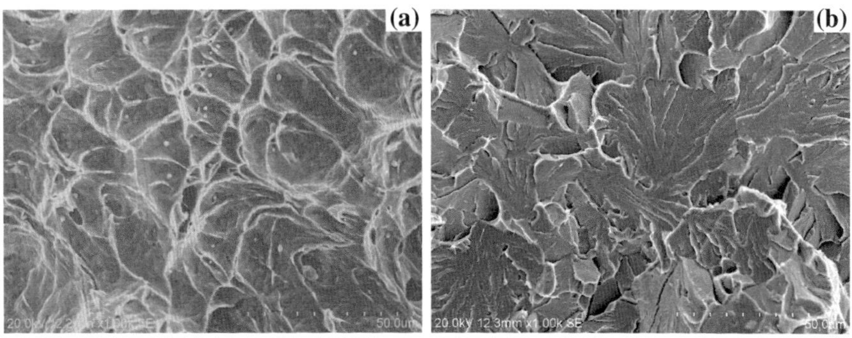

Fig. 5.19 Impact fractography of the broken steel piece tempered at 750 °C. **a** −40 °C, **b** −60 °C. With kind permission from Springer Science+Business Media (Zhang et al. 2012, Fig. 7)

steel tempered at 650 °C gives brittle cleavage fracture at both room temperature and −20 °C. The steel tempered at 700 °C displays ductile dimple fracture at room temperature, but brittle cleavage fracture at −20 °C (Zhang et al. 2012). The steel tempered at 750 °C does not exhibit brittle cleavage fracture until the temperature is decreased to −60 °C. When the temperature is above −40 °C, the surface of the impact fracture shows ductile dimple fracture. In the dimples on the fracture surface, many big size particles can be seen.

5.7.3 Effect of Nitride Precipitation on Yield Strength

The strength of the steel shows normal response to the increasing tempering temperature, i.e. the strength decreases with the increase of tempering temperature. However, it is noticeable that the room temperature yield strength decreases much quicker when tempering temperature is increased from 700 to

750 °C than from 650 to 700 °C, as described in Sect. 5.7.2. The (more) complete precipitation of nitrides should be responsible for the quicker decrease. It is known that the formation of precipitates consumes dislocations. The nitride precipitation will decrease much the number of dislocations in the matrix, resulting in weakening of dislocation strengthening. On the other hand, the formation of nitrides consumes the dissolved nitrogen which could provide strong solid solution strengthening. Therefore, although the nitride precipitation could produce precipitation strengthening, it is not enough to compensate for the loss of dislocation strengthening and nitrogen solid solution strengthening.

However, the high temperature yield strength of 600 °C does not show an obviously accelerated decrease when tempering temperature is increase to 750 °C. It could be interpreted from two views. The first one is that, at the high temperature of 600 °C, the dislocation is easier to move and annihilate. The advantage of high dislocation density is no longer obvious. The second view is that the more movable dislocation would become easier to reproduce because of nitride precipitation in the steel tempered at 750 °C. Therefore, the precipitation strengthening would mainly compensate for the loss of nitrogen solid solution strengthening. Hence, accelerated reduction is not observed in the high temperature yield strength of 600 °C.

5.7.4 Dependence of DBTT on Tempering Temperature

The DBTT decreases from above room temperature to −50 °C when the tempering temperature increases from 650 to 750 °C. In order to reach a clear understanding on DBTT of the steel, it is critical to take an investigation on the yield strength. That is the reason that a detailed discussion on the effect of tempering temperature on the yield strength has been made above.

It is widely believed that the cleavage fracture stress changes with temperature. Thus, it is easy to understand, in a phenomenological way, that the DBTT will be reduced with the decrease of yield strength when the tempering temperature is increased from 650 to 750 °C (Sawada et al. 2003). As discussed previously, the nitride precipitation will lead to the decrease of yield strength. Thus, it actually can be interpreted that the DBTT decrease is really associated with the nitride precipitation. The nitride precipitation has improved the impact toughness and decreased the DBTT by toughening the steel.

In summary of Sect. 5.7, the nitride precipitation in the steel reaches its peak when the tempering temperature is increased to 750 °C. Tempering at 650 or 700 °C could not induce the nitride precipitation, at least not to effective levels. The steel could achieve a martensitic microstructure strengthened by only nitrides after tempering at 750 °C. This microstructure is expected to have good thermal stability and high creep strength. The steel tempered at 750 °C could achieve comparable mechanical properties with the commercial P92 at both room temperature and 600 °C. The room temperature impact toughness of the steel is greatly enhanced from several Joules to nearly 100 J by increasing tempering temperature from 650 to 750 °C. The DBTT shows a great dependence on the tempering temperature. It could be reduced from above room temperature to −50 °C when the tempering temperature is increased from 650 to 750 °C.

References

Abe F, Taneike M, Sawada K (2007) Alloy design of creep resistant 9Cr steel using a dispersion of nano-sized carbonitrides. Int J Press Vessels Pip 84:3–12. doi:10.1016/j.ijpvp.2006.09.003

Baluc N, Gelles DS, Jitsukawa S, Kimura A, Klueh RL, Odette GR, van der Schaaf B, Yu J (2007) Status of reduced activation ferritic/martensitic steel development. J Nucl Mater 367–370:33–41. doi:10.1016/j.jnucmat.2007.03.036

Blach J, Falat L, Ševc P (2009) Fracture characteristics of thermally exposed 9Cr-1Mo steel after tensile and impact testing at room temperature. Eng Fail Anal 16:1397–1403. doi:10.1016/j.engfailanal.2008.09.003

Cipolla L, Danielsen HK, Venditti D, Di Nunzio PE, Hald J, Somers MAJ (2010) Conversion of MX nitrides to Z-phase in a martensitic 12% Cr steel. Acta Mater 58:669–679. doi:10.1016/j.actamat.2009.09.045

Danielsen HK, Hald J (2004) Z-phase in 9–12%Cr steels. In: Viswanathan R, Gandy D, Coleman K (eds) Proceedings of the 4th international conference on advances in materials technology for fossil power plants. ASM International, Materials Park, OH, pp 999–1012

Danielsen HK, Hald J (2006) Behaviour of Z phase in 9–12%Cr steels. Energ Mater 1:49–57. doi:10.1179/174892306X99732

Danielsen HK, Hald J (2007) A thermodynamic model of the Z-phase Cr(V, Nb)N. Calphad 31:505–514. doi:10.1016/j.calphad.2007.04.001

Ghassemi-Armaki H, Chen RP, Maruyama K, Yoshizawa M, Igarashi M (2009) Static recovery of tempered lath martensite microstructures during long-term aging in 9–12% Cr heat resistant steels. Mater Lett 63:2423–2425. doi:10.1016/j.matlet.2009.08.024

Hald J (2008) Microstructure and long-term creep properties of 9–12% Cr steels. Int J Pressure Vessels Pip 85:30–37. doi:10.1016/j.ijpvp.2007.06.010

Hesabi ZR, Simchi A, Reihani SMS (2006) Structural evolution during mechanical milling of nanometric and micrometric Al_2O_3 reinforced Al matrix composites. Mater Sci Eng A 428:159–168. doi:10.1016/j.msea.2006.04.116

Hu X, Xiao N, Luo X, Li D (2009) Effects of delta-ferrite on the microstructure and mechanical properties in a tungsten-alloyed 10% Cr ultra-supercritical steel. Acta Metall Sinica 45:553–558

Hu P, Yan W, Deng L, Sha W, Shan Y, Yang K (2010) Nitride-strengthened reduced activation ferritic/martensitic steels. Fusion Eng Des 85:1632–1637. doi:10.1016/j.fusengdes.2010.04.066

Jitsukawa S, Tamura M, van der Schaaf B, Klueh RL, Alamo A, Petersen C, Schirra M, Spaetig P, Odette GR, Tavassoli AA, Shiba K, Kohyama A, Kimura A (2002) Development of an extensive database of mechanical and physical properties for reduced-activation martensitic steel F82H. J Nucl Mater 307–311:179–186. doi:10.1016/S0022-3115(02)01075-9

Jun HJ, Kang JS, Seo DH, Kang KB, Park CG (2006) Effects of deformation and boron on microstructure and continuous cooling transformation in low carbon HSLA steels. Mater Sci Eng A 422:157–162. doi:10.1016/j.msea.2005.05.008

Kimura K, Toda Y, Kushima H, Sawada K (2010) Creep strength of high chromium steel with ferrite matrix. Int J Press Vessels Pip 87:282–288. doi:10.1016/j.ijpvp.2010.03.016

Klueh RL (2005) Elevated temperature ferritic and martensitic steels and their application to future nuclear reactors. Int Mater Rev 50:287–310. doi:10.1179/174328005X41140

Li Y, Huang Q, Wu Y (2006) Study on impact and tensile properties of CLAM steel. Nucl Phys Rev 23:151–154

Li Y, Nagasaka T, Muroga T (2010) Long-term thermal stability of reduced activation ferritic/martensitic steels as structural materials of fusion blanket. Plasma Fusion Res 5:S1036. doi:10.1585/pfr.5.S1036

Lu Z, Faulkner RG, Riddle N, Martino FD, Yang K (2009) Effect of heat treatment on microstructure and hardness of Eurofer 97, Eurofer ODS and T92 steels. J Nucl Mater 386–388:445–448. doi:10.1016/j.jnucmat.2008.12.152

Maruyama K, Sawada K, Koike JI (2001) Strengthening mechanisms of creep resistant tempered martensitic steel. ISIJ Int 41:641–653. doi:10.2355/isijinternational.41.641

Mungole MN, Sahoo G, Bhargava S, Balasubramaniam R (2008) Recrystalised grain morphology in 9Cr 1Mo ferritic steel. Mater Sci Eng A 476:140–145. doi:10.1016/j.msea.2007.04.105

Pešička J, Kužel R, Dronhofer A, Eggeler G (2003) The evolution of dislocation density during heat treatment and creep of tempered martensite ferritic steels. Acta Mater 51:4847–4862. doi:10.1016/S1359-6454(03)00324-0

Poulachon G, Dessoly M, Lebrun JL, Le Calvez C, Prunet V, Jawahir IS (2002) Sulphide inclusion effects on tool-wear in high productivity milling of tool steels. Wear 253:339–356. doi:10.1016/S0043-1648(02)00122-9

Reith M, Schirra M, Falkenstein A, Graf P, Heger S, Kempe H, Lindau R, Zimmermann H (2003) In: EUROFER 97. Tensile, charpy, creep and structural tests. Wissenschaftliche Berichte FZKA 6911

Ryu SH, Lee YS, Kong BO, Kim JT, Kwak DH, Nam SW et al (2006). In: Proceedings of the 3rd international conference on advanced structural steels. The Korean Institute of Metals and Materials, pp 563–569

Sawada K, Kubo K, Abe F (2001) Creep behavior and stability of MX precipitates at high temperature in 9Cr-0.5Mo-1.8W-VNb steel. Mater Sci Eng A 319–321:784–787. doi:10.1016/S0921-5093(01)00973-X

Sawada K, Kimura K, Abe F (2003) Mechanical response of 9% Cr heat-resistant martensitic steels to abrupt stress loading at high temperature. Mater Sci Eng A 358:52–58. doi:10.1016/S0921-5093(03)00326-5

Sawada K, Taneike M, Kimura K, Abe F (2004) Effect of nitrogen content on microstructural aspects and creep behavior in extremely low carbon 9Cr heat-resistant steel. ISIJ Int 44:1243–1249. doi:10.2355/isijinternational.44.1243

Sawada K, Kushima H, Kimura K, Tabuchi M (2007) TTP diagrams of Z phase in 9–12% Cr heat-resistant steels. ISIJ Int 47:733–739. doi:10.2355/isijinternational.47.733

Shen YZ, Kim SH, Han CH, Cho HD, Ryu WS (2009) TEM investigations of MN nitride phases in a 9% chromium ferritic/martensitic steel with normalization conditions for nuclear reactors. J Nucl Mater 384:48–55. doi:10.1016/j.jnucmat.2008.10.005

Sklenička V, Kuchařová K, Svoboda M, Kloc L, Buršík J, Kroupa A (2003) Long-term creep behavior of 9–12% Cr power plant steels. Mater Charact 51:35–48. doi:10.1016/j.matchar.2003.09.012

Taneike M, Sawada K, Abe F (2004) Effect of carbon concentration on precipitation behavior of $M_{23}C_6$ carbides and MX carbonitrides in martensitic 9Cr steel during heat treatment. Metall Mater Trans A 35A:1255–1262. doi:10.1007/s11661-004-0299-x

Tanigawa H, Shiba K, Möslang A, Stoller RE, Lindau R, Sokolov MA, Odette GR, Kurtz RJ, Jitsukaw S (2011) Status and key issues of reduced activation ferritic/martensitic steels as the structural material for a DEMO blanket. J Nucl Mater 417:9–15. doi:10.1016/j.jnucmat.2011.05.023

Yamada K, Igarashi M, Muneki S, Abe F (2003) Effect of Co addition on microstructure in high Cr ferritic steels. ISIJ Int 43:1438–1443. doi:10.2355/isijinternational.43.1438

Yan W, Shan YY, Yang K (2007) Influence of TiN inclusions on the cleavage fracture behavior of low-carbon microalloyed steels. Metall Mater Trans A 38A:1211–1222. doi:10.1007/s11661-007-9161-2

Yan W, Hu P, Deng L, Wang W, Sha W, Shan Y, Yang K (2012) Effect of carbon reduction on the toughness of 9CrWVTaN steels. Metall Mater Trans A 43A:1921–1933. doi:10.1007/s11661-011-1046-8

Yllmaz Ş, Ipek M, Celebi GF, Bindal C (2005) The effect of bond coat on mechanical properties of plasma sprayed Al_2O_3 and Al_2O_3–13 wt%TiO_2 coatings on AISI 316L stainless steel. Vacuum 77:315–321. doi:10.1016/j.vacuum.2004.11.004

Yong Q (2006) The second phase in steels. Metallurgical Industry Press, Beijing

Yoshizawa M, Igarashi M (2007) Long-term creep deformation characteristics of advanced ferritic steels for USC power plants. Int J Press Vessels Pip 84:37–43. doi:10.1016/j.ijpvp.2006.09.005

Zhang L, Thomas BG (2003) State of the art in evaluation and control of steel cleanliness. ISIJ Int 43:271–291. doi:10.2355/isijinternational.43.271

Zhang L, Thomas BG, Wang X, Cai K (2002) Evaluation and control of steel cleanliness-review. In: 85th Steelmaking conference proceedings. ISS-AIME, Warrendale, PA, pp 431–452

Zhang W, Yan W, Sha W, Wang W, Zhou Q, Shan Y, Yang K (2012) The impact toughness of a nitride-strengthened martensitic heat resistant steel. Sci China Technol Sci 55:1858–1862. doi:10.1007/s11431-012-4903-9

Zhou Q, Zhang W, Yan W, Wang W, Sha W, Shan Y, Yang K (2012) Microstructure and mechanical properties of a nitride-strengthened reduced activation ferritic/martensitic steel. Metall Mater Trans A 43A:5079–5087. doi:10.1007/s11661-012-1311-5

Chapter 6
Ultra High-Strength Maraging Steel

Abstract Maraging steels are high-strength steels combined with good toughness. They are used particularly in aerospace and tooling applications. Maraging refers to the ageing of martensite, a hard microstructure commonly found in steels. Research on the kinetics of precipitate formation and austenite reversion in maraging steels has received great attention due to their importance to steel properties. Judging from the literature in recent years, research into maraging steels has been very active, mainly extending to new types of steels, for new applications beyond the traditional strength requirements. This chapter provides an in-depth overview of the literature in this area. It includes a state-of-the-art review of ultra high-strength steels, and discussions on types of maraging steels, and microstructure and precipitates in maraging steels.

6.1 State-of-the-Art of Ultra High-Strength Steels

Steels combining properties of ultrahigh-strength (UHS) and good ductility are of great importance in automotive, aerospace, nuclear, gear, bearing and other industries. They are the future key materials for lightweight engineering design strategies and corresponding CO_2 savings. Driven by the development of metallurgy and in response to commercial demands, both academia and industry have made great efforts to develop a large variety of steel grades and processing technology, and hence achieved continuous improvement for some decades.

Conventionally produced bulk high-strength steels are known as high-strength low-alloy (HSLA) or microalloyed steels (Chap. 2). This family of steels usually has a strength not exceeding 700–800 MPa and a microstructure of fine-grained ferrite that has been strengthened with carbon and/or nitrogen precipitates of titanium, vanadium or niobium. These steels can be manufactured under relatively well-established processing conditions and have been widely applied for weight reduction in automotive and general construction applications. In order to further enhance the combination of strength and ductility, more alloying elements have been added and more sophisticated alloy systems have been designed employing various mechanisms including dual phase (DP) steels, transformation induced

W. Sha, *Steels*, DOI: 10.1007/978-1-4471-4872-2_6,

plasticity (TRIP) steels, twinning induced plasticity (TWIP) steels and maraging steels. Typical strength-ductility profiles of those steel families are shown in Fig. 6.1. Dual phase steels have a microstructure of mainly soft ferrite, with islands of hard martensite dispersed throughout. The strength level of these grades is related to the amount of martensite in the microstructure along with its distribution and morphology. TRIP steels are multiphase grades that involve special alloying and heat treatments to stabilise some amount of austenite at room temperature embedded in a primary ferritic matrix. During plastic deformation and straining, the retained austenite progressively transforms to martensite with increasing strain. This leads to a volume and shape change within the microstructure, which accommodates the strain and increases the ductility. Unlike DP and TRIP steels, TWIP steels generally have a fully austenitic microstructure at room temperature. The formation of mechanical twins during deformation generates high strain hardening, preventing necking and thus maintaining a very high strain capacity and achieving a better combination of strength and ductility. Another important category of UHS steels is maraging steels, in which martensite is formed upon quenching following solution treatment and subsequently further strengthened by the formation of various precipitates such as Cu clusters, Ni_3Ti, NiAl and other types of intermetallics at a moderate temperature of about 500 °C. The superior properties of maraging steels, such as ultra high strength, high ductility, good hardenability, good weldability, simple heat treatment without deformation steps, sometimes in combination with good corrosion resistance, have led to widespread application of maraging steels for demanding applications. The composition and mechanical properties of some successful maraging steel grades are summarised in Table 6.1. All steels in this table except 300 M, Aermet 100 and Low Ni are stainless grades.

In applications of UHS steels, such as in the automotive, aerospace and nuclear industries, the materials are usually subjected to extreme mechanical loads and harsh environmental conditions in which corrosion is an important issue. However, the most commonly used steels for these applications, such as maraging

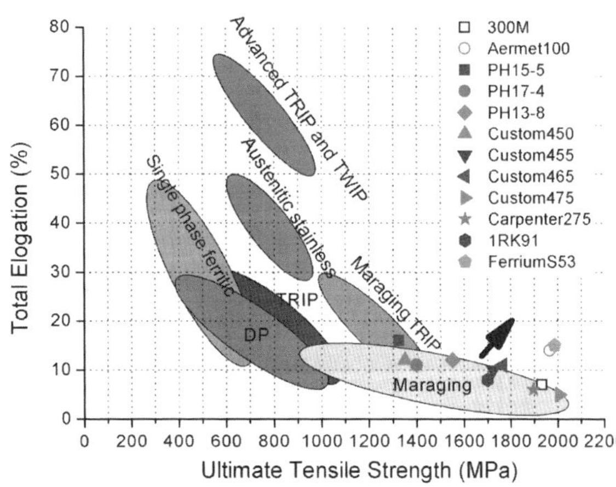

Fig. 6.1 The strength-ductility profile of various steels. The symbols indicate the properties of existing high-end maraging steel grades

Table 6.1 Compositions in weight per cent, yield strength (YS), ultimate tensile strength (UTS) and elongation (El) of some grades of maraging steels

Steel	C	Cr	Ni	Mo	Cu	Mn	Si	Others	YS (MPa)	UTS (MPa)	El
300 M	0.4	0.8	1.8	0.4	–	0.8	1.6	V (0.05)	1586	1995	0.10
Aermet 100	0.23	3.10	11.1	1.2	–	–	–	Co (13.4)	1724	1965	0.14
PH15-5	<0.07	14–15.5	3.5–5.5	–	2.5–4.5	<1.0		–	1228	1325	0.16
PH17-4	<0.07	15.5–17.5	3.0–5.0	–	3.0–5.0	<1.0	<1.0	–	1275	1399	0.11
PH13-8	<0.05	12.3–13.3	7.5–8.5	2.0–2.5	–	<0.1	–	Al (0.90-1.35)	1448	1551	0.12
Custom450	<0.05	14.0–16.0	5.0–7.0	0.5–1.0	1.3–1.8	<1.0	<1.0	–	1296	1351	0.12
Custom455	<0.05	11.0–12.5	7.5–9.5	–	1.5–2.5	<0.5	<0.5	Ti (0.8-1.4), Nb (0.1-0.5)	1689	1724	0.10
Custom465	<0.02	11.0–12.5	10.8–11.3	0.75–1.25	–	<0.25	–	Ti (1.5-1.8)	1707	1765	0.11
Custom475	<0.01	10.5–11.5	7.5–8.5	4.5–5.5	–	<0.5	–	Co (8.0-9.0), Al (1.0-1.5)	1972	2006	0.05
Carpenter275	<0.02	11.0–12.5	10.8–11.3	0.75–1.25	–	<0.25	–	Ti (1.55-1.80), Nb (0.15-0.30)	1758	1896	0.06
1RK91	0.01	12.2	9.0	4.0	1.95	0.32	0.15	Ti (0.87)	1500	1700	0.08
FerriumS53	0.20	10.0	5.5	2.0	–	–	–	Co (14.0), W (1.0), V (0.3)	1517	1986	0.15
Low Ni	0.046	0.20	12.94	1.01	–	<0.01	<0.05	Al (1.61), Nb (0.23)	–	1594	–

300 M for aircraft landing gear, and DP/TRIP in automotive industry, are non-stainless steel grades. Consequently, a specialised and costly coating or plating process has to be employed for corrosion protection. From manufacturing, environmental and reliability considerations, a stainless steel with equivalent mechanical properties but not requiring a corrosion protective coating would be an attractive alternative. Among existing stainless steels, the majority are austenitic, ferritic or duplex (austenite and ferrite) grades. Therefore, much effort for developing ultra high-strength stainless steels have been undertaken, employing martensite as the matrix and strengthening further by various precipitates, while a high Cr content is imposed to ensure corrosion resistance. Some successful grades of these maraging stainless steels are shown in Table 6.1. Notwithstanding the efforts to develop high-strength and high ductility stainless steels, Table 6.1 shows that few stainless steels meet the properties of their non-stainless counterparts in terms of strength and ductility, because of the nature of ferrite and austenite.

6.2 Types of Maraging Steels

Precipitation hardening is one of the most effective mechanisms for increasing the strength of metallic high performance materials. The chemical and crystallographic relations of particles and matrix determine the precipitation behaviour and consequently the material's properties. Precipitation hardening was discovered, a century ago, in aluminium alloys. The principle of particle strengthening is hindering of dislocation movement by the precipitation of nanometre-sized particles. In steels, particle strengthening can either be caused by the precipitation of carbides, e.g. in so-called secondary hardening low-alloy steels, or intermetallic phases, e.g. in so-called maraging steels. Secondary hardening steels are also the major steel grade for today's tool steels. Generally, secondary hardening steels exhibit a complex microstructure depending on alloy composition and the heat treatment applied. The heat treatment consists of hardening, which denotes austenitising in the gamma phase field followed by quenching, and tempering, usually several times. The precipitation of the nanometre-sized carbides has been the focus of numerous studies, and is still a subject of research activities.

The second group of particle-strengthened steels, the maraging steels, was developed over the past 70 years. The term 'maraging' refers to the ageing of martensite, but specifically a martensite that is easily obtained at *low* cooling rates due to a high Ni content in the steel. In addition, these steel grades have very low carbon content and as implied above can belong to the wider group of ultra high-strength steels. Their development started in the 1940s in the USA, when it was noticed that magnetic Fe–Ni-Ti–Al alloys could be hardened significantly by heat treatment. After an initial development period and the addition of Co and Mo, the first applications of maraging steels were set out in the 1960s. Then, the primary goal was to develop high-strength steels for submarine hulls, but maraging steels proved unsuitable for these applications. However, besides a market developed for specialised aerospace and military applications of these first maraging grades, applications for tools and dies became common. Due to the sharp drop in availability and rising cost of Co in the late 1970s, alternatives to cobalt containing maraging steels were in demand. Thus, much effort was made on developing Co-free maraging steels with appropriate mechanical properties (Sha and Guo 2009). This resulted in a variety of steels with different precipitating elements, e.g. Al, Ti and Cu, and additionally, alloys with lowered Ni contents. One example of these developments is the PH13-8Mo types (Sha and Guo 2009). These cobalt-free alloys generally have inferior properties to the cobalt containing grades, but their properties appear satisfactory for their designed applications, and the absence of cobalt is a distinctive cost advantage.

Intermetallic phases have many specific features as they differ from carbides and thus materials that are strengthened by intermetallic precipitates are of great interest for industrial applications. Specifically, they differ from carbides in the following ways:

(1) Intermetallic phases are formed during primary crystallisation without eutectic transformation, because of which they distribute much more evenly in the cast steel and consequently also in the deformed steel than carbides do. Because of the primary crystallisation we already have a homogeneous distribution in the as-cast state, which is not the case for eutectic solidification. When such a material is deformed the distribution is still homogeneous. In contrast, eutectic carbides, which actually form a network, become more or less homogeneously distributed by deformation (forging or rolling) which is always worse. Intermetallic particle size is also smaller, up to 2–3 μm in diameter. Because of this, intermetallic phases have a less detrimental effect on the strength and ductility and on the variations of these properties depending upon the degree of deformation.
(2) Intermetallic phases precipitate from supersaturated solution during ageing to result in precipitation hardening. The composition of these precipitated phases does not differ from that of the phases mentioned above, i.e. appearing during solidification and present in hot rolling, etc. Precipitation of intermetallic phases is characterised by the following essential features.

(i) The dispersion of intermetallic particles being formed is significant: at the maximum hardness, their size is less than 5–20 nm and the distances between particles around 100 nm, which is lower than the respective values for carbides.
(ii) The distribution of precipitated particles is relatively uniform, since they form in low-carbon or carbon-free martensite (or austenite), especially when Ni is present in the solution. For that reason, the embrittlement effect of the precipitation hardening, even with fine particles, is weaker than that of carbides. However, the embrittling effect of intermetallics does increase with increasing volume fraction.
(iii) The strengthening effect of precipitated intermetallics is very high, greater than with carbides. It is a function of the structure of the matrix,

being greater for a martensitic matrix when compared to an austenitic one. Precipitation hardening increases the hardness by 20–40 HRC for carbon-free martensite (against 3–10 HRC with carbides) and by ~30 HRC for austenite. This is due to the more refined dispersion of intermetallic phases precipitated, with higher volume fractions, compared with carbides.

(3) The temperatures of precipitation hardening which produce the maximum rise in hardness depend on the composition of the metallic matrix and the type of precipitated intermetallic compounds. For martensitic steels, they are lower when intermetallic compounds $(Fe,Ni,Co)_7(Mo,W)_6$ and $(Fe,Cr)_3(Ti,Al)$ precipitate, around 500–550 °C, and higher for the intermetallic phase $(Fe,Co)_7(W,Mo)_6$, around 580–650 °C. In austenitic alloys, these processes are shifted towards still higher temperatures, 750–800 °C and higher. For that reason, steels based on carbon-free martensite can acquire a very high hardness, up to 68–69 HRC, and an increased thermal stability, up to 600–720 °C, depending on the type of intermetallic phase. Austenitic steels have a still higher thermal stability, but a lower hardness.
(4) During precipitation hardening of nickel-containing steels, their hardness rises intensively during the first 10–15 min, but then attains its maximum only after a long holding, 5–10 h, instead of 30–40 min required in carbide strengthening.
(5) Coagulation of precipitated phases and, therefore, reduction in hardness, occur at correspondingly higher temperatures of ageing.

Generally, there are different ways as to how steel grades can be classified, e.g. according to their composition, or strength levels. Such a classification is also possible for maraging steels. A classification according to the main alloy system gives the best overview for maraging steels. The Fe–Ni-Mo system is briefly described in the following.

The first alloys, e.g. Fe-28Ni-4Ti-4Al identified and further work with the Co-Mo hardening system led to the development of the well-known maraging steels based on the Fe–Ni-Mo system, namely the well-known 18Ni (200), 18Ni (250) and 18Ni (300) alloys (Table 6.2). The numbers in parenthesis refer to the nominal yield strengths in ksi (pound-force per square inch, lbf/in^2) in the aged condition. Titanium was also added to this group of alloys as a supplemental hardener. In these alloys, hardening is produced by the combination of Co and Mo as Co

Table 6.2 Nominal compositions (wt%) and the respective yield strength (YS) of commercial maraging steels (Inco)

Alloy designation	Ni	Mo	Co	Ti	Al	YS (MPa)
18Ni (200)	18	3.3	8.5	0.2	0.1	1400
18Ni (250)	18	5.0	8.5	0.4	0.1	1700
18Ni (300)	18	5.0	9.0	0.7	0.1	2000
18Ni (350)	18	4.2	12.5	1.6	0.1	2400
18Ni (cast)	17	4.6	10.0	0.3	0.1	1650

lowers the solubility of Mo and thus increases the amount of Mo-rich precipitates that form during ageing. A number of studies have been conducted on the precipitation behaviour of these maraging steels. Overall, the following precipitation reactions take place during ageing. Strengthening in the underaged conditions is caused by Mo-rich zones. With further ageing, the metastable orthorhombic Ni_3Mo forms which transforms into $Fe_2(Mo,Ti)$ hexagonal Laves phase after longer ageing. The formation of Ni_3Mo is accelerated by Co. However, it is suggested that higher Ti contents (alloy 350) lead to the formation of Ni_3Ti instead of Ni_3Mo.

As mentioned above, around 1978–80, development of cobalt-free grades was promoted (Sha and Guo 2009). Considering that the goal was the development of high fracture toughness in cobalt-free maraging steel, various elements that might successfully substitute for Co were considered. An analysis of a number of variants revealed that the alloy should contain 3 % Mo and 1.4 % Ti to achieve the desired yield strength of 250 ksi (1700 MPa) with good transverse ductility. Ni_3Ti was found to be the major precipitation-hardening phase in cobalt-free maraging steels, whereas the achieved yield strength rises with increased Mo and Ti contents (up to 2400 MPa). The mechanism of strengthening and toughening is the same in low and high Mo and Ti containing alloys. However, the high Mo and Ti alloys additionally contain a large number of coarse particles of type Fe_2Ti and $Fe_2(Mo,Ti)$ embedded in the martensite lath boundary or within the lath. Those particles produce a detrimental effect on fracture toughness and ductility.

Other maraging systems include broadly Fe–Ni-Cr and Fe–Ni-Mn, which will be discussed in the following sections.

6.3 Microstructure and Precipitates in Maraging Steels

6.3.1 PH13-8Mo Maraging Steels

The microstructural constituents that were observed in PH13-8Mo maraging steels include the following:

(1) δ ferrite
(2) cubic martensite
(3) retained and reverted austenite
(4) nanometre-sized precipitates of intermetallic type
(5) Laves phase.

A phase that is sometimes present in PH13-8Mo maraging steels is δ ferrite, the high temperature bcc ferrite. The presence of this phase is associated with the alloy composition and production route. Although the solidification path of the PH13-8Mo steel grades is known to be fully ferritic via the δ phase, the transformation of $\delta \rightarrow \gamma$ strongly depends on cooling rate. Thus, large cast sizes and an appropriate composition can provoke the incomplete transformation of $\delta \rightarrow \gamma$.

PH13-8Mo maraging steels receive their excellent combination of strength and ductility by a heat treatment which consists of a first-step solution anneal in the temperature range 900–1000 °C and subsequent air-cooling to room temperature. When quenched from the austenitic single-phase field after solution annealing, the austenite transforms into a soft, but heavily dislocated Ni-martensite, exhibiting a cubic structure. The martensite in this steel grade is of lath type and it forms a packet that consists of many similar-sized laths arranged parallel to each other. Each prior austenite grain includes several lath packets.

Depending mainly upon the Ni content, but also on all other alloying elements, a certain amount of austenite can be retained after quenching to room temperature. Besides this retained austenite, which is the result of a martensite finish temperature below room temperature, a so-called 'reverted austenite' can also be present in maraging steels. The occurrence of this reverted austenite also depends upon chemical composition, but additionally, the applied ageing temperature and time play decisive roles. The reverted austenite emerges from martensite during ageing at temperatures below the global $\alpha \rightarrow \gamma$ transformation temperature. This is the temperature where the entire material transforms, which is significantly higher than the local temperature because of Ni-enrichment.

Ageing, typically carried out in a temperature range of 400–600 °C, leads to the precipitation of the strengthening nanometre-sized intermetallic particles.

Laves phases formed at higher ageing temperatures have also been observed in this steel grade.

However, the nanometre-sized precipitates and the austenite phase fraction have the strongest effect on the mechanical properties of maraging steels and, thus, a more detailed survey of the literature for these aspects is given in the following.

6.3.2 Precipitates

The precipitation behaviour and strengthening mechanisms in cobalt-free maraging steels have been extensively studied by employing a variety of characterisation techniques. It has been shown that maraging steels containing Ni and Al are strengthened by the formation of the ordered β'-NiAl phase with B2 (CsCl) superlattice structure (Sha and Guo 2009). This structure consists of two interpenetrating primitive cubic cells, where the Al atoms occupy the cube corners of the first sublattice and the Ni atoms occupy the cube corners of the second sublattice. The lattice constant of the stoichiometric composition is 0.2887 nm. This is very close to the lattice constant for ferrite, which has a value of 0.28664 nm. Due to this fact, the NiAl precipitates are coherent with the matrix and even remain coherent after long ageing times. Precipitation takes place immediately after heating the material to the ageing temperature. Sha and Guo (2009) show Ni- and Al-rich clusters after just 6 min of ageing at 593 °C and after 40 min of ageing at 510 °C. However, the composition of the precipitates, even after longer ageing times, is far from stoichiometric NiAl phase, the precipitates containing a significant amount of Fe. The shape of the precipitates is a matter of debate in the literature. In general, the NiAl phase is believed to be of spherical shape, but some authors assume a change of morphology from spherical- to needle- and plate-shaped forms at longer ageing times. However, it is generally thought that their formation proceeds via solute-rich clusters within the martensitic matrix. Out of these nuclei, the coherent NiAl precipitates are formed and distributed uniformly within the matrix (Leitner et al. 2010).

In contrast, it has been reported that strengthening in Ti-containing maraging steels is caused by the precipitation of the η-phase ($Ni_3(Ti,Al)$) (Leitner et al. 2010). The η-phase exhibits a hexagonal lattice with $a = 0.255$ nm and $c = 0.42$ nm. In the literature, some debate exists on the mechanism of formation of the η-phase. In most studies, heterogeneous nucleation on dislocations is proposed, with subsequent growth taking place via pipe diffusion (Dutta et al. 2001). Others advance the theory that formation of coherent zones on dislocations in the martensitic matrix takes place first, acting as nucleation sites for the η precipitates (Leitner et al. 2010). There is also some discrepancy on the dominating strengthening mechanism in this kind of maraging steels. While the majority of studies attributed strengthening to the formation of Ni_3Ti, others proposed an additional contribution from some B2-type ordering of Fe and Ni atoms in the matrix (Leitner et al. 2010).

A further type of precipitate phase, called G-phase, was found in Ti-containing maraging steels alloyed with Si. The G-phase exhibits the chemical composition $Ti_6Si_7Ni_{16}$ and precipitates primarily on grain boundaries. Depending upon the chemical composition of the alloy, the G-phase ($Ti_6Si_7Ni_{16}$) and the η-phase (Ni_3Ti) can precipitate either simultaneously or separately. The shape of the Ni_3Ti-phase was thought to be rod-like, whereas the G-phase shows a spherical morphology. Later studies revealed that both the spherical G-phase and the rod-shaped η-phase are formed independently out of an undefined precursor phase, which is responsible for strengthening up to peak hardness. Atom probe tomography (APT) was used to follow the precipitation sequence in an Si-free Fe–Cr–Ni–Al–Ti stainless steel during ageing at 525 °C (Schober et al. 2009). In that work, instead of splitting into G-phase ($Ti_6Si_7Ni_{16}$) and η-phase (Ni_3Ti), the formation of spherical NiAl particles and elongated $Ni_3(Ti,Al)$ particles out of an undefined precursor phase in the early stages was found.

A further age-hardener in maraging steels is copper. It is used in PH15 5 and PH17–4 alloys as a precipitating hardening element, but also in alloy systems such as 1RK91 (Sandvik in-house grade designation) and C455 (Custom455, a registered trademark of Carpenter Technology Corporation) (for compositions see Table 6.1) as nucleation site for the precipitation of the strengthening causing phase (Schnitzer et al. 2010a).

In the case of 1RK91, the precipitation sequence starts with the formation of an Ni, Ti and Al-rich phase nucleating on Cu-rich clusters (Schnitzer et al. 2010a). During further ageing, $Ni_3(Ti,Al)$ precipitates develop adjacent to Cu-rich precipitates. The C455, which contains less Al and no Mo, forms clusters of Cu, Ti and Ni at the early stages, and they are thought to be a precursor phase of η-Ni_3Ti (Schnitzer et al. 2010a). An extended ageing treatment then leads to the separation of Ni-rich (η-Ni_3Ti) and Cu-rich precipitates. The effect of Cu on the precipitation evolution in PH13-8Mo type maraging steel, which exhibits NiAl and η-phase precipitates, was recently investigated by Schnitzer et al. (2010a). This study revealed the formation of NiAl out of a Cu-containing precursor phase and the nucleation of η-phase on independent Cu clusters.

There exists almost a comprehensive picture of precipitation behaviour in PH13-8Mo type maraging steels. Only the influence of Cu on the formation of G-phase is missing.

Additionally to the strengthening precipitates, the formation of carbides has been observed in maraging, although the carbon content is generally very low in such steels. These carbides precipitate adjacent to or inside reverted austenite because of the higher carbon solubility in austenite. The carbides were found to be of type $Cr_{23}C_6$ and $(Cr,Mo)_2C$.

6.4 Reverted Austenite and Mechanical Properties

6.4.1 Reverted Austenite

In maraging steels, partial reversion from martensite to austenite can take place during ageing below the global $\alpha \rightarrow \gamma$ temperature (Schober et al. 2009). The amount of reverted austenite present after ageing depends on the ageing temperature as well as ageing time (Sha and Guo 2009). The formation of reverted austenite is reported to be the result of local Ni-enrichment, which arises during ageing. As Ni, one of the major alloying elements in maraging steels, is an austenite stabilising element, the local transition temperature of martensite to austenite is lowered. In this area, the applied ageing temperature is, therefore, higher than the transition temperature from martensite to austenite. Due to this Ni enrichment, this austenite remains stable even when the material is cooled back to room temperature. This type of austenite has to be distinguished from retained austenite, which can also be present in maraging steels after cooling from the solution annealing temperature. Depending on the alloy composition and the applied heat treatment, different morphologies of reverted austenite have been observed in maraging steels, shown in Fig. 6.2. The reverted austenite morphologies are classified into three types: (i) matrix austenite, (ii) lath-like austenite and (iii) recrystallised austenite. Matrix austenite (Fig. 6.2a) grows either from retained austenite with the same orientation, or nucleates on prior austenite grain boundaries and then forms a single grain. Lath-like austenite (Fig. 6.2b) can develop within the martensite lath or grow along the lath boundaries of martensite. The growing direction of the austenite laths is parallel to the martensite lath resulting in a lamellar structure of austenite and martensite. The elongated shape is supposed to be caused by a growth impediment of the adjacent martensite laths. Recrystallised austenite (Fig. 6.2c) forms at higher ageing temperatures or longer ageing times, and is characterised by a very low density of defects and dislocations (Viswanathan et al. 2005).

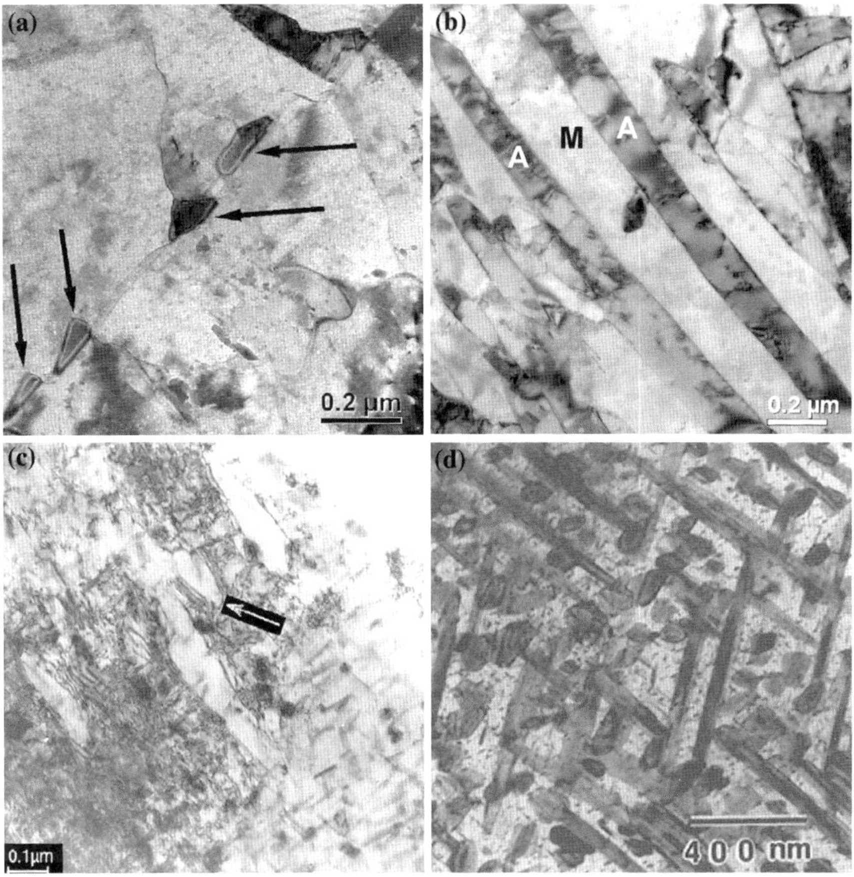

Fig. 6.2 Morphologies of reverted austenite in maraging steels **a** matrix austenite (Schnitzer et al. 2010b), **b** lath-like austenite (A = austenite, M = martensite) (Schnitzer et al. 2010b), **c** recrystallised austenite (Viswanathan et al. 2005), **d** Widmanstätten austenite (Kim and Wayman 1990)

In addition, a fourth morphology found in maraging steels, is Widmanstätten (Fig. 6.2d), only evidenced in high Ni-alloyed and Ti-containing maraging steels (Schnitzer et al. 2010b). The morphology of reverted austenite developed depends on the applied heating rate and on the ageing temperature (Schnitzer et al. 2010c).

Independently of morphology, the orientation relationship between austenite and martensite was found to be either of type Kurdujumov–Sachs, with $(110)_{bcc}//(111)_{fcc}$ and $[111]_{bcc}//[110]_{fcc}$, or Nishiyama-Wassermann with $(110)_{bcc}//(111)_{fcc}$ and $[100]_{bcc}//[110]_{fcc}$ (Schnitzer et al. 2010b).

An important finding for the understanding of the formation of reverted austenite is the fact that the regions of reverted austenite are free of the precipitates causing strengthening. Only carbides precipitate inside the austenite or adjacent to it, due to the higher solubility for carbon in austenite compared to martensite.

The formation mechanism of reverted austenite in maraging steels is supposed to be a diffusion-controlled process. It is assumed that the formation of reverted austenite is connected to the dissolution of precipitates, resulting in a local enrichment of austenite stabilising elements (Sha and Guo 2009). This was also reported for 18 % Ni steels, where the formation of reverted austenite was encouraged by the dissolution of $Ni_3(Mo,Ti)$ precipitates. The formation of reverted austenite can be accompanied by the precipitation of Fe_2Mo. This leads to the assumption that the Ni of the dissolved $Ni_3(Mo,Ti)$ precipitates contributes to the formation of reverted austenite, while the released Mo is used to form Fe_2Mo precipitates. Others exclude the theory that the formation of reverted austenite is related to the dissolution of precipitates. For example, Ni diffuses to dislocations and other defects, resulting in microsegregation of the austenite stabilising elements in localised areas. Hsiao et al. (2002) reported that in Cu-containing maraging alloys, such as PH17-4, fine Cu precipitates act as nucleation sites for austenite formation since both copper and austenite exhibit the same crystallographic structure with similar lattice parameter. Kim and Wayman (1990) suggested that the formation of lath like austenite in high Ni-containing maraging alloys is shear dominated, but assisted by a diffusion-controlled process. Schnitzer et al. (2010b) investigated the formation of reverted austenite in PH13-8Mo type maraging steel. The interpretation of the experimental data was supported by thermodynamic and kinetic calculations. It was shown that the formation of NiAl precipitates and reverted austenite starts from the very beginning of ageing widely independent of each other, i.e. the formation of both, austenite and NiAl nuclei, is thermodynamically possible in the virgin martensitic matrix. Thus, dissolution of precipitates is not necessarily the initial driving factor for the formation of reverted austenite. Consequently, a formation mechanism was proposed as follows: beginning from lath martensite with high dislocation density after solution annealing, NiAl precipitates and reverted austenite nucleate independently. The NiAl precipitates are homogeneously distributed in the martensitic matrix and reverted austenite forms preferentially on prior austenite grain boundaries or on martensite lath boundaries. Growth of reverted austenite is related to dissolution of adjacent NiAl precipitates, whereas the reverted austenite incorporates Ni stemming from the dissolved precipitates. In contrast, the released Al content is transferred to the NiAl precipitates left in the matrix.

6.4.2 Mechanical Properties

Maraging steels belong to the group of high-strength steels, i.e. tensile strength >1000 MPa. The achieved strength values range from 1000 MPa up to 3000 MPa. The stress–strain behaviour is characterised by a yield strength that is above 80 % of the ultimate tensile strength, i.e. the work hardening is small. The reason for this behaviour is proposed to be as follows. The coherent precipitates introduce a friction stress acting on dislocations, which raises the level of the stress–strain curve without markedly altering the dislocation interactions leading to work hardening.

The following microstructural features have an influence on the mechanical properties, in different ways/extents:

(1) Nanometre-sized precipitates
(2) Retained or reverted austenite
(3) Prior austenite grain size.

The required strength and toughness is reached after an ageing treatment, which conforms to the application (Sha and Guo 2009). Of course, the chemical composition, but also the applied ageing temperature and time influence types, size and volume fraction of precipitates as well as the amount of austenite and thus the mechanical properties. In maraging alloys, the main strengthening contribution stems from the precipitation of nanometre-scale precipitates. The high-strength of maraging steels, thereby, is because of dense and homogenously distributed precipitates in the martensitic matrix. Depending upon their size and coherency, they are either sheared or circumvented by dislocations.

Retained and reverted austenite also shows a strong influence on the mechanical properties of maraging steels. Reverted austenite is free of precipitates and possesses a much lower yield strength compared to the martensitic matrix (Sha and Guo 2009). Reverted austenite influences the mechanical properties. Viswanathan et al. (2005) observed a pronounced decrease of the yield strength with increasing amount of reverted austenite. Reverted austenite is supposed to increase the ductility and impact toughness. Therefore, for high toughness requirements, increased amounts of reverted austenite are beneficial and the material is subjected to higher ageing temperatures. On prolonged ageing, coarsening of precipitates results also in an increase of ductility but loss of strength. However, it was reported that the influence of austenite reversion is much more pronounced on softening than the influence of precipitate coarsening (Schnitzer et al. 2010b). In the literature, there are also investigations that report a detrimental effect of reverted austenite. Viswanathan et al. (2005) showed that the ductility is only increased in the early stages of ageing. Severe embrittlement was noticed in samples subjected to prolonged ageing. Reverted austenite in a Co-containing maraging steel leads to deterioration in ductility at low temperatures as fracture occurs along the boundaries of the martensite packets. However, it is known that the austenite phase in maraging steels is not stable during deformation (Schnitzer et al. 2010c), which results in transformation to martensite (so-called transformation induced plasticity, TRIP effect). This transformation behaviour also strongly influences the mechanical properties of maraging steels.

Normally, the amount of austenite is associated with the size of precipitates. Schnitzer et al. (2010c) performed a study on PH13-8 Mo-type maraging steel to judge the individual influence of precipitates and reverted austenite on mechanical properties. Calculation revealed that approximately 40 % of the loss in yield strength during ageing could be attributed to the influence of higher amounts of reverted austenite. The remaining amount of the observed decrease in strength is due to coarsening of NiAl precipitates.

A further microstructure constituent in maraging steels, which may affect the mechanical properties, is the martensitic matrix and thus, the prior austenite grain size. The grain size depends on the solution treatment temperature as well as time, whereas the martensite lath spacing in maraging steels is independent of the solution treatment temperature (Schnitzer et al. 2010c). The influence of the grain size on mechanical properties depends on the nucleation site of the precipitates. When precipitates nucleate homogenously in the martensitic matrix, no influence of grain size was observed (Schnitzer et al. 2009). In contrast, the relationship between yield strength and grain size follows the Hall–Petch equation after ageing when the precipitates nucleate on grain boundaries.

6.5 Evolution of Precipitates and the Overall Process

A research monograph on maraging steels (Sha and Guo 2009) has described recent theoretical research on the two main types of phase transformation in maraging steels, *precipitation* and *austenite reversion*. Before either of these transformations happens in these steels (during heating), there is a third transformation, which is the martensitic transformation (during cooling). Precipitation and austenite reversion occur in this martensite matrix, the former generally desirable as long as it is not too complete and the latter usually undesirable. In simple terms, precipitation leads to hardening and austenite reversion leads to softening. Although martensitic transformation is a prerequisite of the functioning of maraging steels, it is easily achievable and its details do not strongly determine the final steel properties, at least to a far lesser extent than precipitation and austenite reversion. For this reason, these sections will not consider the martensitic transformation in maraging steel.

Section 6.5.1 will give a brief discussion of the evolution of precipitates in maraging steels, with particular reference to its complexity in inter-particle spacing. This will be followed by the brief Sect. 6.5.2 on the overall transformation process.

6.5.1 Calculation of Inter-Particle Spacing

When precipitation fraction increases, the particle size should not be ignored when calculating the inter-particle spacing L. Assuming that the precipitates are spherical, the relationship between precipitate fraction f, inter-particle spacing and particle radius r is given by Sha and Guo (2009), which also contains the other equation in this section:

$$L = \left(1.23\sqrt{2\pi/(3f)} - 2\sqrt{2/3}\right) r \tag{6.1}$$

When precipitation fraction increases, the L/r ratio decreases dramatically (Fig. 6.3). L and r become very much comparable when f increases to about 0.1. The simple formula significantly overestimates the inter-particle spacing when the precipitation fraction is large. In addition to Eq. 6.1 for spherical precipitates,

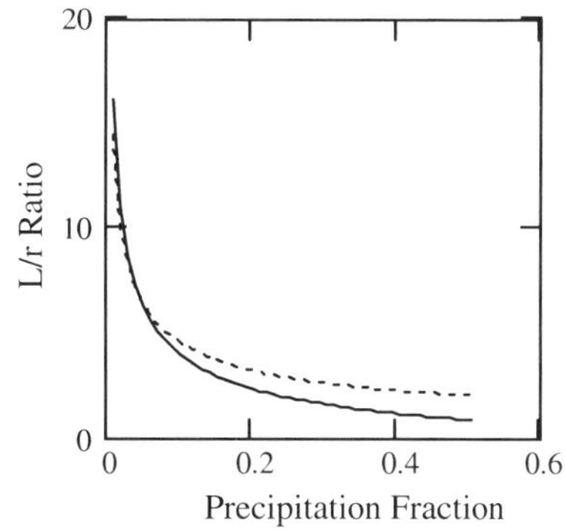

Fig. 6.3 Comparison of *L*/*r* ratio as a function of precipitation fraction, bwteeen Eq. 6.1 (solid line) and the simple formula $L/r = \sqrt{2\pi/3f}$ (dotted line)

there are equations for plate-like particles, which is the case for Ni_3Ti-type precipitates in maraging steels. If the volume fraction can be estimated using the Johnson–Mehl–Avrami (JMA) equation, then particle spacing as a function of time will be known. Such a procedure allows the quantification of age hardening to be carried out accurately.

The size increment of particles during the early ageing period may follow a slower procedure than the classical parabolic growth law would allow, regardless of whether the phase separation follows classical nucleation or spinodal decomposition. Although sometimes it is difficult to determine the controlling step, and in turn the operating coarsening law, Lifshitz-Slyozov-Wagner (LSW) theory (Lifshitz and Slyozov 1961; Wagner 1961) may be used for practical purposes. The LSW theory is the first statistical mechanical formulation of the kinetics of precipitate ageing, for diffusion-limited as well as interface-limited precipitate coarsening. The prediction of LSW theory that the cube of the average length scale of particles increases linearly with time was validated by numerous experiments, even in cases where there was a finite volume fraction of the dispersed phase. In LSW theory, the coarsening rate of a precipitate particle is given by the linear form

$$\frac{\mathrm{d}r}{\mathrm{d}t} = \frac{K_{\mathrm{LSW}}}{r}\left(\frac{1}{r^*} - \frac{1}{r}\right) \tag{6.2}$$

where r and r^* are the radius of the particle, and the critical radius of the precipitate population, respectively. K_{LSW} is the kinetic coefficient and is independent of the volume fraction of the precipitates.

6.5.2 The Overall Process

The overall transformation curve can be recorded using dilatometry (Fig. 6.4) measured for a cobalt-free maraging steel as an example. The steel composition is Fe-18.9Ni-4.1Mo-1.9Ti (wt %). Uniform expansion continues until 510 °C when a small contraction starts to occur, indicating the start of precipitation at this

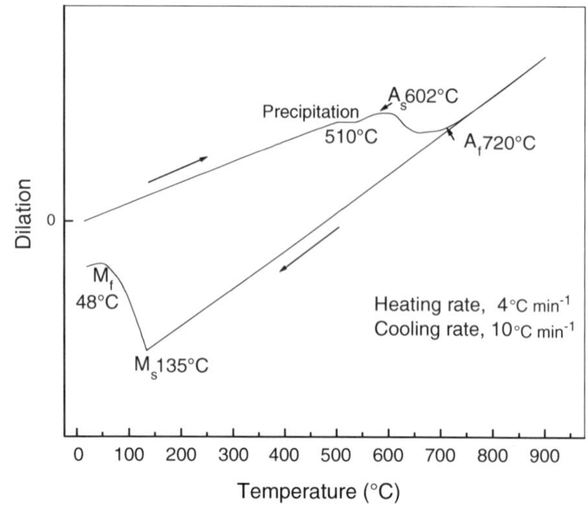

Fig. 6.4 Dilatometry curve of the 2000 MPa grade cobalt-free maraging steel

temperature. This is followed by a small period of linear expansion with increasing temperature. When the temperature is raised to 602 °C, a large contraction appears in the dilatometry curve, which can be taken as the start of austenite formation (A_s). At 660 °C, this contraction starts to slow down, until 720 °C when the curve resumes linear expansion. Therefore, the austenite transformation ends at around 720 °C (A_f). Complete solution is ensured by heating continuously to 900 °C and holding at this temperature for 30 min. During cooling to room temperature, there is drastic expansion at approximately 135 °C, due to the sudden start of rapid transformation from austenite to martensite, corresponding to the martensite start temperature, M_s. Martensite finish temperature (M_f) is about 48 °C. Thus, the cobalt-free maraging steel should have a single-phase structure upon cooling to room temperature that is martensite.

A discrepancy in dilatation at room temperature is apparent in Fig. 6.4. In dilatometry curves for T300 (Fe-18Ni-2.4Mo-2.2Ti, wt%) and C350 (Fe-18.77Ni-10.8Co-4.2Mo-1Ti, wt%) recorded in the literature (He et al. 2003), the cooling curve normally does not return to zero but to below the original point. This residual strain may be induced by the martensite phase transformation, due to both thermal and transformation changes. The strength level (2400 MPa) of the commercial cobalt-free T300 steel is the same as the experimental cobalt-free maraging steel used in Fig. 6.4, although the experimental steel, due to its very low impurity levels, has better toughness. The C350 steel is also ultra-purified steel, reaching a strength level of 2800 MPa. The transformation-induced plasticity in the free dilatometry cycle is seen in the martensitic transformation in many experiments (He et al. 2003; Coret et al. 2004). However, the exact mechanism is not clear. It is possible that the lattice has been sheared and it could not return to the original position.

Cobalt raises the martensitic phase transformation temperature. In general, the martensitic transformation temperature in 18Ni maraging steels is 50–100 °C higher than in their cobalt-free counterparts. However, controlling correctly the amount of nickel and molybdenum in cobalt-free maraging steels can still ensure complete martensitic transformation upon cooling to room temperature after solution treatment.

6.6 Research Trends

Although research in maraging steels usually involves the quantification of phase transformation kinetics, this section will expand beyond this theme, to the general developments in this type of steel. These can be roughly divided into the following areas.

6.6.1 Ultrafine-Grained Fe–Ni–Mn Steels, Novel Maraging Steels and Atom Probe Tomography

The research includes the study of behaviour during isothermal ageing (Nedjad et al. 2009a), the effects of chromium addition (Meimandi et al. 2008) and further alloying (Nedjad et al. 2008a, 2009b) on microstructure and mechanical properties including tensile behaviour (Nedjad et al. 2009c), and the correlation between intergranular brittleness and precipitation reactions (Nedjad et al. 2008b). Transmission electron microscopy studies have been carried out on grain boundary precipitation (Nedjad et al. 2006, 2008c). X-ray diffraction including peak profile analysis (Garabagh et al. 2008a) and microstructural study (Mobarake et al. 2008) have been carried out on a nanostructured 18Ni maraging steel prepared by equal-channel angular pressing (Garabagh et al. 2008b), aimed at better understanding of the deformation process and deformed structure. This equal-channel angular pressing is a natural extension after studying the extent and mechanism of nanostructure formation during cold rolling and ageing of lath martensite (Nedjad et al. 2008d).

Characterisation using advanced equipment and techniques has been carried out, the most recent example being of the precipitation behaviour of intermetallic NiAl particles in Fe-6at.%Al-4at.%Ni by small angle neutron scattering (SANS) and three-dimensional atom probe (3DAP) (Schober et al. 2010). Transformation during continuous-to-isothermal ageing applied on a maraging steel has been studied (Primig and Leitner 2010), as has the effect of Cu on the evolution of precipitation in a Fe–Cr-Ni–Al–Ti maraging steel (Schnitzer et al. 2010a) and the splitting phenomenon (Leitner et al. 2010). In addition, reverted austenite in a PH13-8Mo maraging steel has been investigated (Schnitzer et al. 2010b), in particular its influence on static and dynamic mechanical properties (Schnitzer et al. 2010c). Modelling work has also been carried out on the yield strength of stainless maraging steel (Schnitzer et al. 2010d). The latest techniques of atom probe tomography have been used in this field (Pereloma et al. 2004, 2009; Shekhter et al. 2004).

6.6.2 Computer-Aided Alloy Design Approach to New Steel Compositions

The design is always followed by characterisation of the steels manufactured on a laboratory scale. Studies include real-time martensitic transformation kinetics in maraging steel under high magnetic fields (San Martin et al. 2010). A novel ultra high-strength stainless steel, strengthened by various coexisting and multi-species (Xu et al. 2008) nanoprecipitates including Ni_3Ti intermetallic, has been designed (Xu et al. 2010a) based on a genetic algorithm and thermokinetics, and has been characterised (Xu et al. 2010b). This genetic computational design of novel ultra high-strength maraging precipitation hardenable stainless steels has incorporated alloy composition as well as austenitisation and ageing temperatures as combined optimisation parameters (Xu et al. 2009a, 2009b); the model description and first experimental validation have been given (Xu et al. 2009c).

Traditionally, new alloys have been developed through an empirical 'trial-and-error' approach, which is based on systematic experimental investigations. This semi-empirical approach is very dependent on intuition, experience and, to some extent, luck. Depending on the matrix type and the strengthening system, modern high-end UHS steels possess intentionally, and with specified amounts, a range of elements such as C, Cr, Ni, Al, Ti, Mo, V, Mn, Nb, Co, Cu, W, Si, B and N in wide ranges of concentrations. The interactions of alloying elements display complex cross effects. To understand the competition, conflict and synergy among various alloying elements experimentally, it would be necessary to systematically produce a large number of alloys with different compositions. Apart from the alloy composition, the heat treatment plays a paramount role in determining mechanical properties as well. After producing trial alloys, processing parameters need to be optimised via a tremendous amount of experimental work to achieve the best mechanical properties. Therefore, the conventional trial-and-error approach has been recognised to have low success rate, and be very costly and time-consuming.

With the improved understanding of processing/structure/property/performance relationships, and with the advent of more powerful and reliable computer hard- and software, alloy design philosophies are now increasingly oriented in the goal/means direction as demonstrated by the chain design model by Olson (1997). The goal/means oriented alloy design approach starts with defining the target properties required by the applications. This first link, between performance and properties, is mainly based on mechanical engineering. In the second step, the target microstructures are subsequently designed to obtain the desired properties, according to metallurgical and mechanical principles. The last link, attaining the tailored microstructures, is the key and most sophisticated process in the goal/means alloy design strategy. The alloy composition and heat treatment parameters are designed and optimised, employing various approaches, to obtain the desired microstructures.

Following fully or partially the goal/means oriented alloy design methodology, computational methods of different types have become a more efficient and powerful tool in providing guidance for designing new alloys as well as for process development. Thermodynamics aided design approaches have been applied, in different degrees, to various systems such as maraging. Another popular computational approach is artificial neural networks (ANN) which extracts empirical trends from exploring large databases. The neural networks are often coupled with genetic algorithms (GA) for alloy design to fulfil the multi-objective optimisation goals. More recently, atomic level ab initio calculations are also applied to provide theoretical guidance in selecting and optimising composition of Ti-based alloys, stainless steels and UHS steels. With respect to the process development, the experimental approach can also be better guided by extracting information from computational thermodynamics and artificial neural networks.

References

Coret M, Calloch S, Combescure A (2004) Experimental study of the phase transformation plasticity of 16MND5 low carbon steel induced by proportional and non-proportional biaxial loading paths. Eur J Mech A-Solid 23:823–842. doi:10.1016/j.euromechsol.2004.04.006

Dutta B, Palmiere EJ, Sellars CM (2001) Modelling the kinetics of strain induced precipitation in Nb microalloyed steels. Acta Mater 49:785–794. doi:10.1016/S1359-6454(00)00389-X

Garabagh MRM, Nedjad SH, Shirazi H, Mobarekeh MI, Ahmadabadi MN (2008a) X-ray diffraction peak profile analysis aiming at better understanding of the deformation process and deformed structure of a martensitic steel. Thin Solid Films 516:8117–8124. doi:10.1016/j.tsf.2008.04.019

Garabagh MRM, Nedjad SH, Ahmadabadi MN (2008b) X-ray diffraction study on a nanostructured 18Ni maraging steel prepared by equal-channel angular pressing. J Mater Sci 43:6840–6847. doi:10.1007/s10853-008-2992-4

He Y, Liu K, Yang K (2003) Effect of solution temperature on fracture toughness and microstructure of ultra-purified 18Ni(350) maraging steel. Acta Metall Sin 39:381–386

Hsiao CN, Chiou CS, Yang JR (2002) Aging reactions in a 17–4 PH stainless steel. Mater Chem Phys 74:134–142. doi:10.1016/S0254-0584(01)00460-6

Kim SJ, Wayman CM (1990) Precipitation behavior and microstructural changes in maraging Fe-Ni-Mn-Ti alloys. Mater Sci Eng A 128:217–230. doi:10.1016/0921-5093(90)90230-Z

Leitner H, Schober M, Schnitzer R (2010) Splitting phenomenon in the precipitation evolution in an Fe-Ni-Al-Ti-Cr stainless steel. Acta Mater 58:1261–1269. doi:10.1016/j.actamat.2009.10.030

Lifshitz IM, Slyozov VV (1961) The kinetics of precipitation from supersaturated solid solutions. J Phys Chem Solids 19:35–50. doi:10.1016/0022-3697(61)90054-3

Meimandi SH, Nedjad SH, Yazdani S, Ahmadabadi MN (2008) Effect of chromium addition on the microstructure and mechanical properties of Fe-10Ni-6Cr-2Mn maraging steels. In: New developments on metallurgy and applications of high strength steels, TMS, Warrendale, pp 1151–1157

Mobarake MI, Nili-Ahmadabadi M, Poorganji B, Fatehi A, Shirazi H, Furuhara T, Parsa H, Nedjad SH (2008) Microstructural study of an age hardenable martensitic steel deformed by equal channel angular pressing. Mater Sci Eng A 491:172–176. doi:10.1016/j.msea.2008.02.034

Nedjad SH, Ahmadabadi MN, Mahmudi R, Furuhara T, Maki T (2006) Analytical transmission electron microscopy study of grain boundary precipitates in an Fe-Ni-Mn maraging alloy. Mater Sci Eng A 438:288–291. doi:10.1016/j.msea.2006.02.097

Nedjad SH, Garabagh MRM, Ahmadabadi MN, Shirazi H (2008a) Effect of further alloying on the microstructure and mechanical properties of an Fe-10Ni-5Mn maraging steel. Mater Sci Eng A 473:249–253. doi:10.1016/j.msea.2007.05.093

Nedjad SH, Ahmadabadi MN, Furuhara T (2008b) Correlation between the intergranular brittleness and precipitation reactions during isothermal aging of an Fe-Ni-Mn maraging steel. Mater Sci Eng A 490:105–112. doi:10.1016/j.msea.2008.01.070

Nedjad SH, Ahmadabadi MN, Furuhara T (2008c) Transmission electron microscopy study on the grain boundary precipitation of an Fe-Ni-Mn maraging steel. Metall Mater Trans A 39A:19–27. doi:10.1007/s11661-007-9407-z

Nedjad SH, Ahmadabadi MN, Furuhara T (2008d) The extent and mechanism of nanostructure formation during cold rolling and aging of lath martensite in alloy steel. Mater Sci Eng A 485:544–549. doi:10.1016/j.msea.2007.08.008

Nedjad SH, Ahmadabadi MN, Furuhara T (2009a) Annealing behavior of an ultrafine-grained Fe-Ni-Mn steel during isothermal aging. Mater Sci Eng A 503:156–159. doi:10.1016/j.msea.2007.12.054

Nedjad SH, Teimouri J, Tahmasebifar A, Shirazi H, Ahmadabadi MN (2009b) A new concept in further alloying of Fe-Ni-Mn maraging steels. Scr Mater 60:528–531. doi:10.1016/j.scripta mat.2008.11.046

Nedjad SH, Meimandi S, Mahmoudi A, Abedi T, Yazdani S, Shirazi H, Ahmadabadi MN (2009c) Effect of aging on the microstructure and tensile properties of Fe-Ni-Mn-Cr maraging alloys. Mater Sci Eng A 501:182–187. doi:10.1016/j.msea.2008.09.062

Olson GB (1997) Computational design of hierarchically structured materials. Sci 277:1237–1242. doi:10.1126/science.277.5330.1237

Pereloma EV, Shekhter A, Miller MK, Ringer SP (2004) Ageing behaviour of Fe-20Ni-1.8Mn-1.6Ti-0.59Al (wt %) maraging steel: clustering, precipitation and hardening. Acta Mater 52:5589–5602. doi:10.1016/j.actamat.2004.08.018

Pereloma EV, Stohr RA, Miller MK, Ringer SP (2009) Observation of precipitation evolution in Fe-Ni-Mn-Ti-Al maraging steel by atom probe tomography. Metall Mater Trans A 40A:3069–3075. doi:10.1007/s11661-009-9993-z

Primig S, Leitner H (2010) Transformation from continuous-to-isothermal aging applied on a maraging steel. Mater Sci Eng A 527:4399–4405. doi:10.1016/j.msea.2010.03.084

San Martin D, van Dijk NH, Jiménez-Melero E, Kampert E, Zeitler U, van der Zwaag S (2010) Real-time martensitic transformation kinetics in maraging steel under high magnetic fields. Mater Sci Eng A 527:5241–5245. doi:10.1016/j.msea.2010.04.085

Schnitzer R, Zickler GA, Zinner S, Leitner H (2009) Structure-properties relationship of a PH 13–8 Mo maraging steel. In: Beiss P, Broeckmann C, Franke S, Keysselitz B (eds) Tool steels—deciding factor in worldwide production, Proceedings of the 8th International Tooling Conference, vol I. Aachen, Germany, pp 491–503

Schnitzer R, Schober M, Zinner S, Leitner H (2010a) Effect of Cu on the evolution of precipitation in an Fe-Cr-Ni-Al-Ti maraging steel. Acta Mater 58:3733–3741. doi:10.1016/j.actamat.2010.03.010

Schnitzer R, Radis R, Nöhrer M, Schober M, Hochfellner R, Zinner S, Povoden-Karadeniz E, Kozeschnik E, Leitner H (2010b) Reverted austenite in PH 13–8 Mo maraging steels. Mater Chem Phys 122:138–145. doi:10.1016/j.matchemphys.2010.02.058

Schnitzer R, Zickler GA, Lach E, Clemens H, Zinner S, Lippmann T, Leitner H (2010c) Influence of reverted austenite on static and dynamic mechanical properties of a PH 13–8 Mo maraging steel. Mater Sci Eng A 527:2065–2070. doi:10.1016/j.msea.2009.11.046

Schnitzer R, Zinner S, Leitner H (2010d) Modeling of the yield strength of a stainless maraging steel. Scr Mater 62:286–289. doi:10.1016/j.scriptamat.2009.11.020

Schober M, Schnitzer R, Leitner H (2009) Precipitation evolution in a Ti-free and Ti-containing stainless maraging steel. Ultramicroscopy 109:553–562. doi:10.1016/j.ultramic.2008.10.016

Schober M, Lerchbacher C, Eidenberger E, Staron P, Clemens H, Leitner H (2010) Precipitation behavior of intermetallic NiAl particles in Fe-6 at.%Al-4 at.%Ni analyzed by SANS and 3DAP. Intermetallics 18:1553–1559. doi:10.1016/j.intermet.2010.04.007

Sha W, Guo Z (2009) Maraging steels: Modelling of microstructure, properties and applications. Woodhead Publishing, Cambridge. doi:10.1533/9781845696931

Shekhter A, Aaronson HI, Miller MK, Ringer SP, Pereloma EV (2004) Effect of aging and deformation on the microstructure and properties of Fe-Ni-Ti maraging steel. Metall Mater Trans A 35A:973–983. doi:10.1007/s11661-004-0024-9

Viswanathan UK, Dey GK, Sethumadhavan V (2005) Effects of austenite reversion during overageing on the mechanical properties of 18 Ni (350) maraging steel. Mater Sci Eng A 398:367–372. doi:10.1016/j.msea.2005.03.074

Wagner C (1961) Theory of the ageing of precipitates by redissolution (Ostwald maturing). Z Elektrochem 65:581–591

Xu W, del Castillo PEJRD, van der Zwaag S (2008) Computational design of UHS stainless steel strengthened by multi-species nanoparticles combining genetic algorithms and thermokinetics. In: New developments on metallurgy and applications of high strength steels, TMS, Warrendale, pp 1167–1181

Xu W, Rivera-Díaz-del-Castillo PEJ, van der Zwaag S (2009a) Computational design of UHS maraging stainless steels incorporating composition as well as austenitisation and ageing temperatures as optimisation parameters. Philos Mag 89:1647–1661. doi:10.1080/14786430903019081

Xu W, del Castillo PEJR, van der Zwaag S (2009b) A combined optimization of alloy composition and aging temperature in designing new UHS precipitation hardenable stainless steels. Comput Mater Sci 45:467–473. doi:10.1016/j.commatsci.2008.11.006

Xu W, del Castillo PRD, Yang K, Yan W, van der Zwaag S (2009c) Genetic computational design of novel ultra high strength stainless steels: model description and first experimental validation. In: TMS 2009 138th annual meeting and exhibition - supplemental proceedings, vol 2. Materials characterization, computation and modeling, TMS, Warrendale, pp 319–326

Xu W, Rivera-Díaz-del-Castillo PEJ, Yan W, Yang K, San Martín D, Kestens LAI, van der Zwaag S (2010a) A new ultrahigh-strength stainless steel strengthened by various coexisting nanoprecipitates. Acta Mater 58:4067–4075. doi:10.1016/j.actamat.2010.03.005

Xu W, Rivera-Díaz-del-Castillo PEJ, Wang W, Yang K, Bliznuk V, Kestens LAI, van der Zwaag S (2010b) Genetic design and characterization of novel ultra-high-strength stainless steels strengthened by Ni_3Ti intermetallic nanoprecipitates. Acta Mater 58:3582–3593. doi:10.1016/j.actamat.2010.02.028

Chapter 7
Low Nickel Maraging Steel

Abstract The kinetics of precipitate formation are analysed using the Johnson–Mehl–Avrami (JMA) theory. Nano-sized precipitates are formed homogeneously during the ageing process, which results in high hardness. As the ageing time is prolonged, precipitates grow and hardness increases. As-forged Fe-12.94Ni-1.61Al-1.01Mo-0.23Nb (wt%) steel has mixed ductile and brittle fracture and has good toughness. Relationships among heat treatment, microstructure and mechanical properties are discussed. Austenitisation with lower temperature and intercritical annealing are introduced in the treatment of the maraging steel. Intercritical annealing treatments do not increase hardness either before or after ageing. The impact toughness in aged condition is enhanced after austenitisation at 950 °C. Reverted austenite is suspected after austenitisation at 950 °C followed by ageing at 600 °C.

7.1 Precipitation Kinetic Theory in Fe–12Ni–6Mn

This section will concentrate on the quantification or modelling of the precipitation transformation in Fe-12Ni-6Mn maraging steel. Aimed at developing cheaper alternatives than the classical 18Ni maraging steels (Table 6.2), efforts were were made in Russia and Japan, studying maraging in a series of experimental Fe–Ni-Mn alloys. Research with an Fe-12Ni-6Mn alloy showed that age hardening is achieved due to precipitation of θ-NiMn in the lath martensite, the quenched structure before ageing, as observed using electron microscopy.

7.1.1 Theoretical Analysis of the Early Stage Ageing Process

Coherent zones, possibly with a bcc structure, form at the early stages of ageing. After these, e.g. 0.2 h at 450 °C, θ-NiMn forms well before peak hardness. Deformation occurs by dislocations cutting through these coherent zones (or precipitates). Because the increase in yield stress is proportional to the increase in hardness (ΔH) for maraging steels and applying a coherency-strengthening mechanism, one may get:

$$\Delta H = Ar^{1/2} f^{1/2} \tag{7.1}$$

W. Sha, *Steels*, DOI: 10.1007/978-1-4471-4872-2_7,

where r is radius of the particle, f is volume fraction of the transformed particles and A is a coefficient relating the increase in hardness to precipitate size and fraction:

$$A = \frac{M_T}{q}(\kappa\varepsilon)^{3/2}\mu\left(\frac{3}{2\pi b}\right)^{1/2} \tag{7.2}$$

where M_T is the Taylor factor, q is a conversion constant between Vickers hardness and yield strength, κ is a numerical constant between 3 and 4, taken as 3.5, ε is a strain energy constant, μ is the shear modulus of the matrix, taken as 81 GPa and b is the dislocation Burgers vector.

Since κ, ε, μ and b are all material constants and can be obtained from the literature, the quantification of ΔH is now switched to the determination of the particle size r and precipitation fraction f.

The relationship between ageing time t and radius r of zone or precipitate (assumed spherical) is given by Zener's parabolic relationship:

$$r = \alpha\,(Dt)^{1/2} \tag{7.3}$$

where α is a constant related to the solid solubility of the precipitate and the matrix and the composition of the alloy, and D is the diffusion coefficient.

When time t is small for the early ageing stage, α can be considered as a constant and calculated following Christian's suggestion for a spherical precipitate and small degree of supersaturation (Christian 2002):

$$\alpha = \frac{2^{1/2}(c_0 - c_\alpha)^{1/2}}{(c_\theta - c_0)^{1/2}} \tag{7.4}$$

where c_0 is concentration of the precipitating elements in the matrix before ageing, taken as the sum of Ni and Mn, equal to the composition of the alloy, c_α is the solid solubility of the controlling elements in the parent phase and c_θ is the concentration of the elements in the new phase, that is, the θ-NiMn precipitate.

The Johnson–Mehl–Avrami (JMA) equation can be used to describe the relationship between transformation fraction and time at a certain temperature:

$$\frac{f}{f_{eq}} = 1 - \exp\left[-(kt)^m\right] \tag{7.5}$$

where f_{eq} is the equilibrium fraction of the precipitation (temperature dependent), k is a reaction rate constant, and m is the Avrami index.

At an early ageing stage, that is, when $kt \ll 1$, the above equation reduces to the following:

$$f = f_{eq}(kt)^m \tag{7.6}$$

Combining Eqs. (7.1), (7.3) and (7.6), one obtains:

$$\Delta H = A(\alpha f_{eq})^{1/2} D^{1/4} k^{m/2} t^{(m/2+1/4)} = (Kt)^n \tag{7.7}$$

where n is the time exponent in the relationship between the increase in hardness and ageing time in the early stage of ageing, $n = (2\,m + 1)/4$, and K is a temperature-dependent rate constant.

Assuming that K follows the Arrhenius type of equation:

$$K = K_0 \exp\left(-\frac{Q}{RT}\right) \tag{7.8}$$

where K_0 is a pre-exponential term, R is the gas constant, T is temperature (in Kelvin) and the activation energy Q for the precipitation process during ageing can be calculated. Using Eq. (7.8) to replace K in Eq. (7.7) and then taking the natural logarithms of both sides, one may get:

$$\ln \Delta H = n\left(\ln K_0 - \frac{Q}{RT} + \ln t\right) \tag{7.9}$$

For a constant increase in hardness ΔH_0 at different temperatures, one will have:

$$\ln t = \frac{Q}{RT} + \frac{\ln \Delta H_0}{n} - \ln K_0 \tag{7.10}$$

Assuming that n is constant, the activation energy Q can be obtained by plotting $\ln t$ versus $1/T$, where t is the time to reach ΔH_0 at temperature T. The slope of the straight line is Q/R as can be seen from the above equation. K_0 can be obtained simultaneously from the interception of the straight line with the $\ln t$ axis. When the activation energy Q and K_0 are known, ΔH during the age hardening can be calculated using Eq. (7.7), in conjunction with Eq. (7.8).

7.1.2 Overall Ageing Process

There are a few assumptions in the above theory that may prohibit its application to the overall ageing process:

(A1) coherent precipitate strengthening mechanism;
(A2) spherical precipitate (zone or particle);
(A3) constant activation energy;
(A4) $kt \ll 1$, which allows the simplification from Eq. (7.5) to Eq. (7.6);
(A5) constant α for precipitate growth.

The strengthening is due to coherent precipitates before peak hardness is reached, and the particles do not change to platelets until overageing takes place. These allow assumptions A1 and A2 to be applied to describe the overall ageing period. As for the activation energy, assumption A3, this may change during the ageing process. However, when ΔH_0 takes different values, the calculations to be shown later imply that the activation energy does not vary very much. Therefore, the activation energy will be considered as a constant value in the current model.

Assumptions A4 and A5 will not be acceptable when ageing proceeds beyond the early stage. The growth constant α in Eq. (7.3) is not constant any longer when the precipitates continue to form and grow, because the composition of the matrix changes significantly when more precipitates form from the matrix. Consequently, a more accurate way to calculate r is formulated below, considering α as a function of t through the change of c_0, the composition of the matrix. Before ageing takes place, c_0 is the concentration of the precipitating elements in the alloy. Both are considered as the total amount of Ni and Mn.

From Eq. (7.3), the growth rate of particles is given by:

$$\frac{\mathrm{d}r}{\mathrm{d}t} = \frac{1}{2}D^{1/2}\alpha(t)t^{-1/2} + D^{1/2}\frac{\mathrm{d}\alpha(t)}{\mathrm{d}t}t^{1/2} \tag{7.11}$$

so

$$r = \int D^{1/2}\left(\frac{1}{2}\alpha(t)t^{-1/2} + \frac{\mathrm{d}\alpha(t)}{\mathrm{d}t}t^{1/2}\right)\mathrm{d}t \tag{7.12}$$

where α can still be calculated from Eq. (7.4). However, it should be noted that the concentration in the matrix c_0, taken as the sum of Ni and Mn, changes during the ageing process simultaneously with the fraction of the precipitation f, denoted as $c_0{}'$, therefore:

$$c_0{}' = c_0 - c_\theta f \tag{7.13}$$

Thus, for the overall ageing process, we have:

$$\Delta H = AD^{1/4}f_{\mathrm{eq}}^{1/2}\left(\int\left(\frac{1}{2}\alpha(t)t^{-1/2} + \frac{\mathrm{d}\alpha(t)}{\mathrm{d}t}t^{1/2}\right)\mathrm{d}t\right)^{1/2}\left(1-\exp\left(-(kt)^m\right)\right)^{1/2} \tag{7.14}$$

where A is as defined in Eq. (7.2).

In Eq. (7.14), if the transformation fraction f is known (i.e., k is known), the precipitate size r can be calculated (assuming D is known), through the calculation of $c_0{}'$ and α. As a result, ΔH as a function of time and temperature can be quantitatively described.

7.2 Parameter Determination

7.2.1 Strengthening, Activation Energy and Avrami Index

In Eq. (7.14), A can be calculated when parameters M_T, q, κ, ε, μ, b in Eq. (7.2) are known. The strain energy constant ε for precipitation of θ-NiMn in the α-iron matrix is:

$$\varepsilon = \frac{3K_\theta(1+\nu_\alpha)\delta}{3K_\theta(1+\nu_\alpha)+2E_\alpha} \tag{7.15}$$

Table 7.1 Values of parameters involved in the calculation of precipitation strengthening in Fe-12Ni-6Mn maraging-type alloy

Parameter	M_T	q (MPa/HV)	κ	Ω_α (nm^3)	Ω_θ (nm^3)	δ	ε	μ (GPa)	b (nm)
Value	2.75	2.5	3.5	0.01176	0.01224	0.01333	0.0084	81	0.248

where K_θ is the bulk modulus of the θ-NiMn precipitate, ν_α is Poisson's ratio of ferrite in pure iron, taken as 0.282, E_α is the Young modulus of ferrite in pure iron, taken as 206 GPa (Ledbetter and Reed 1973), and δ is the linear strain accompanying precipitation from the matrix:

$$\delta = \frac{2(\Omega_\theta - \Omega_\alpha)}{3(\Omega_\theta + \Omega_\alpha)} \tag{7.16}$$

where Ω_α and Ω_θ are the atomic volumes of the ferrite (0.01176 nm^3/atom) and the θ-NiMn precipitate (0.01224 nm^3/atom), respectively. Alternatively, δ can be determined from $(1 + \delta)^3 = \Omega_\theta/\Omega_\alpha$, which gives a value close to that from Eq. (7.16). The above parameters are summarised in Table 7.1. The Taylor factor M_T is taken as 2.75 (Hosseini and Kazeminezhad 2009) for body-centred-cubic materials, and κ is assigned as 3.5.

The activation energy Q is calculated using Eq. (7.10), based on early ageing data. When ΔH_0 takes different values 70, 100, 150 HV, Q is determined as 141, 133 and 125 kJ mol^{-1}, respectively. It is reasonable to treat it as a constant. In the following calculation, Q is taken as 133 kJ mol^{-1}, the average of the above three values. Growth-related constants n and m are 0.475 and 0.45, respectively.

7.2.2 Reaction Rate Constant in the Johnson–Mehl–Avrami Equation

Experimental research shows that peak hardness is reached either by holding at 400 °C for 16 h, at 450 °C for 1.4 h, or at 500 °C for 0.24 h. Assuming that the precipitate fraction corresponding to peak hardness at 400 or 450 °C is known, f/f_{eq} is known, one can obtain the value of the pre-exponential term k_0, if the reaction rate constant in the JMA equation k is assumed to follow an Arrhenius type equation with the same activation energy as K:

$$k = k_0 \exp\left(-\frac{Q}{RT}\right) \tag{7.17}$$

Combining Eqs. (7.17) and (7.5), one will obtain:

$$k_0 = \frac{\left(-\ln(1 - f/f_{eq})\right)^{1/m}}{t_p} \exp\left(\frac{Q}{RT}\right) \tag{7.18}$$

where t_p is the time to reach peak hardness.

The percentage of the precipitate formed at peak hardness will be determined later. As will also be shown later, the precipitation fraction corresponding to peak hardness at 400 °C is higher than that at 450 °C, which partially contributes to the stronger age hardening effect at 400 °C than at 450 °C.

7.2.3 Growth and Diffusion Constants

Equation (7.4) shows that α is a function of c_0, c_θ and c_α. In the calculations below, as discussed earlier, c_0 will be replaced by c_0', calculated from Eq. (7.13). Both c_θ and c_α are to be considered as the sum of Ni and Mn in atomic fraction. Since NiMn only contains Ni and Mn, c_θ always equals 1. The values of c_α at different temperatures are calculated using thermodynamic software (Table 7.2, together with the equilibrium amount of NiMn precipitate). The value for c_0 is 0.1728 in atomic fraction, which is the equivalence of 0.1765 in weight fraction, that is the sum of 11.9 wt% Ni and 5.75 wt% Mn, the precise composition of the alloy studied.

At 450 °C, the average diameter of precipitates is around 3 nm after 0.2 h ageing, and 6 nm after 2 h (peak hardness). Based on these, the diffusion coefficient can be estimated. One may assume that D follows an Arrhenius-type equation with the same activation energy as K:

$$D = D_0 \exp\left(-\frac{Q}{RT}\right) \tag{7.19}$$

Then, the pre-exponential term D_0 can be obtained by combining Eqs. (7.19) and (7.12):

$$D_0 = \left(\frac{r}{\int \left(\frac{1}{2}\alpha(t)t^{-1/2} + \frac{\mathrm{d}\alpha(t)}{\mathrm{d}t}t^{1/2}\right)\mathrm{d}t}\right)^2 \exp\left(\frac{Q}{RT}\right) \tag{7.20}$$

Table 7.2 Data calculated with thermodynamics related to precipitation in Fe-12Ni-6Mn maraging-type alloy

Temperature (°C)	350	375	400	425	450
Ni in ferrite (at.%)	6.6	6.7	6.8	7.0	7.2
Mn in ferrite (at.%)	0.4	0.5	0.7	0.9	1.1
c_α (at.%)	7.0	7.2	7.5	7.9	8.3
f_{eq} (vol%)	11.4	11.1	10.8	10.5	9.7

7.2.4 *Critical Nucleus Size and Precipitation Fraction at Peak Hardness*

Critical nucleus size (R_c) is the size of the precipitate above which the nucleus is stable and able to grow. It can be calculated as (Porter and Easterling 1981):

$$R_c = \frac{-2\sigma N_A \Omega_\theta}{\Delta G_\nu} \tag{7.21}$$

where σ is the interface energy per unit area between precipitate and matrix, taken as 0.2 J m^{-2}, N_A is Avogadro's number, and the Gibbs energy difference between the precipitate and ferrite, ΔG_ν, can be calculated using thermodynamics software.

Rivera-Díaz-del-Castillo and Bhadeshia (2001) calculated the critical nucleus size taking into account the Gibbs–Thomson capillarity effect, which influences the equilibrium compositions at the particle/matrix boundary:

$$R_c = \frac{2c_\alpha \Gamma}{c_0 - c_\alpha} \tag{7.22}$$

where the capillarity constant Γ is given by:

$$\Gamma = \frac{\sigma N_A \Omega_\theta}{RT} \frac{1 - c_\alpha}{c_\theta - c_\alpha} \tag{7.23}$$

ΔG_ν and R_c calculated from both methods at various temperatures are listed in Table 7.3. When calculating the incubation period, the larger R_c values calculated from Eq. (7.21) are used. It should be noted that even for sizes smaller than the critical radius, a parabolic growth law described by Eq. (7.3) is assumed.

The determination of precipitation fraction f at peak hardness, f_p, as a percentage of the equilibrium fraction f_{eq} at different temperatures involves an optimisation procedure. Using the various transformation fraction values, the age hardening curves at various temperatures can be obtained. One may therefore obtain the transformation fraction value to best fit the calculated hardness curves with the experimental curves (Table 7.4). In general, peak hardness increases with decreasing temperature due to greater supersaturation giving a greater volume fraction of precipitate.

Table 7.3 Driving force and the critical nucleus radius of precipitation in Fe-12Ni-6Mn maraging-type alloy

Temperature (°C)	350	375	400	425	450
Driving force ($-\Delta G_\nu/RT$)	0.97	0.86	0.77	0.67	0.59
R_c (nm) (Eq. 7.21)	1.05	1.10	1.16	1.24	1.34
R_c (nm) (Eq. 7.22) (Rivera-Díaz-del-Castillo and Bhadeshia 2001)	0.39	0.39	0.41	0.43	0.46

Table 7.4 Values of parameters obtained by best fitting

Temperature (°C)	k_0 (s^{-1})	D_0 (m^2 s^{-1})	f_p/f_{eq}	f_{eq} (vol%)	f_p (vol%)	r_p (nm)
400	2.392×10^5	2.532×10^{-10}	0.56	10.8	6.1	3.9
450	2.392×10^5	2.532×10^{-10}	0.44	9.7	4.3	3.0
500	2.392×10^5	2.532×10^{-10}	0.39	8.7	3.4	2.2

7.3 Basic Measurements

7.3.1 Phase Transformation Determinations

The composition of the steel used in Sects. 7.3–7.5 is included at the bottom of Table 6.1. The phase transformation temperatures labelled in the dilation curves are shown in Figs. 7.1, 7.2. The standard way of measuring A_{c3} is used by San Martín et al. (2008) and Gomez et al. (2009), where A_{c3} is defined as the temperature at which expansion begins again to depend linearly on temperature once the sample is fully austenitic.

Koistinen and Marburger have shown for athermal martensite that the volume fraction transformed, y, at temperature T may be related to the martensite start temperature M_s by an expression of the form (Chong et al. 1998):

$$y = 1 - \exp\left[\alpha\left(M_s - T\right)\right] \tag{7.24}$$

where $\alpha =$ constant, < 0, unit K^{-1}. Rearranging and taking logarithms gives

$$\ln\left(1 - y\right) = \alpha\left(M_s - T\right) \tag{7.25}$$

The dilatometer curve for the as-forged steel analysed according to the Koistinen and Marburger (K–M) analysis is shown in Fig. 7.3, in the same way as shown with Fe-15 %Ni (Chong et al. 1998; Wilson and Medina 2000). The portion of the K–M curve between 378 and 340 °C probably corresponds to grain boundary massive ferrite. The remainder of the K–M curve should correspond to martensite inside the prior austenite grains. The K–M analysis can reveal different transformation processes and products, as shown by Wilson and Medina (2000), where multiple linear fitting lines were demonstrated, corresponding to different transformations in different temperature ranges during one continuous cooling experiment. That is not the case for the present maraging steel, because the vast portion of the transformation fits to one linear line.

7.3.2 Hardness

There is a close relation between hardness and precipitation in maraging steels. The hardness, aged at 550 °C for 10 h and air cooled is 488–499 HV30. The hardness data in Fig. 7.4 show age hardening curves of the maraging steel at different

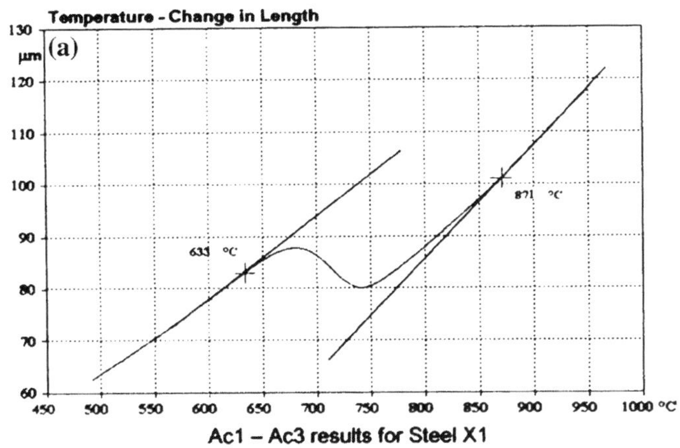

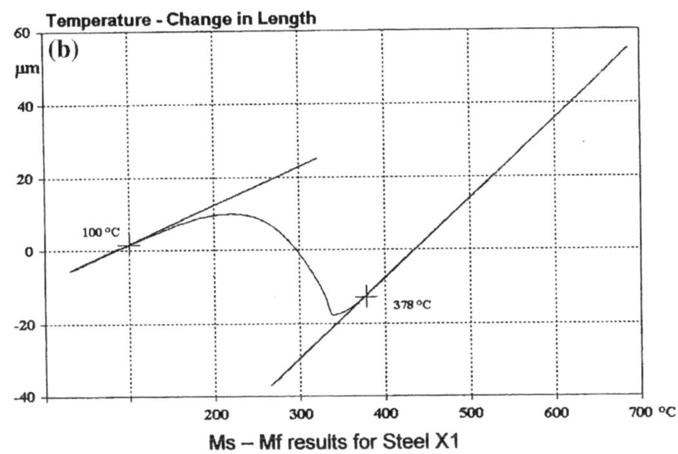

Fig. 7.1 Dilation curves of as-forged low nickel steel. **a** A_{c1}–A_{c3}; **b** M_s–M_f. (From Sha et al. (2011), www.maney.co.uk/journals/mst and www.ingentaconnect.com/content/maney/mst)

ageing temperatures. When aged at the lowest temperature of 450 °C, the steel can attain the hardness of 401 HV2 after 1 h of ageing. Both 550 and 600 °C ageing temperature have led to rapid hardening responses. At 600 °C ageing temperature, the hardness reaches its peak when the ageing time is 0.25 h, with the maximum hardness of 467 HV2, followed by slow reduction to 301 HV2 after ageing for 257 h. It takes 2 h for the steel at 550 °C ageing temperature to reach the peak hardness of 496 HV2. The hardness increase rate is marginally slower than at 600 °C ageing temperature. Ageing at 450 °C gives the lowest hardness increase rate, and requires the longest time of 66 h to reach the peak hardness 500 HV2. When aged at 450 °C, the hardness keeps increasing up to and likely after the longest ageing time used at this temperature. Moreover, at the higher ageing temperature 500 °C, the maximum ageing hardness is the same as the maximum hardness at 450 °C, within error ranges. The peak hardness is 501 HV2 at 17.35 h.

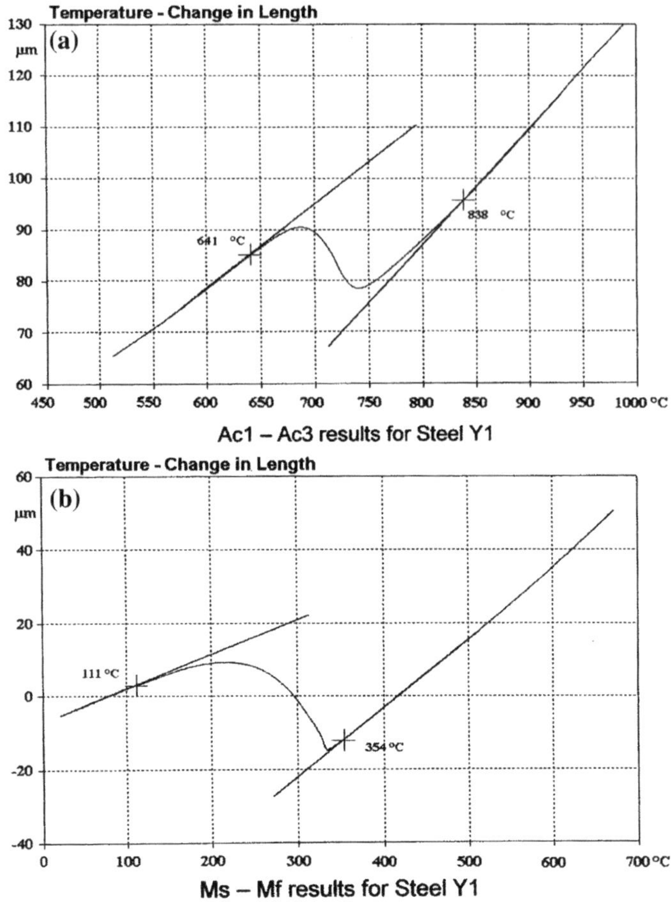

Fig. 7.2 Dilation curves of as-forged steel after cooling to −196 °C. **a** A_{c1}–A_{c3}; **b** M_s–M_f (From Sha et al. (2011), www.maney.co.uk/journals/mst and www.ingentaconnect.com/content/maney/mst)

Similar peak hardness can be reached at all four ageing temperatures, but the time of reaching the peak hardness is different. This permits the presumption that similar microstructures with the peak hardness can be achieved by ageing at these four temperatures for different times. Precipitation plays a dominant role in hardness of maraging steels. Other factors are retained or reverted austenite content, and dislocation density change. In this steel, no austenite is detected by X-ray diffraction (XRD) throughout the ageing process (see Sect. 7.4.2), so this factor has no contribution. The ageing is in a temperature range of 450–600 °C. Therefore, if we assume that the dislocation density change is comparable at different temperatures when the peak hardness is reached (i.e., shorter time at higher temperatures and longer time at lower temperatures), we would lead ourselves to the belief

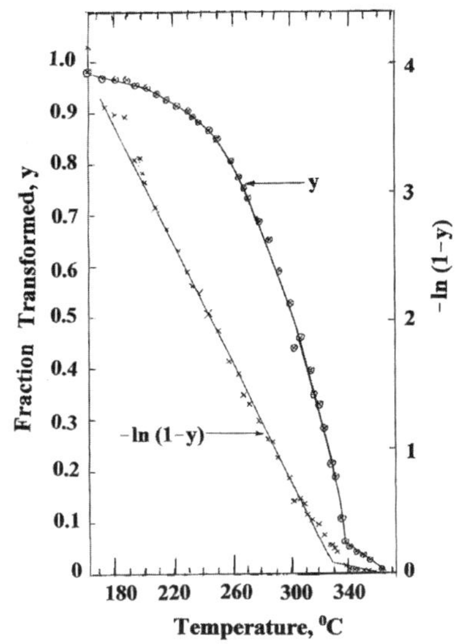

Fig. 7.3 K–M analysis of the dilatometer curve for the as-forged steel. (From Sha et al. (2011), www.maney.co.uk/journals/mst and www.ingentaconnect.com/content/maney/mst)

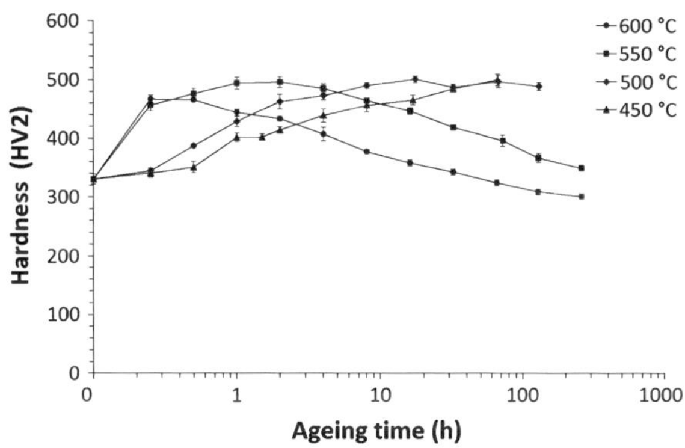

Fig. 7.4 Age hardening curves, showing the variation of hardness, with ageing time, in the maraging steel aged at 450–600 °C. (From Sha et al. (2011), www.maney.co.uk/journals/mst and www.ingentaconnect.com/content/maney/mst)

that similar size precipitates are similarly distributed in the matrix, when the peak hardness is achieved.

It should also be noted that the hardness after ageing at 550 and 600 °C temperatures decreases not long after it reaches the peak. Moreover, the hardness keeps increasing up to and likely after the longest ageing time at 450 °C. The hardness after ageing at 500 °C temperature displays a large plateau around the peak hardness and shows almost no decrease. These interesting behaviours may be explained from two points of view. On the one hand, ageing temperatures of 450 and 500 °C are not sufficiently high for the precipitates in the steel to grow quickly, while 550 °C is high enough for them to grow to large sizes after long ageing times. Therefore, overageing happens to the steel relatively early when aged at the temperatures of 550 and 600 °C than at the temperatures of 450 and

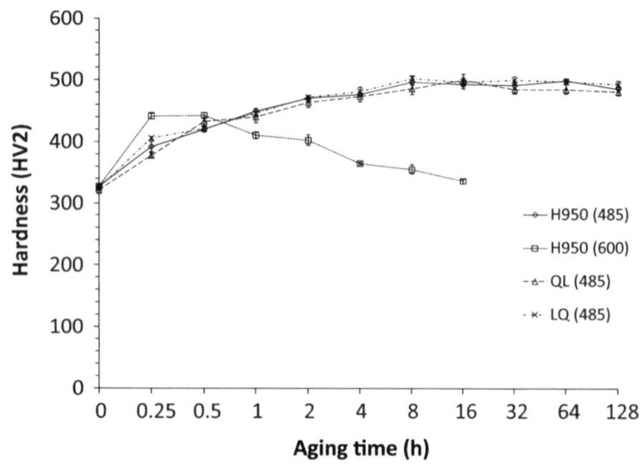

Fig. 7.5 Age hardening curves at two ageing temperatures (485 and 600 °C). Reprinted from Sha et al. (2012), with permission from Elsevier

500 °C. On the other hand, when the ageing temperature is as high as 550 °C, after a relatively long ageing time, some amount of reverted austenite may form around some nickel rich precipitates (Sha and Guo 2009). They may also be responsible for the decrease of the hardness.

The age hardening after different heat treatments, i.e. H950 (485), H950 (600), QL (485) and LQ (485) treatments, are shown in Fig. 7.5. H950 (485) treatment includes heating at 950 °C for 1 h followed by air cooling and then ageing at 485 °C. H950 (600) uses the ageing temperature of 600 °C. Q refers to half an hour at 950 °C, followed by water quenching and L refers to 2 h at 750 °C, followed by air cooling. The steel after the H950 (600) treatment reaches the peak hardness within a shortest time of about 30 min, with the lowest peak hardness value of 442 HV2, comparing with the other treatments. During H950 (600) treatment, the hardness decreases significantly after the peak hardness. H950 (600) and H950 (485) treatments only have a difference in ageing temperatures, namely ageing at 600 °C and 485 °C, respectively. When the ageing time is 64 h, H950 (485) treatment reaches its peak hardness with a value of 498 HV2. Hence, the 485 °C ageing temperature has resulted in the slower hardening response, while having a significant increase in peak hardness. QL (485) treatment takes 16 h to reach the peak hardness value of 500 HV2 and LQ (485) treatment has a similar peak hardness value of 503 HV2 when the ageing time is 8 h. Additionally, H950 (485), QL (485) and LQ (485) treatments lead to the similar increasing rate and the hardness reaches a large plateau after ageing for 8 h. At a range of 8–64 h, the hardness is close and can be considered as the same when combining standard deviation. Considering the scatter of the results, there are no differences between the hardness between 8 and 128 h.

Summarising the above, despite the several pre-ageing treatments, the only important variable is ageing temperature. Within experimental error all data aged at 485 °C superimpose and the steel aged at 600 °C exhibits a well-known response of reaching a peak in hardness at shorter times. There appears to be no effect of the different pre-ageing treatments.

7.3.2.1 Precipitation

We now continue to discuss the effect of the heat treatments on the hardness due to precipitation (this section) and overageing (Sect. 7.3.2.2). The hardness after ageing shows a pronounced increase compared to before ageing, for all treatments including the H950 treatment. Conventionally, precipitates formed during ageing should be responsible for the hardening during ageing (Askeland et al. 2010). The type of precipitates is analysed in Sect. 7.4.3.

Thermodynamic calculations indicate that Nb_2C and NbC form during austenitisation (950 °C) and intercritical annealing (750 °C) (see Sect. 7.4.3). Other precipitates such as NiAl whose formation temperature is lower than 950 and 750 °C can form during subsequent cooling.

7.3.2.2 Overageing

The existence of austenite cannot be absolutely excluded owing to the limitation of the conventional XRD. Guo et al. (2004) stated that when the amount of austenite is lower than 2 %, conventional XRD cannot detect the austenite. Hence, even though no austenite is detected for this material, the amount of retained or reverted austenite should be considered not higher than this threshold. Additionally, for L treatment, there is no indication in XRD profile of retained austenite either. Hence, most likely, the amount of Ni element in this material is not enough to generate the reverted austenite which is formed at the expense of martensite during ageing.

The clear sign of overageing is the decrease in hardness. A possible reason of hardness decrease is precipitate coarsening (Lach et al. 2010). However, the image after H950 (600) treatment (Sect. 7.4.1) shows needle-like structures. These clear and bright microstructures are suspected to be reverted austenite which is considered to be predominately responsible for the toughness of steel (Xiang et al. 2011), but the amount of reverted austenite is too small to detect by XRD. Hence, the pronounced drop in hardness could be attributed to the reverted austenite. Accurate characterisation of the reverted austenite must be carried out to validate the hypothesis of its link with the drop of hardness.

7.4 Microstructure

7.4.1 Microscopy

The microstructure of the steel after etching (Fig. 7.6) consists of the martensite laths and the grain boundaries. After ageing for a long time at 550 or 600 °C, some dark areas are revealed, probably related to overageing products.

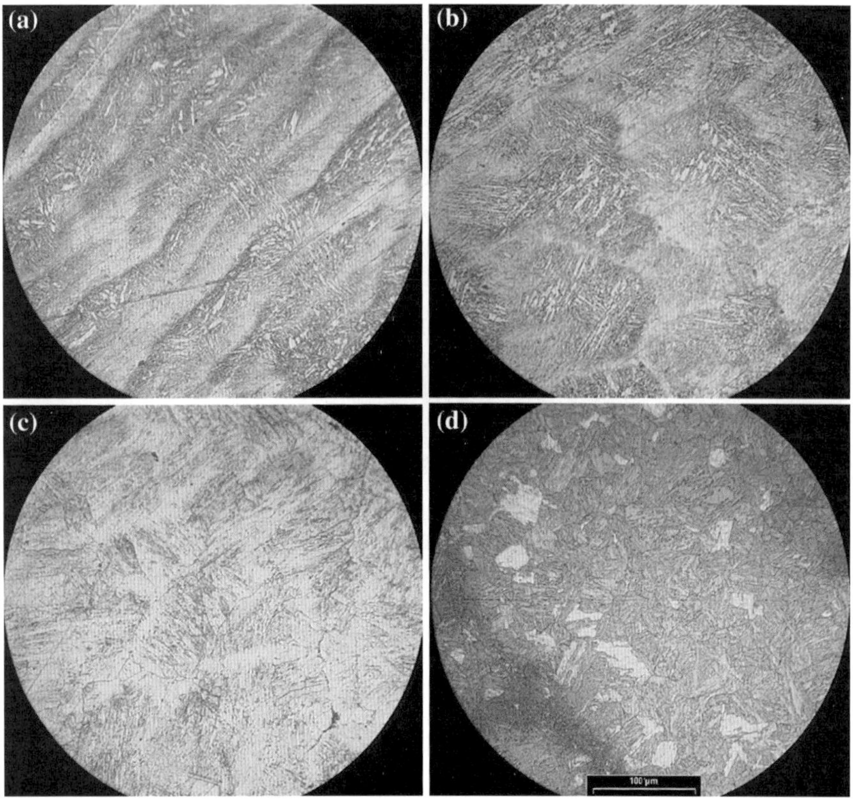

Fig. 7.6 Optical micrographs before and after ageing. **a** As-forged; **b** 500 °C, 0.25 h; **c** 500 °C, 0.5 h; **d** 550 °C, 72 h. The four micrographs have a same magnification. (From Sha et al. (2011), www.maney.co.uk/journals/mst and www.ingentaconnect.com/content/maney/mst)

The martensite laths and the grain boundaries are more clearly revealed in SEM images (Fig. 7.7). Very fine precipitates are also seen, homogeneously lying on the surface. Under the TEM, the microstructure consists of martensite laths with many precipitates in the lath matrix randomly (Fig. 7.8). With increasing ageing time, small precipitates become large very slowly.

Table 7.5 compares the size of precipitates measured at different ageing times on SEM micrographs. The measurements on the SEM micrographs could have three sources of errors. Because some particles might have more than half of their volume embedded in the matrix and the sectioning nature of the imaging, the measured diameter might be the diameter of the exposed spherical cap instead of the diameter of the particle sphere, assuming spherical particles. This would result in underestimation of precipitate size in SEM measurements. On the other hand, the SEM might not have picked up all of the particles given the scale of the observations. This would lead to overestimation of the precipitate size due to not including the smaller particles in the counts. The two error sources would give opposite

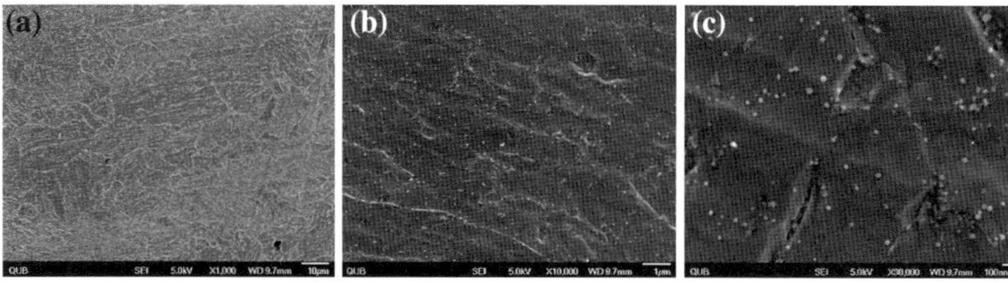

Fig. 7.7 Scanning electron micrographs after ageing. (**a**) 500 °C, 32 h; (**b**, **c**) 450 °C, 66 h. (From Sha et al. (2011), www.maney.co.uk/journals/mst and www.ingentaconnect.com/content/maney/mst)

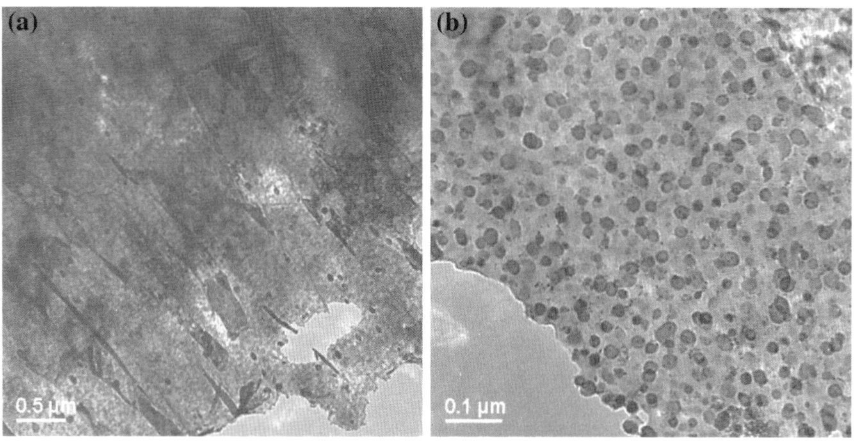

Fig. 7.8 Transmission electron micrographs after ageing. **a** 500 °C, 8 h; **b** 550 °C, 256 h. (From Sha et al. (2011), www.maney.co.uk/journals/mst and www.ingentaconnect.com/content/maney/mst)

Table 7.5 Comparison of size of precipitates after ageing at 500 °C measured in SEM micrographs

Ageing time (h)	Average size (nm)
1	18 ± 4
8	19 ± 5
32	19 ± 5
67	22 ± 9

errors, and so they might cancel each other out to some extent. However, the error values shown, using standard deviations, should be mostly from the real scatter of precipitate sizes.

Another source of error in SEM at such high magnifications is the local sampling. The technique cannot easily reveal any possible long-range variation of precipitate size and density, due to, for example, inhomogeneous chemical composition.

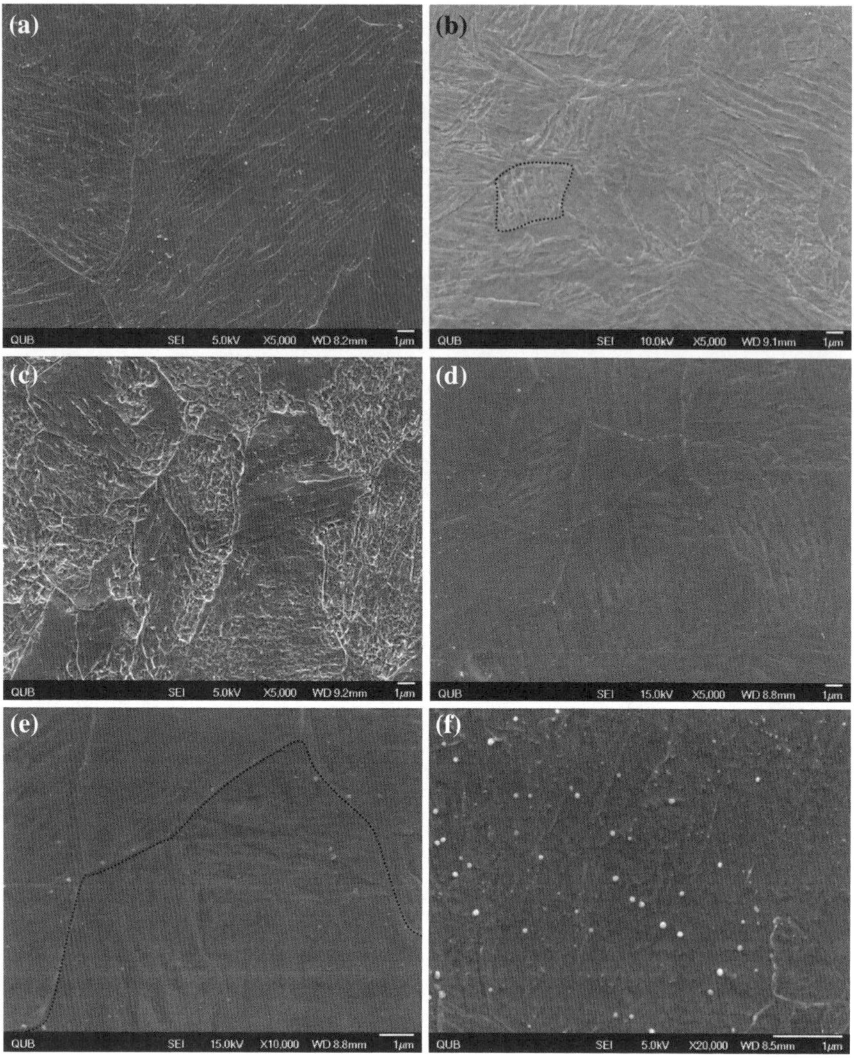

Fig. 7.9 SEM images after different heating treatments without ageing. **a** Q; **b** H950; **c** QL; **d**, **e**, **f** LQ. A lath-like martensite pack is masked in **b**. A grain is masked in **e**. Reprinted from Sha et al. (2012), with permission from Elsevier

Figure 7.9 shows the microstructures after Q, H950, QL and LQ treatments prior to ageing. Many lath-like martensite packs (one example is masked in Fig. 7.9b) are clearly visible in the grain (one example is masked in Fig. 7.9e).

In comparison with before ageing, the microstructures after QL (Fig. 7.10b) and LQ (Fig. 7.10c) treatments comparing to that after H950 treatment (Fig. 7.10a) show that martensite matrix is more complex, in that it does not show clear, long laths, and heterogeneous.

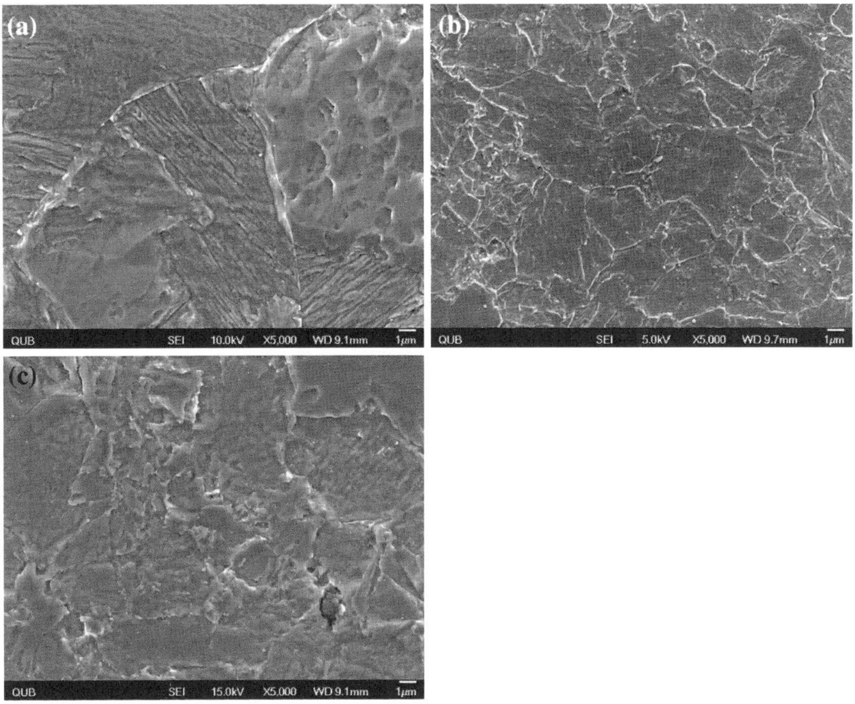

Fig. 7.10 SEM images after different heating treatments with ageing. **a** H950 (485) for 8 h; **b** QL (485) for 16 h; **c** LQ (485) for 8 h. Reprinted from Sha et al. (2012), with permission from Elsevier

The overaged microstructure after H950 (485), LQ (485) and QL (485) for 128 h, and after H950 (600) for 4 h, is given in Fig. 7.11. The matrix, martensitic structure that appears, at such high magnifications, should be treated with caution. There may be contribution from the effect of the sample preparation process, including scratch marks and etching, which is magnified with the high magnifications. Thus, not all fine features are real matrix features. Similar attention should be given for Figs. 7.9, 7.10 when magnifications at these levels are used.

7.4.2 X-ray Diffraction Analysis

XRD analysis is used to detect retained or reverted austenite, but no reflection peaks for fcc austenite are in the patterns after ageing at any temperature. The darkening in Fig. 7.6d does not seem to be related to austenite. Otherwise, such an austenite fraction would have been detected by XRD analysis.

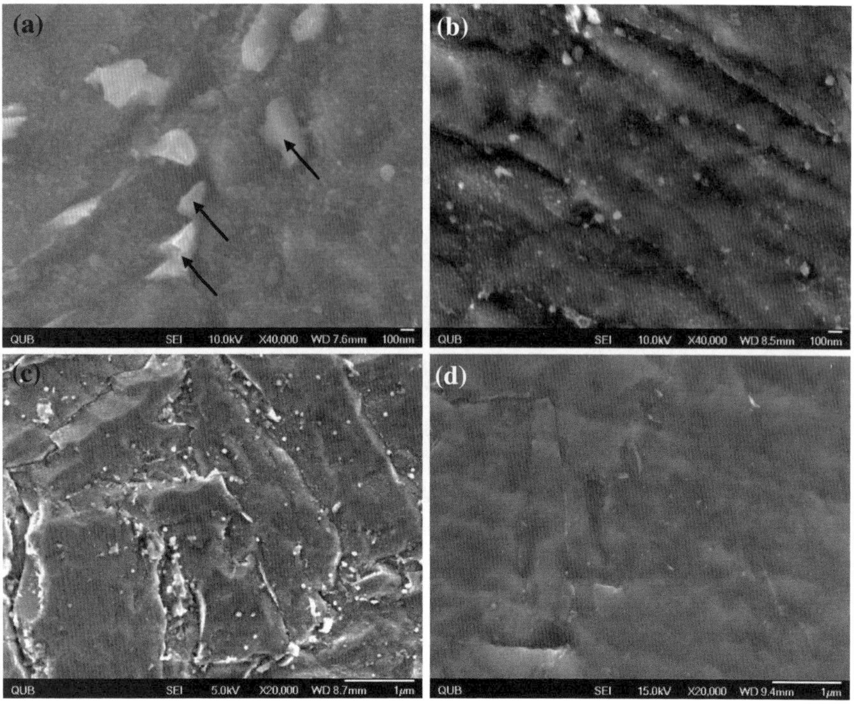

Fig. 7.11 SEM images after different heating treatments with prolonged ageing. **a** H950 (600) for 4 h; **b** H950 (485) for 128 h; **c** QL (485) for 128 h; **d** LQ (485) for 128 h. Three needle-like structures are indicated by arrows in (**a**), although there are more in the picture. Reprinted from Sha et al. (2012), with permission from Elsevier

Only two peaks are in the X-ray diffraction patterns, where a sharp peak occurs at 44.8° position and a small peak is at 64.9° position, in Fig. 7.12 using the example after QL treatment before and after ageing. Guo et al. (2004) quantified the precipitate fraction in maraging steel, which also gave diffraction patterns showing the existence of austenite. By comparing with Guo et al. (2004), there is no indication of peaks for retained or reverted austenite (γ) in Fig. 7.12, namely no peaks occur at 43°, 50° and 74°, the primary positions of austenite peaks.

7.4.3 Thermodynamic Calculations

Thermodynamic calculations of equilibrium phases, phase fractions and their compositions in the Fe-12.94Ni-1.61Al-1.01Mo-0.23Nb-0.0046C (wt%) system are made at 450, 500, 550 and 600 °C. The phases in the calculations are liquid, fcc, bcc, hcp, diamond, graphite, σ, Laves, R, P-phase, μ, χ, MoNi , $MoNi_4$ β, $MoNi_3$ γ, $Al_{13}Fe_4$, Al_2Fe, Al_5Fe_2, Ti_3Al, TiAl, cementite, ξ carbide, $M_{23}C_6$, M_7C_3, M_6C,

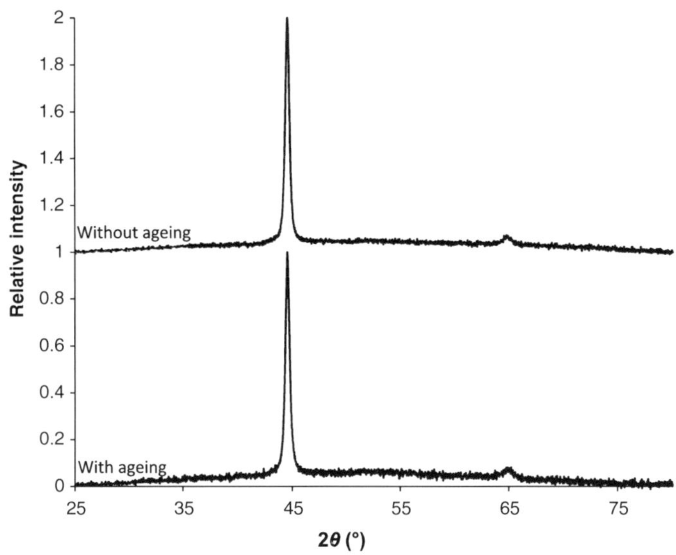

Fig. 7.12 XRD profiles after QL treatment without ageing and aged at 485 °C for 128 h

Table 7.6 Equilibrium phase mole fractions when austenite is entered in the thermodynamic calculations

Phase	450 °C	500 °C	550 °C	600 °C
NiAl B2	0.024	0.007	–	–
bcc	0.723	0.631	0.497	0.297
fcc	0.247	0.354	0.495	0.696
NbC	0.002	0.003	0.003	0.003
Mo_2C	0.003	–	–	–
M_6C	–	0.006	0.005	0.005

M_3C_2, V_3C_2, MC η, M_5C_2, Al_3Ni, Al_3Ni_2, AlNi B2, AlCu θ and FeCN χ. The components in each of the equilibrium phases will be given further below when showing the calculation results.

The equilibrium phases and their mole percentages are given in Table 7.6. In this table, the fcc phase is the austenite. X-ray results show, however, that this phase does not form even after the steel is significantly overaged, for 257 h. These results, although showing the real equilibrium state, have no practical value because, in the actual ageing process of this steel at these temperatures, austenite does not form due to its slow kinetics. The reason for this is that, due to the virtually zero carbon in the bcc, the difference between austenite and bcc compositions are mainly nickel contents. Hence, the formation of austenite is controlled by the diffusion of nickel, plus the driving force for the formation of difference phases. Among the precipitation phases in Table 7.6, NiAl has a very fast kinetics (Sha and Guo 2009), as well as the carbides, because carbon has a faster diffusion rate than the substitutional elements. Because the results in Table 7.6 have no practical value, we will not continue to show and discuss the compositions of each phase in detail.

The austenite phase is now excluded from the calculations, which is more relevant to the practical heat treatment of maraging steels (Table 7.7). NiAl is found to be the main precipitation phase, in agreement with an atom probe work characterising precipitates in an aluminium-containing precipitation hardening (PH) steel (Sha and Guo 2009). The calculated composition of the NiAl B2 phase is $Ni_{53}Al_{47}$. The compositions of the FeNi hcp phase and all other equilibrium phases are given in Table 7.8.

The calculated equilibrium phase mole fractions at 750 and 950 °C are given in Table 7.9. The main phase is austenite which accounts for more than 99 %, whether at 750 or 950 °C. At 950 °C, another phase is NbC accounting for 0.044 %. At 750 °C, two other phases are Nb_2C and $Fe_2(Nb,Mo)$ which is a Laves phase.

Table 7.7 Equilibrium phase mole fractions when austenite is not entered in the thermodynamic calculations

Phase	450 °C	500 °C	550 °C	600 °C
NiAl B2	0.066	0.063	0.060	0.054
bcc	0.897	0.919	0.932	0.938
NbC	0.003	0.003	0.003	0.003
FeNi	0.029	0.009	–	–
M_6C	0.006	0.006	0.005	0.005

Table 7.8 Components in the calculations in each phase and their concentration ranges in atomic per cent at 450/500/550/600 °C[a]

Phase	Fe	Ni	Al	Mo	Nb	C
bcc	91.1/90.1/ 89.4/88.9	8.4/9.2/ 9.7/10.0	0.3/0.4/ 0.5/0.8	0.2/0.3/ 0.3/0.3	0.0	0.0
NbC	0.0	0.0	0.0	0.0/0.1/ 0.4/1.3	52.2/51.3/ 50.3/49.1	47.7/48.6/ 49.3/49.6
FeNi[b]	56.9/58.6	42.6/40.8	0.0	0.5/0.5	0.0	0.0
M_6C	28.9/29.4/ 30.5/32.2	0.0	–	56.8/56.3/ 55.2/53.5	–	14.3

aThe four numbers are for at 450, 500, 550 and 600 °C, respectively. If only one number is shown, it is for all the four temperatures, i.e. no change with temperature. *Dash* indicates that the element is not a component in the calculations in the phase

[b] Only exists at 450 and 500 °C, thus only two numbers are shown

Table 7.9 Equilibrium phase mole fractions in thermodynamic calculations

Phase	750 °C	950 °C
fcc	0.99804	0.99956
NbC	–	0.00044
Nb_2C	0.00064	–
Fe_2 (Nb, Mo)	0.00133	–

7.5 Mechanical Properties

7.5.1 Tensile and Impact Properties and Fractography

The tensile properties after ageing treatment at 500 °C for 2 and 6 h are similar, with tensile strength of 1594 and 1577 MPa, respectively. The reduction of area in both cases is 15 %. These testing results show that the steel has good tensile strength.

The 5 × 10 × 55 mm half-size impact energy values in as-forged condition (forged to 25.4 mm thick disc followed by air cooling), before ageing, range 21–36 J, at −196 °C to room temperature. The steel was tough in this condition because of grain refining with Nb and excess C over stoichiometry of around 160 ppm C (Wilson et al. 2008). However, the fracture toughness was low after ageing. The ductile–brittle-transition-temperature (DBTT) of the maraging steel, aged at 550 °C for 10 h, is above room temperature.

The fracture surface of the as-forged impact specimens is characterised by radial ridges (Fig. 7.13a). These shearing dimples are very big and shallow. Figures 7.13b, 7.13c show small and deep tensile dimples, with small precipitates at the bottom of many dimples, confirming that the fracture process was typically ductile. The two kinds of dimples emerging on the fracture surface on the same specimens are due to the complicated stress experienced during the impact. They also show that the steel has good toughness under the as-forged condition.

The properties of the steel may be compared with its most equivalent but more expensive commercial counterpart, the Vascomax (2000) T-250 (Fe-18.5Ni-3Mo-1.4Ti) as follows. Following ageing treatment, T-250 has Rockwell C hardness of 49–52, equivalent to 498–544 HV10, tensile strength of 1793 MPa and reduction of area 58 %. Its room temperature full-size impact energy is 34 J. The low nickel maraging steel has matched the commercial maraging steel in its hardness and strength, but is low on ductility and toughness.

The Charpy impact energy after H950, LQ and QL treatments and then ageing at 485 °C, and after H950 treatment and then ageing at 600 °C, are shown in Table 7.10. H950 (600) treatment gives a pronounced increase in toughness of

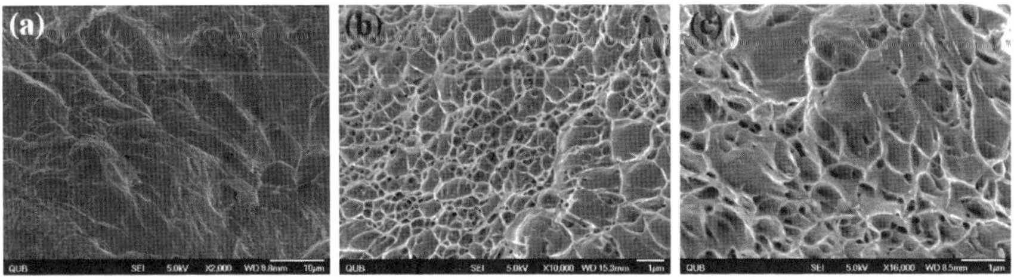

Fig. 7.13 Fractographs of as-forged impact specimens, at increasing magnifications. **a** Big and shallow shearing dimples, tested at –110 °C; **b** small and deep tensile dimples, tested at –196 °C; **c** small and deep tensile dimples, tested at –110 °C. (From Sha et al. (2011), www.maney.co.uk/journals/mst and www.ingentaconnect.com/content/maney/mst)

Table 7.10 Charpy impact toughness after different heat treatments

Heat treatment	Room temperature impact energy (J, half-size)
H950 (485) for 16 h	3.9
H950 (600) for 30 min	19.7
LQ (485) for 16 h	2.8
QL (485) for 16 h	5.4

around 400 % at the expense of hardness of around 12 % comparing with H950 (485) treatment. Consistent with the hardness data in Sect. 7.3.2, all data after ageing at 485 °C are experimentally the same within experimental uncertainty.

7.5.2 Intercritical Annealing and the Dual-Phase Structure

In this section, we attempt to link different aspects together. The discussion with intercritical annealing and the dual-phase structure aims to explain the effect of the QL and LQ treatments on the microstructure.

For L treatment (750 °C), thermodynamically, there is no bcc phase at this temperature (see Sect. 7.4.3). It has a high possibility that the temperature/time of L treatment is not enough to transform martensite into austenite completely. Therefore, in the following cooling, even if no austenite is retained, the total amount of the fresh martensite should be less than what Q treatment would result, where Q for the austenitisation can achieve the complete transformation of martensite into austenite. QL also embodies Q treatment as the first heating step, while the further L treatment is to temper martensite which can soften martensite (Baltazar Hernandez et al. 2011). LQ treatment, where Q treatment is the second heating step comparing with QL treatment, results in more martensite to transform into austenite. Hence after subsequent quenching, more martensite is formed. There are slight differences between Q and H950, in terms of quenching media, but their hardness prior to ageing is the same, when considering standard deviation. No retained austenite has been detected after heating at around 950 °C.

Clear and long martensite laths are seen in Fig. 7.10a, but they look short and disrupted in Figs. 7.10b and 7.10c. This is probably due to the formation of the intercritical annealing product, the 'dual-phase' structure, which is a mixture of tempered and fresh martensite. It is noted that some precipitates form at 750 °C shown in thermodynamic calculations (see Sect. 7.4.3). Large precipitate particles may impede grain growth with ageing time more effectively than small precipitate particles. Hence, those two constituents of dual-phase have different responses to ageing. This also gives a reasonable explanation why this microstructure can only be observed after ageing. As mentioned in Sect. 7.3.1, the A_{c1}/A_{c3} phase transformation temperatures are 635/871 °C. However, the A_{c1}/A_{c3} phase transformation temperatures are measured during continuous heating, while the heat treating is isothermal. Thermodynamic calculations only show one phase, γ, with no bcc, at

750 °C. However, 2 h, the heating time of L treatment, may not be long enough for the transformation of martensite to austenite to complete. If so, effectively, L treatment can successfully heat the steel to an α and γ two-phase region. For two-step heating treatments, the first heating step causes the transformation of ferrite into austenite and then austenite transforms into martensite during quenching. At the second heating step, the main portion of martensite transforms into reverted austenite and the remainder of martensite is tempered and relaxed, denoted as tempered martensite. At subsequent quenching, reverted austenite transforms completely into martensite, due to no austenite detected after XRD, where the martensite is denoted as fresh martensite. Tempered martensite and fresh martensite compose a dual-phase structure, where tempered martensite is soft but lean and fresh martensite is hard but rich. The difference between QL and LQ treatments is that QL treatment sets up austenitisation as first heating step, which can transform ferrite into austenite completely. Hence, more martensite is tempered after QL treatment eventually.

This dual-phase microstructure should be responsible for the fine-grained microstructure, shown in Figs. 7.10b and 7.10c compared to Fig. 7.10a. Guo et al. (2003) gave an explanation as to why the two-step martensitic transformation of dual-phase results in the refinement of grain size. Owing to substitutional alloying elements such as Ni, Mo and Al replacing a portion of the position of iron atoms in fcc structure, there are two kinds of austenite formed, namely low-alloy and high-alloy phases, leading to two different quenching responses. The atomic concentrations in the austenite formed at high temperature (during Q treatment) and at low temperature (during L treatment) are not the same, because of the austenite composition changes as a function of temperature in the phase diagram (ASM Handbook 1992). At following quenching, the low-alloy phase transforms into martensite first. Then, the high-alloy phase starts to transform. This two-step martensitic transformation is restricted actually. The latter is constrained by surrounding martensite which is formed already. Eventually, the fine-grained microstructure is achieved. In the same paper by Guo et al. (2003), it was reported that LQ treatment nearly has no effect on grain refinement, but significant refinement can be obtained if iterating these heating procedures, namely LQLQ treatment (for QL treatment, it should be QLQL treatment).

7.5.3 Toughness

We now complete the discussion with the effect of the heat treatments on the toughness. The primary difference shown in the data in Table 7.10 is that the material aged at 600 °C is tougher than the material aged at 485 °C. Since the material is aged at each temperature to the peak hardness shown in Fig. 7.5, then it is obvious that the higher temperature ageing produces a material with lower hardness (e.g., 440 HV at 600 °C vs. 500 HV at 485 °C). Thus, the CVN data support the well-known observation that most materials within a given microstructural class exhibit higher toughness with a decrease in strength (Kim et al. 2008). The steel is expected to be used at around peak hardness, provided that the toughness is satisfactory.

For the steel after ageing at 550 °C for 10 h, without initial heat treatment, at room temperature, the half-size Charpy impact energy is around 4 J, and the hardness is about 455 HV2. The hardness after H950 (600) for 30 min is close, whereas the Charpy impact toughness is significantly enhanced. Therefore, low austenitisation temperature (950 °C) can achieve the significant enhancement in Charpy impact toughness compared with steel without initial heat treatment.

7.5.4 Summary

In summary, there is rapid age hardening across the ageing treatment temperatures of 450–600 °C. There are no significant changes of grain size and precipitates. There is no reverted austenite detected by X-ray diffraction. The steel is tough before ageing, but extremely brittle at room temperature after ageing, with high tensile strength and hardness. Thermodynamic calculations show that NiAl is the major precipitate phase at 450–600 °C.

There is different age hardening behaviour after austenitisation at 950 °C and intercritical annealing treatments of the Fe-12.94Ni-1.61Al-1.01Mo-0.23Nb (wt%) maraging steel, the steel with reduced nickel. There is no reverted austenite detected by XRD. However, some suspected reverted austenite, with a needle-like microstructure, is formed after austenitisation at 950 °C followed by ageing at 600 °C for 4 h. Laves phase, $Fe_2(Nb,Mo)$, is formed at 750 °C. The toughness after austenitisation with lower temperature is significantly increased. However, intercritical annealing treatments do not increase hardness either before or after ageing, and do not improve the toughness after ageing.

References

Askeland DR, Fulay PP, Wright WJ (2010) The science and engineering of materials, 6th edn. Cengage Learning, Stamford

ASM Handbook (1992), Alloy phase diagrams, vol 3. ASM International, Metals Park, OH

Baltazar Hernandez VH, Nayak SS, Zhou Y (2011) Tempering of martensite in dual-phase steels and its effects on softening behavior. Metall Mater Trans A 42A:3115–3129. doi:10.1007/s11661-011-0739-3

Chong SH, Sayles A, Keyse R, Atkinson JD, Wilson EA (1998) Examination of microstructures and microanalysis of an Fe-9% Ni alloy. Mater Trans, JIM 39:179–188

Christian JW (2002) The theory of transformations in metals and alloys. Pergamon, Oxford

Gomez G, Pérez T, Bhadeshia HKDH (2009) Air cooled bainitic steels for strong, seamless pipes. Part 1—alloy design, kinetics and microstructure. Mater Sci Technol 25:1501–1507. doi:10.1179/174328408X388130

Guo Z, Sha W, Wilson EA, Grey RW (2003) Improving toughness of PH 13–8 stainless steel through intercritical annealing. ISIJ Int 43:1622–1629. doi:10.2355/isijinternational.43.1622

Guo Z, Li D, Sha W (2004) Quantification of precipitate fraction in maraging steels by X-ray diffraction analysis. Mater Sci Technol 20:126–130. doi:10.1179/026708304225010398

Hosseini E, Kazeminezhad M (2009) Dislocation structure and strength evolution of heavily deformed tantalum. Int J Refract Met Hard Mater 27:605–610. doi:10.1016/j.ijrmhm.2008.09.006

Kim BC, Park SW, Lee DG (2008) Fracture toughness of the nano-particle reinforced epoxy composite. Compos Struct 86:69–77. doi:10.1016/j.compstruct.2008.03.005

Lach E, Schnitzer R, Leitner H, Redjaimia A, Clemens H (2010) Behaviour of a maraging steel under quasi-static and dynamic compressive loading. Int J Microstruct Mater Prop 5:65–78. doi:10.1504/IJMMP.2010.032502

Ledbetter HM, Reed RP (1973) Elastic properties of metals and alloys, I. Iron, nickel, and iron-nickel alloys. J Phys Chem Ref Data 2:531–617. doi:10.1063/1.3253127

Porter DA, Easterling KE (1981) Phase transformations in metals and alloys. Van Nostrand Reinhold, New York

Rivera-Díaz-del-Castillo PEJ, Bhadeshia HKDH (2001) Theory for growth of spherical precipitates with capillarity effects. Mater Sci Technol 17:30–32. doi:10.1179/026708301101509089

San Martín D, de Cock T, García-Junceda A, Caballero FG, Capdevila C, de Andrés CG (2008) Effect of heating rate on reaustenitisation of low carbon niobium microalloyed steel. Mater Sci Technol 24:266–272. doi:10.1179/174328408X265640

Sha W, Guo Z (2009) Maraging steels: modelling of microstructure, properties and applications. Woodhead Publishing, Cambridge. doi:10.1533/9781845696931

Sha W, Li Q, Wilson EA (2011) Precipitation, microstructure and mechanical properties of low nickel maraging steel. Mater Sci Technol 27:983–989. doi:10.1179/17432847 10Y.0000000019

Sha W, Ye A, Malinov S, Wilson EA (2012) Microstructure and mechanical properties of low nickel maraging steel. Mater Sci Eng A 536:129–135. doi:10.1016/j.msea.2011.12.086

Vascomax® (2000) Nickel maraging alloys technical data sheet, Allvac, Monroe

Wilson EA, Medina SF (2000) Application of Koistinen and Marburger's athermal equation for volume fraction of martensite to diffusional transformations obtained on continuous cooling 0.13%C high strength low alloy steel. Mater Sci Technol 16:630–633. doi:10.1179/026708300101508397

Wilson EA, Ghosh SK, Scott PG, Hazeldine TA, Mistry DC, Chong SH (2008) Low cost grain refined steels as alternative to conventional maraging grades. Mater Technol 23:1–8. doi:10.1179/175355508X266908

Xiang S, Wang JP, Sun YL, Yan YY, Huang SG (2011) Effect of ageing process on mechanical properties of martensite precipitation-hardening stainless steel. Adv Mater Res 146–147:382–385. doi:10.4028/www.scientific.net/AMR.146-147.382

Part II
Steels in Structural Engineering

Chapter 8
Concrete Structures

Abstract Research has proved that the use of stainless steel has definite benefits but that it must be used intelligently and with careful planning and assessment of economic advantage for the future. The interest in the material is apparent but additional measures need to be instituted to further the knowledge of the engineer about the practical benefits stainless steel reinforcement presents. The second part of the chapter is concerned with differential scanning calorimetry (DSC). The thermal stability of ground granulated blast-furnace slag (ggbs) powder is examined. The third part of the chapter involves DSC of the thermal behaviour of hydration products in ggbs. The two-step loss of water from calcium silicate hydrates and dehydroxylation of calcium hydroxide ($Ca(OH)_2$) contributes, respectively, to the two major peaks in the DSC curves. Peaks due to the formation of ettringite and Fe_2O_3 solid solution are also present. The crystallisation peaks from amorphous phases in the ggbs are significant in the DSC thermograms.

8.1 Is Stainless Steel Reinforcement a Viable Option?

The use of stainless steel reinforcement has developed considerably over the past years. Stainless steel can be used in any situation where carbon steel reinforcement is normally used, although its use is particularly in precast concrete, repair of structures damaged by reinforcement corrosion or structures with enhanced design life.

Stainless steels used for reinforcement generally have maximum carbon contents of 0.07 % or below. In the presence of air or any other oxidising environment, a chromium-rich film forms spontaneously on the surface of the steel. This film is invisible, inert and tightly adherent to the metal. If it is damaged it will reform immediately in the presence of oxygen. Although the film is very thin it is both stable and non-porous and so prevents the steel from reacting further with the atmosphere. It is therefore called a passive layer and its stability increases with increasing chromium content. However, aggressive environments can give rise to local or widespread breakdown of this passive layer resulting in corrosion of the unprotected surface. Stainless steels are considered as homogeneous;

W. Sha, *Steels*, DOI: 10.1007/978-1-4471-4872-2_8,

i.e. their properties do not vary through their thickness. This has major advantage over coated carbon steel as the benefit of the coating can be easily destroyed by a scratch or damage on site. Stainless steels show an increasing strength advantage with temperature while carbon steel reinforcement shows a significant drop.

Both the general environment in which the structure is situated and the direct surroundings of the reinforcement should be considered. The corrosiveness of an environment is governed by a number of variables such as the humidity, air temperature, presence of chemicals and their concentration and oxygen content. Steel reinforcement corrodes most severely in areas such as bridge joints, splash zones, infrequently saturated zones, support structures and column heads. If stainless steel reinforcement is used in these areas corrosion rates can be reduced and cracking can be controlled.

In the past, stainless steel has been regarded by the building industry as a desirable but prohibitively expensive material. Emphasis has been placed on its corrosion resistance to the exclusion of other properties that have a considerable influence on cost-effectiveness. Although the initial cost of stainless steel reinforcement is several times higher than carbon steel, the intrinsic properties of stainless steel permit the use of designs requiring less material. The initial costs are also reducing due to changes in the manufacturing processes. It is not only the initial costs that need to be considered however, but also the total life-cycle costs (LCCs). If the life cycle costs of the structure are taken into account then stainless steel is often a viable solution due to the longer service life and zero maintenance and repair requirements (Table 8.1). The cost of cutting, bending, transporting and fixing stainless steel reinforcement is the same as for carbon steel.

The coefficient of thermal expansion of stainless steel is 50 % greater than that of carbon steel, whereas its thermal conductivity is only 30 % of carbon steel. This means that stainless steels conduct heat more slowly that other steels and steeper temperature gradients can build up causing unequal expansion and distortion. Excessive distortion will lead to failure to meet fabrication tolerances as well as causing fitting problems during erection and post-weld straightening may be necessary. Thermal distortion can be minimised in all metals by keeping the heat input and passage temperatures down, designing to accommodate wider dimensions and ensuring that good fitting and alignment are obtained prior to welding.

Table 8.1 Comparison of the costs involved in the use of stainless steel and carbon steel reinforcements

Stainless steel	Carbon steel
Materials costs	Materials costs
Installation costs	Installation costs
	Maintenance costs
	Lost production costs
	Replacement costs
	Additional operating costs
	Unexpected costs

8.2 Slag Powder

Granulated blast-furnace slag (ggbs) is a waste product in the manufacture of pig iron, the amounts of iron and slag obtained being of the same order. ggbs is not a pozzolan. Formed by rapid quenching of the molten blast-furnace slag from iron production, it comprises calcium or magnesium silicates and aluminosilicates. The slag is a mixture of lime, silica, and alumina, the same oxides that make up Portland cement, but not in the same proportion. Thus, it has many similarities to Portland cements, but the proportion of the constituents differs. Some hydraulic reaction takes place in water but it is slow. Reactions are accelerated to useful levels when combined with hydroxyl ions, e.g. cement plus water. Because of its inherent potential hydraulic reaction, higher percentage combinations with Portland cement are possible than with fly ash whilst retaining structurally useful strengths. In combinations of the Portland cement and slag, the cement component hydrates in the normal manner and it appears that the calcium hydroxide ($Ca(OH)_2$) thus liberated gives the correct alkalinity needed to provide a 'starter' for the hydration of the granulated slag. Depending on the proportion of ggbs in the cement, common applications include reduced heat of hydration in large pours of concrete, improved sulphate resistance, and reduced chloride ingress compared with normal Portland cement.

The slag powder composition is given in Table 8.2. The specific surface area is 440 m^2/kg. Figure 8.1 shows a differential scanning calorimetry (DSC) thermogram from pure ggbs powder, prior to any mixing. The peak temperatures and enthalpy values are shown in Table 8.3. ggbs has a glass content of 93 %. The two exothermal peaks, at 612 °C and 958 °C (smaller), respectively, are likely crystallisation peaks, of the amorphous phases, and the high glass content resulted in the very large peak at 612 °C. Other glass crystallising around this temperature is for example $Zr_{36}Ni_{64}$ (591 °C). It is interesting to compare with a study of metakaolin (Sha and Pereira 2001a). A crystallisation peak in metakaolin powder is at 992 °C, close to the temperature for the smaller exothermic peak from ggbs powder, 958 °C. The corresponding enthalpy for metakaolin powder is 98 J/g. It may also be noted that there is a larger peak in a similar temperature in the fly ash powder (Sha and Pereira 2001b). Again, the peaks in the fly ash and ggbs could be from a similar source. Comparison of the compositions of the two materials may

Table 8.2 Chemical compositions of the slag

Compound	Percentage
SiO_2	35
Al_2O_3	11
Fe_2O_3	1
CaO	41
MgO	8
SO_3	0.1
S	0.9

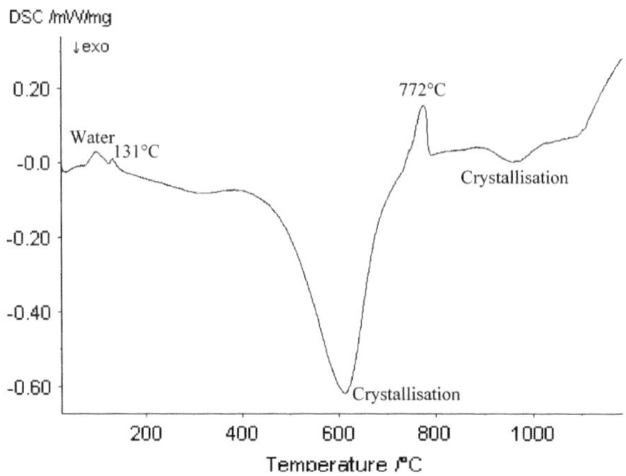

Fig. 8.1 DSC thermogram of ggbs powder. Reprinted from Cement and Concrete Research (Sha and Pereira 2001c), with permission from Elsevier

Table 8.3 DSC peak temperatures and enthalpy values during heating of ground granulated blast-furnace slag powders

Peak temperature (°C)	Nature	Enthalpy (J/g)
97	Endothermic	5.3
131	Endothermic	1.2
612	Exothermic	493
772	Endothermic	22
958	Exothermic	19

give some hint in the identification of this source. Analysis by some quenching technique by X-ray diffraction (XRD) is needed to identify the phases.

It is not clear as regard to the origins for the endothermic peaks at 131 and 772 °C. The latter peak appears to be at a temperature too high for the decarbonation of $CaCO_3$, should there be any residual amount of this in the powder. The average temperature for the decarbonation of $CaCO_3$ in fly ash is 730 ± 20 °C. XRD and chemical analyses of quenched samples either side of peak temperatures can help in understanding the mechanisms involved.

8.3 Hydrated Ground Granulated Blast-Furnace Slag

DSC is used to characterise the hydration products and the hydration process in Ordinary Portland Cement (opc) paste. Various hydration product phases are identified corresponding to the peaks in the DSC curves. In this section, this is extended to the study of the hydration products of ground granulated blast-furnace slag.

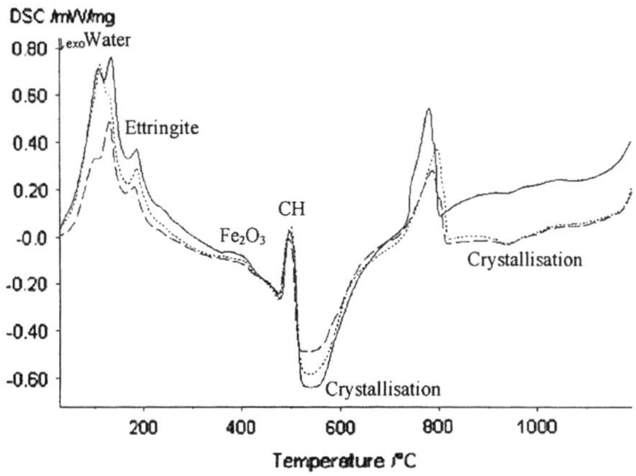

Fig. 8.2 DSC curves of ggbs hydrated at the ages of 26 (*solid line*), 29 (*long dashed line*) and 35 (*short dashed line*) days. Reprinted from Cement and Concrete Research (Sha and Pereira 2001c), with permission from Elsevier

Table 8.4 Peak temperatures and enthalpy in differential scanning calorimeter curves

Peak	Temperature (°C)	Enthalpy (J/g of sample)
First water loss from calcium silicate hydrates	106 ± 4	–
Second water loss from calcium silicate hydrates	129 ± 4	–
Ettringite	183 ± 3	13 ± 2
Fe_2O_3 solid solution	368 ± 20	–
Dehydroxylation of $Ca(OH)_2$	496 ± 2	59 ± 12
Crystallisation (low temperature)	537 ± 3	414 ± 80
Crystallisation (high temperature)	938 ± 2	15 ± 11

Comparing with DSC runs from pure ggbs powder before mixing (Fig. 8.1), Fig. 8.2 is from the part of the cement mix containing ggbs, showing the hydration products from an almost pure ggbs mix. ggbs itself has some hydration capability, especially in the presence of hydroxyl ions released during the hydration of opc.

There are a number of peaks in the DSC curves, in Fig. 8.2. The peak temperatures and enthalpy values are summarised in Table 8.4. Most of the peaks also appear in DSC curves from opc paste, thus their identifications are the same. The curves at three different ages, 26, 29 and 35 days, are very similar as can be seen from Fig. 8.2, with no sign of significant progressive development with increasing age. In Table 8.4, the values are the averages from the three different ages.

The most significant reaction from the presence of ggbs in the mix is the large exothermic peak at around 537 °C. This peak is also in unmixed ggbs powder, and is from the crystallisation of an amorphous phase in ggbs (Sect. 8.2). There is an overlap between the peak corresponding to the dehydroxylation of $Ca(OH)_2$ and the crystallisation peak, which are in opposite directions as the former reaction absorbs heat while the latter releases heat. The overlap needs to be taken into account when calculating the enthalpy for crystallisation. There is a second crystallisation peak at a higher temperature, around 938 °C, which may be due to a different amorphous phase. This peak is also present in the DSC curve of ggbs powder (Sect. 8.2). Peaks in similar positions are also in DSC runs of fly ash and metakaolin powders (Sha and Pereira 2001a, b), which are also from the crystallisation of amorphous phases.

The quantity of $Ca(OH)_2$ can be quantified using the heat of dehydroxylation of $Ca(OH)_2$, which is 1021 J/g as estimated from published calorimetric data (Ray et al. 1996). This is similar to the value of 856 J/g calculated from the Gibbs energy of formation of $Ca(OH)_2$, CaO, and H_2O. Therefore, a measured enthalpy of 59 J/g of the original 80 mg DSC sample gives a total heat of $59 \times 0.08 = 4.72$ J, which corresponds to the decomposition of 4.6 mg of $Ca(OH)_2$. If assuming that the initial DSC sample was a mixture of only water and ggbs in the ratio of 1:3, the percentage of $Ca(OH)_2$ is 7.7 % of the ggbs originally present. This is below the level from the hydration of normal opc mixes. There is quite possibly some contribution from the hydration of opc from elsewhere in the mix as a result of migration, thus the level of $Ca(OH)_2$ generated by hydration of ggbs should be even lower.

There is a large peak at 786 ± 8 °C in DSC curves, with an enthalpy (absorption of heat) of 102 ± 13 J/g of the entire DSC sample. The temperature range of this peak is not much above the limit for that for the decarbonation of $CaCO_3$, but its size is much larger than what would be expected from the residual $CaCO_3$ in the system.

References

Ray I, Gupta AP, Biswas M (1996) Physicochemical studies on single and combined effects of latex and superplasticiser on portland cement mortar. Cem Concr Compos 18:343–355. doi:10.1016/0958-9465(96)00025-X

Sha W, Pereira GB (2001a) Differential scanning calorimetry study of ordinary Portland cement paste containing metakaolin and theoretical approach of metakaolin activity. Cem Concr Compos 23:455–461. doi:10.1016/S0958-9465(00)00090-1

Sha W, Pereira GB (2001b) Differential scanning calorimetry study of normal Portland cement paste with 30 % fly ash replacement and of the separate fly ash and ground granulated blast-furnace slag powders. In: Venturino M (compiled) Proceedings of the seventh CANMET/ACI international conference on fly ash, silica fume, slag and natural pozzolans in concrete, Supplementary volume. ACI, Detroit, pp 295–309

Sha W, Pereira GB (2001c) Differential scanning calorimetry study of hydrated ground granulated blast-furnace slag. Cem Concr Res 31:327–329. doi:10.1016/S0008-8846(00)00472-5

Chapter 9
Cold-Formed Steel Portal Frame

Abstract Cold-formed steel portal frames are a popular form of construction for low-rise commercial, light industrial and agricultural single-storey buildings of spans of up to 30 m. Such buildings typically use cold-formed steel channel-sections for the columns and rafters, with joints formed through back-to-back gusset plates bolted to the webs of the channel-sections. This chapter investigates effects of frame topography on the frame weight and cost per metre length of building. An optimisation technique that uses a real-coded genetic algorithm is applied to search for the optimum topography of steel portal frame for a building, to minimise the cost of the main frame of such buildings. The key decision variables considered in this algorithm consist of both the spacing and pitch of the frame as continuous variables, as well as the discrete section sizes. A routine taking the structural analysis and frame design for cold-formed steel sections is embedded into the genetic algorithm. The real-coded genetic algorithm handles effectively the mixture of design variables, with high robustness and consistency in achieving the optimum solution. All wind load combinations according to Australian code are considered in this research. Also, frames with knee braces are included for which the optimisation achieved even larger savings in cost.

9.1 Design Optimisation of Portal Frame Using Genetic Algorithms

Many methods have been proposed during the past decade to solve optimisation problems. These methods can be divided into two main categories: deterministic and stochastic. The methods in the first category, known as mathematic programming, are more difficult to be implemented as they depend on the gradient information about the objective function. Compared with deterministic optimisation techniques, stochastic methods have many advantages because they do not require differential information other than the objective function. Furthermore, heuristic search techniques search simultaneously by using a population of decision

W. Sha, *Steels*, DOI: 10.1007/978-1-4471-4872-2_9,

variable sets, and not a single solution as in a deterministic way. With the ability of determining the optimum, or near-optimum solution, particularly if the optimisation function is non-continuous, non-differentiable and with any kind of constraints, stochastic methods are a powerful tool in non-conventional optimisation problems (Eid et al. 2010). In general, some examples of stochastic methods are adaptive random search, completive evolution, controlled random search, simulated annealing, genetic algorithms, differential evolution and particle swarm optimisation (Tsoulos 2008). Among stochastic techniques presented, genetic algorithm (GA) is a common method applied to solve optimisation problems in the field of structural design optimisation.

In recent years, GAs have been successfully applied in structural optimisation of hot-rolled steel portal frames, using elastic analysis. First, the optimum design of nonlinear hot-rolled steel portal frames was carried out using binary-coded GA. The binary string for each design variable being the cross-section areas of members is used to represent the sequence number from the standard universal beam sections set (Kameshki and Saka 2001). Thereafter, the design of pitched roof hot-rolled steel frames with haunched rafters was also optimised by binary GAs. The key discrete variables in this research are hot-rolled steel sections selected from standard steel sections table and haunch sizes (Saka 2003). To enhance the performance of GAs in finding the best optimum solution, a distributed genetic algorithm (DGA) that uses a number of population groups and implements genetic operations in parallel is applied. Then the best populations of each group are allowed to migrate to another group to make the quicker convergence (Issa and Mohammad 2010). However, applying binary-coded GA will not be suitable when optimisation problems involve continuous design variables other than sections, for instance pitch of frame or frame spacing. This is because the higher precision that is required, the larger is the string length, thereby increasing the computational complexity of the algorithm (Deb 2001). In this case, a real-coded GA is an alternative method that has demonstrated its powerful capacity in solving the optimisation problem. This application allows us to obtain precise results without the effect of the length of binary string. A real-coded GA is used in this chapter.

9.2 Real-Coded Genetic Algorithm and its Application

In a real-coded GA, genetic operators are directly applied on real numbers and decision variables can be directly used to compute the fitness values. The characteristic of real-coded GAs is that genetic operators are directly applied to the design variables without coding and decoding as with binary GAs. Solving optimisation problems using real-coded GAs is, therefore, less cumbersome when compared to the binary-coded GAs. Since the reproduction operator works with fitness value, any reproduction operator used with binary-coded GAs can also be applied in real-coded GAs. In real-parameter GAs, the encountered difficulty is how to use a pair of real-parameter decision variable vectors to produce a new pair of offspring vectors or how to perturb a decision variable vector with a mutated vector in a meaningful manner (Deb 2001). The algorithm used in this chapter randomly generates a set of solutions known as initial population. From this population, the next generation of solutions is evolved by conducting three genetic operations: selection, crossover and mutation. In the chapter, we use simulated binary crossover (SBX) and polynomial mutation (Deb 2001; Deb and Gulati 2001) operators to create the new individuals for the next generation. The flowchart of the real-coded GA used in this chapter is shown in Fig. 9.1.

9.2.1 Selection and Elitism Strategy

In tournament selection operator, the process is conducted by picking at random two solutions from the current population to compare their fitness values. The solution with a better fitness value is selected for the next operation. The process

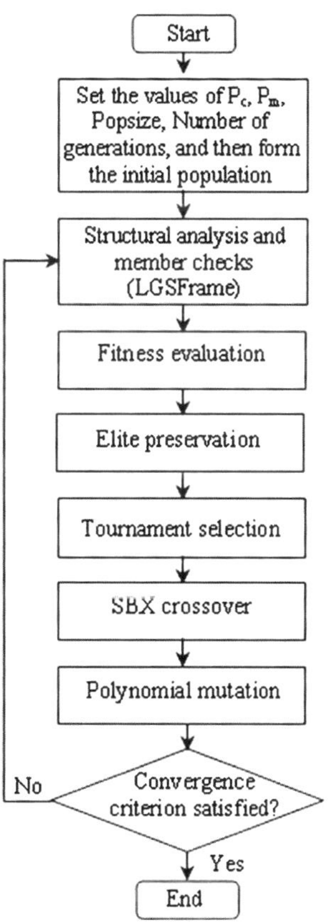

Fig. 9.1 Flowchart of the real-coded GA from Phan et al. (2013), reprinted by permission of the publisher (Taylor & Francis Ltd, http://www.tandf.co.uk/journals)

of random selection ensures that the best solutions in the population will not dominate the mating pool, as in the proportional selection method. The diversity of the population is, thus, preserved to increase the exploration component of the algorithm.

The best individuals in the population, depending on the adopted percentage of population, are retained and carried forward unchanged to the next generation. The rest of the new population is created by the three genetic operators of selection, crossover and mutation applied to the entire current population including elite individuals.

9.2.2 Real-Coded Crossover and Mutation Operators

The SBX operator picks at random two solutions in the current population, known as parents to create two offspring symmetrically about the parent solutions for next generation to avoid a bias towards any particular parent solution in a single crossover operation. The formulation used for SBX is as follows:

$$\begin{aligned} x_i^{(1,t+1)} &= 0.5[(1+\beta)x_i^{(1,t)} + (1-\beta)x_i^{(2,t)}] \\ x_i^{(2,t+1)} &= 0.5[(1-\beta)x_i^{(1,t)} + (1+\beta)x_i^{(2,t)}] \end{aligned} \tag{9.1}$$

where $\beta(\eta_c)$ is the probability distribution function for offspring after crossing over with η_c being distribution index for crossover, $x_i^{(1,\,t)}$ and $x_i^{(2,\,t)}$ are the parent solutions, and $x_i^{(1,\,t+1)}$ and $x_i^{(2,\,t+1)}$ are the children created for the next generation.

To ensure the new values of the decision variable remain within the range $[x_i^l, x_i^u]$, where x_i^l and x_i^u are the lower and upper bounds, respectively, the probability distribution for the crossover operator has the form:

$$\beta(\eta_c) = \begin{cases} [\alpha u]^{1/(\eta_c+1)} & \text{if} \quad u \le 1/\alpha, \\ [1/2 - \alpha u]^{1/(\eta_c+1)} & \text{if} \quad 1/\alpha < u \le 1 \end{cases} \tag{9.2}$$

where u is a random number between 0 and 1, η_c is the distribution index for crossover, $\alpha = 2 - \chi^{-(\eta_c+1)}$ and χ is calculated as follows:

$$\chi = 1 + \frac{2}{x_i^{(2,t)} - x_i^{(1,t)}} \min\left[(x_i^{(1,t)} - x_i^l), (x_i^u - x_i^{(2,t)})\right]; \text{ assuming } x_i^{(1,t)} < x_i^{(2,t)}.$$

Like in the SBX operator, polynomial mutation also uses probability distribution $\bar{\delta}(\eta_m)$ being a polynomial function to create the child solution in the vicinity of a parent solution. The formulation for the mutation operator has the form as follows (Deb 2001, Deb and Gulati 2001):

$$y_i^{(1,t+1)} = x_i^{(1,t+1)} + (x_i^u - x_i^l)\bar{\delta} \tag{9.3}$$

where x_i^u and x_i^l are the boundaries of decision variables and $y_i^{(1,\,t+1)}$ is a new solution obtained from the mutation operator. η_m is distribution index for mutation.

To ensure that no solution would be created outside the range of x_i^u and x_i^l (Deb and Gulati 2001), the parameter $\overline{\delta}(\eta_m)$ has the form as follows:

$$\overline{\delta} = \begin{cases} \left[2u + (1-2u)(1-\delta)^{\eta_m+1}\right]^{1/(\eta_m+1)} - 1 & \text{if} \quad u \le 0.5, \\ 1 - \left[2(1-u) + 2(u-0.5)(1-\delta)^{\eta_m+1}\right]^{1/(\eta_m+1)} & \text{if} \quad 0.5 < u \le 1 \end{cases} \tag{9.4}$$

where u is a random number between 0 and 1, η_m is the distribution index for mutation, and $\delta = \min\left[(x^{(1,t+1)} - x_i^l), (x_i^u - x^{(1,t+1)})\right]/(x_i^u - x_i^l)$.

In this chapter, $\eta_m = \eta_c = 1$ is used. A technique that rounds off the number in dealing with discrete design variables is used, in case of SBX crossover or polynomial mutation creating the decimal numbers. Constant probabilities are assigned to both crossover and mutation operators to reduce the possibility of destroying the good solutions. A crossover probability P_c of 0.9 was used throughout in this chapter. Premature convergence happens with a low mutation probability. To increase the GA's exploration capacity in the solution space to increase the chance of locating the optimum solution, the mutation probability, P_m, is as high as 0.1.

In general, the GA framework for an optimal design is described in Eid et al. (2010). The termination condition of GA program is a prespecified number of generations.

Since genetic algorithms are used to solve unconstrained optimisation problems, penalty functions are required to define the relationship between the objective function and constraints and to transform a constrained problem to an unconstrained one. The effective fitness function often has the form as in following (Pezenshk et al. 2000):

$$F = W\left[1 + \sum C\right] \tag{9.5}$$

where F is the fitness function; W is the weight (or cost) of frame per unit length of building and C is the penalty value for each violated constraint. In this chapter, penalty values are imposed based on the severity of constraint violation. With the assumption of full lateral restraint, constraint for ultimate limit state (ULS) is checked on column and rafter members of the frame in terms of local capacity check (unity factor).

9.3 Exemplar Frames

9.3.1 Frame Geometries, Frame Loading and Member Checks

In this section, three cold-formed steel portal frames will be considered, namely Frames A, B and C. Each frame has different frame geometry (Table 9.1). Figure 9.2 shows details of the exemplar portal frame, such as frame span of L_f,

Table 9.1 Geometries and section sizes for three exemplar frames

			Section sizes for unit weight (kg/m)			Section sizes for unit cost (AUD/m)		
Frame	Span (m)	Height (m)	Column section	Rafter section	Unit weight	Column section	Rafter section	Unit cost
A	15	3	BBC30030	BBC30030	135.46	BBC30030	BBC30030	384.82
B	20	4	BBC30030	BBC30030	180.60	BBC30030	BBC30030	513.34
C	25	5	BBC35030	BBC35030	269.46	BBC35030	BBC35030	791.60

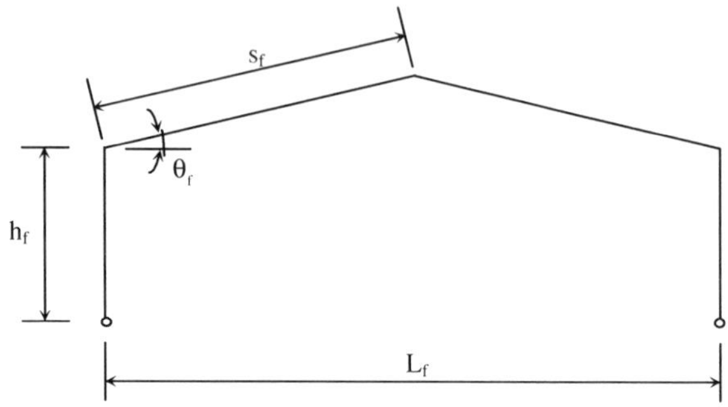

Fig. 9.2 Geometry of the rigid-jointed cold-formed portal frame from Phan et al. (2013), reprinted by permission of the publisher (Taylor & Francis Ltd, http://www.tandf.co.uk/journals)

height to eaves of h_f and frame pitch being θ_f. Pitch angle and frame spacing for these frames are assumed to be 10° and 4 m, respectively. Geometries and optimum section sizes for three exemplar frames are shown in Table 9.1.

It is assumed that frames A, B and C are located in three wind regions in Australia with the wind pressures of 1.45 kN/m^2 (region C), 0.91 kN/m^2 (region W) and 0.68 kN/m^2 (region A1), respectively. Wind pressure is calculated according to Australian code from wind region. The vertical dead load is 0.05 kN/m^2, the live load is 0.25 kN/m^2, cladding load is 0.05 kN/m^2 and ceiling load is 0.05 kN/m^2. In total, 34 ultimate limit state load combinations are considered.

Each frame is designed in accordance with the Australian code (AS/NZS 4600 2005). In order to simplify the problem, it is assumed that there are sufficient purlins and sufficient points of lateral and torsional restraint so that member instability does not need to be checked. With this assumption of sufficient lateral and torsional restraint, only local capacity is checked for the column and rafter members for all ultimate limit state (ULS) load combinations as given in equation below:

$$\frac{N^*}{\phi_c N_c} + \frac{M_x^*}{\phi_b M_{bx}} \leq 1 \tag{9.6}$$

where N_c is the nominal member capacity of the member in compression, M_{bx} is nominal member moment capacity about the x-axis, N^* and M_x^* are design axial compression and bending moment, respectively, and ϕ_c and ϕ_b are capacity reduction factor for compression and bending, respectively (AS/NZS 4600 2005). This local capacity check is applied at every node and in every load combination.

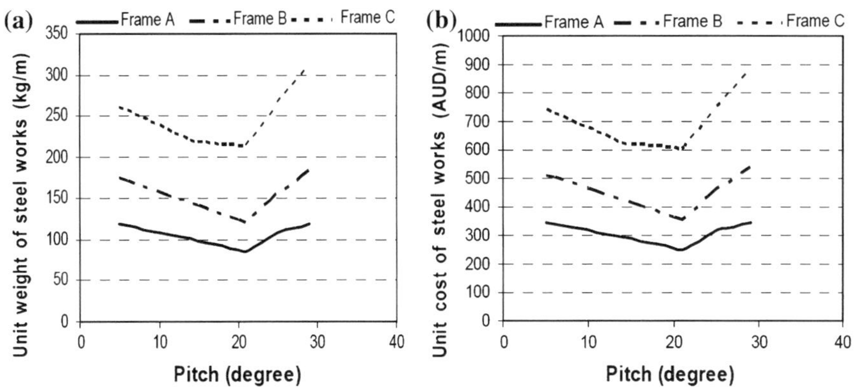

Fig. 9.3 Effect of frame's topography on unit weight and unit cost of three exemplar frames. **a** Pitch-unit weight (kg/m) relationship; **b** pitch-unit cost (AUD/m) relationship. Reprinted from Phan et al. (2011), with permission from Elsevier

9.3.2 Effect of Pitch

To investigate the effect of pitch on the weight and cost per unit length of building, the pitch of the frames will be changed from 5 to 30° with the increment of 2°. For every pitch of frame, section sizes of members are chosen by considering all the different permutations of sections, in conjunction with frame analysis and local capacity checks to determine the minimum weight per unit length of building or the minimum cost per unit length of building, independently. Only 20 channel-section sizes, ranging from C10010 to C35030, manufactured by Lysaght BlueScope will be considered. The unit weight or unit cost calculated will be inversely proportional to ULS factor. The channel-section prices are based on approximate values in the Australian market.

The pitch of frame has a large influence on the unit weight or unit cost of building for the three exemplar frames (Fig. 9.3). The larger wind pressure, as shown in Frames A and B, the clearer this influence can be seen. Interestingly, the lowest value of each graph has the same pitch angle of 21°.

9.4 Topography Optimisation

In the previous section, the effect of roof pitch is investigated. The choice of roof pitch has a large effect on optimum design of the frame. In this section, the optimisation technique will be carried out to search for optimum topographies (i.e. roof pitch and frame spacing) of the three exemplar frames mentioned above using the real-coded genetic algorithm. The structural analysis and optimisation procedure are programmed.

In this section, the objective function is chosen being the cost per unit length of building. The fitness function that combines the objective function and constraints has the form of Eq. (9.5). In the optimisation process presented in this section, as described by Deb (2001), a crossover probability of 0.9, mutation probability of 0.1 and offspring distribution indexes (η_c and η_m) of 10 in both crossover and mutation operators have been used. In real-coded GA, these parameters are called to be exogenous ones that need to be tuned empirically for achieving the effective searching ability. In general, it is expected that the required population size would depend on the problem complexity.

According to the generation histories of the optimisation processes of exemplar frames shown in Fig. 9.4, the number of generations required for convergence is different for different runs with different population sizes. In the same population of solutions, the different runs of GA with different random seeds also lead to the same solution. In general, the number of generations required for convergence reduces as the population size increases.

Table 9.2 shows that real-coded GA results in an optimum pitch that matches with results in the graph of Fig. 9.3. It can also be seen that the optimum frame spacings obtained by real-coded GA are smaller than the 4 m frame spacing used in Sect. 9.3 to achieve a cheaper unit cost. The unit cost in optimum frame topography from Table 9.2 is lower than in the three exemplar frames from Table 9.1 by 23.3–35.5 %. This demonstrates the need of topography optimisation covering both pitch and frame spacing.

9.5 Parametric Study and Design Recommendations

9.5.1 Parametric Study

As can be seen in the previous section, the topography of the frame has the considerable effect on unit cost of cold-formed steel building and the optimum solution can be achieved by applying a real-coded GA. Determining the optimum topography and frame spacing for frames having different spans and column heights located in some typical wind regions can be useful as a design aid for engineers.

Portal frames with four different column heights will be considered: 3, 4, 5 and 6 m. Each column height will be combined with five different frame spans: 5, 10, 15, 20 and 25 m. Each frame is assumed to be located in three typical wind regions in Australia, namely region A1, region W and region C. The other loads are identical to those used in Sect. 9.3.

With the assumption of full lateral restraints, the optimum frame spacing reduces as the span or column height increases (Phan et al. 2011). For instance, optimum frame spacing decreases considerably when frame span increases from 10 to 15 m in the wind region A1 and W, while this reduction in frames having spans more than 15 m is not as steep. In the high pressure wind region (region C),

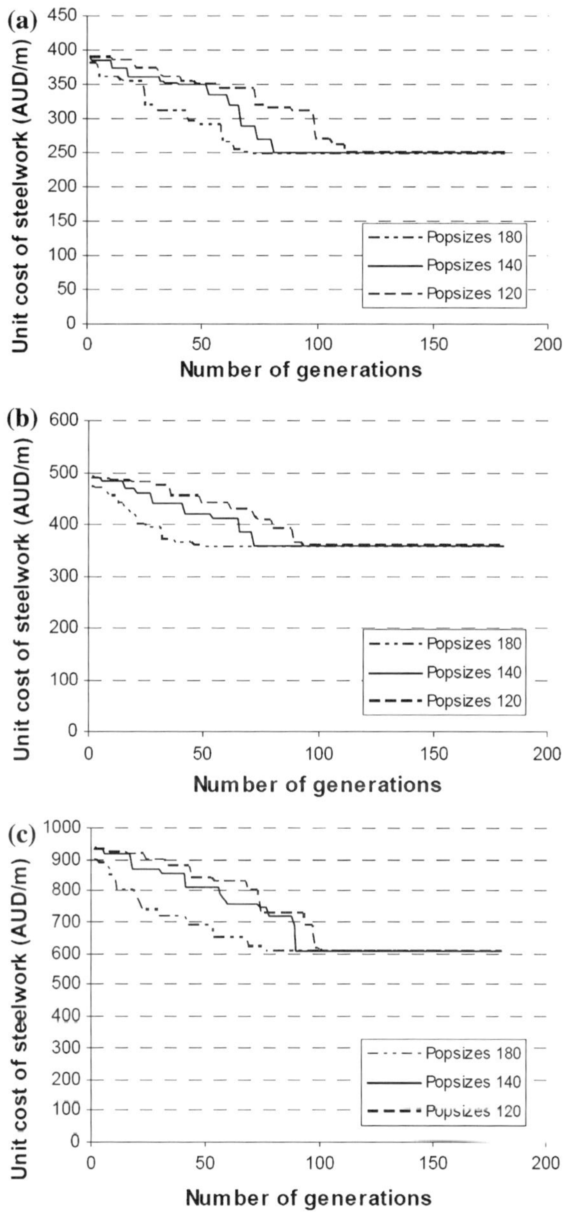

Fig. 9.4 Generation history for three population sizes. **a** Frame A; **b** Frame B; **c** Frame C. Reprinted from Phan et al. (2011), with permission from Elsevier

the contour graph of optimum bay spacings is rather steep in the range of span varying from 5 to 10 m and slightly shallow with frames having spans larger than 10 m. Interestingly, the optimum section determined from the algorithm is the same for column and rafter members in each frame, known as back-to-back channel-section of BBC25024. The minimum unit cost can also be achieved with

Table 9.2 The optimum solutions for three exemplar frames

Frame	Span (m)	Height (m)	Pitch (°)	Bay spacing (m)	Wind pressure (kN/m^2)	Column section	Rafter section	Unit cost (AUD/m)
A	15	3	21	3.17	1.45	BBC25024	BBC25024	248.19
B	20	4	21	2.95	0.91	BBC25024	BBC25024	355.60
C	25	5	20.9	2.16	0.68	BBC25024	BBC25024	606.80

C-single section of C25024 with half frame spacings. In different wind regions, the frame with the same geometry has larger optimum frame spacing in the low pressure wind region than in high pressure wind region.

According to Phan et al. (2011), the optimum pitch of rafter increases as length of span increases from 5 to 25 m. With spans greater than 10 m, changing the optimum pitch is similar in three researched wind regions for frames with different column heights. The slope of rafter is steepest in the frames having a maximum span of 25 m. The result can be related to the behaviour of 'arch effect' in the structure. In contrast to relationship with span, there is a decrease in optimum pitch as column height increases. This reduction is because of the effect of wind pressure on the height to apex of frame.

9.5.2 Design Recommendations for Optimum Topography

From the parametric study, it can be seen that optimum pitch is larger than 10° that is commonly used for practical design. For frame spacings less than 10 m, some comments on optimum frame spacings and optimum section with the objective function being minimum unit cost per unit length are as follows:

- The optimum frame spacings are reasonable for frame spans larger than 12 and 10 m in the wind region A1 and wind region W, respectively (Phan et al. 2011). In contrast, the optimum frame spacing will be more feasible if designer uses C-single section of C25024. With the length of span between 10 and 15 m, the optimum frame spacings can be used for either section of C25024 or BBC25024.
- In the high pressure wind region as in region C, optimum frame spacing is rather small when frame span is larger than 15 m (Phan et al. 2011) with optimum section of BBC25024. Thus, it will be more feasible if the larger back-to-back channel-sections are used in these cases. However, the unit cost will certainly increase compared with minimum unit cost. With the length of span less than 10 m, the optimum frame spacings can be used for either section of C25024 or BBC25024.

In summarising Sects. 9.1–9.5, the optimum topography of a cold-formed portal steel frame can be determined using real-coded GA with assumptions of full lateral restraint and constraint that the frame spacing is less than 10 m.

The optimal solutions from real-coded GA are verified and satisfy all the design constraint. Because of its simplicity of real coding, proposed optimum technique described here can be applied to a wide variety of optimisation problems.

Compared with the popular practical design, say pitch of 10°, it is noticed that the optimum pitch of rafter should be larger than 10°. For optimum bay spacing, minimum cost will be achieved when designer changes frame centres more than 4 m with small frame span, say 15 m, in the low and medium pressure wind regions. For the high pressure wind regions, the feasible frame centres are when frame spans are less than 20 m. Unit cost in optimum frame topography is lower than in common practical designs by 23.3–35.5 %. However, the optimisation will be more reasonable if the optimum design of purlins, cladding and other user requirements is included in the optimisation process. The research shown here can be considered as a first step towards globally optimising topography of steel portal frames. Future work would involve the design of the purlins, side rails and sheeting, and also to take serviceability deflection limits into account.

9.6 More Complicated Frames

9.6.1 Frame Parameters and Geometry

The parameters used to define a rigid-jointed cold-formed steel portal frame have been discussed with Fig. 9.2; a rigid-jointed cold-formed steel portal frame having knee braces at the eaves is shown in Fig. 9.5. These parameters are as follows: span of frame L_f, height to eaves h_f, length of rafter s_f, pitch of frame θ_f, flexural rigidity of members EI, axial rigidity of members EA and frame spacing b_f. For the case of portal frames having knee braces at the eaves, the positions of the knee braces are fixed relative to h_f as shown in Fig. 9.5.

Figure 9.6 shows the dimensions of the cold-formed steel channel-sections used for the primary load-carrying members in portal frame buildings. Table 9.3 shows the cold-formed steel channel-sections used in this chapter. These channel-sections

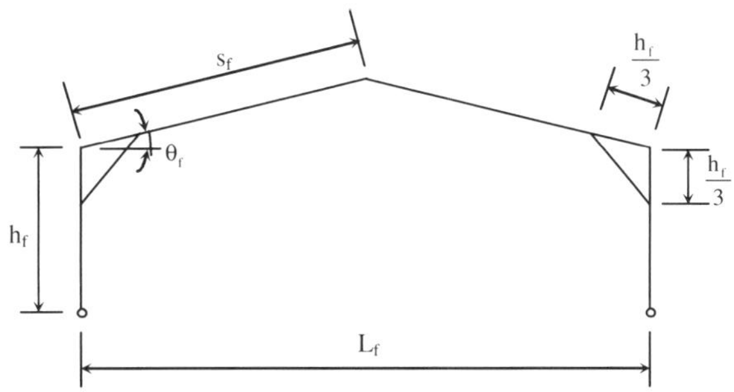

Fig. 9.5 Geometry of the rigid-jointed cold-formed portal frame having knee braces at the eaves. From Phan et al. (2013), reprinted by permission of the publisher (Taylor & Francis Ltd, http://www.tandf.co.uk/journals)

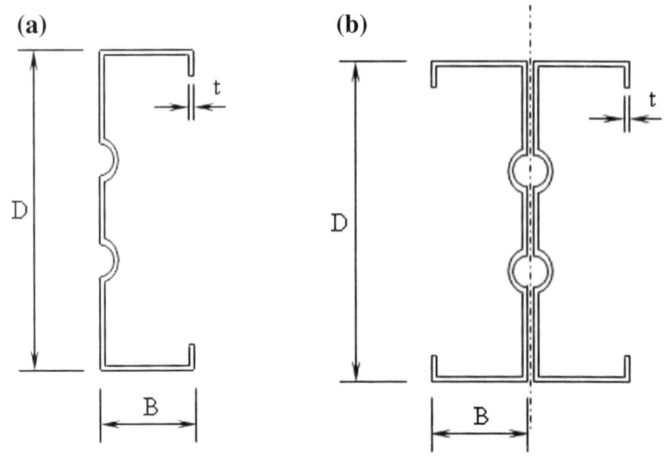

Fig. 9.6 Details of cold-formed steel channel-section. **a** Single channel-section (C-section); **b** back-to-back channel-section (BBC) from Phan et al. (2013), reprinted by permission of the publisher (Taylor & Francis Ltd, http://www.tandf.co.uk/journals)

can either be used singly or back-to-back. The swages on the web of channel-section obviously improve the load carrying capacity of the members. However, for simplifying the checking procedure and obtaining a conservative design, the section properties and member checks are based on plane channel-sections and therefore ignore the benefit of the swages.

In this and the following sections, the design optimisation of two exemplar frames with L_f of 20 m and h_f of 4 m is considered: Frame A without knee braces (Fig. 9.2) and Frame B with knee braces (Fig. 9.5). To minimise the cost per unit length of the building, the decision variables are the pitch, frame spacing and section sizes of the members. It is assumed that the column bases are pinned and that the purlins and side rails positioned within the web of the members are spaced sufficiently close each other to prevent out-of-plane buckling from occurring.

9.6.2 Frame Loading Analysis

The permanent and imposed roof loads (AS/NZS 1170-1 2002) that will be applied to the frames are as follows: permanent load (G) 0.10 kN/m^2 (purlins, rails, cladding) and self-weight of the members (Table 9.3); imposed load (Q) 0.25 kN/m^2.

From the Australian Standard code of practice on wind loading for the design of buildings (AS/NZS 1170-2 2002), the basic wind pressure q_u for the ultimate limit state is calculated from a design wind speed V_{des}, which in turn is calculated

Table 9.3 Dimensions and section properties of cold-formed steel sections

Section	D (mm)	B (mm)	t (mm)	EA ($\times 10^2$) (kN)	EI ($\times 10^6$) (kN.mm^2)	Weight (kg/m)	Cost (A$ /m)
C10010	102	51	1.0	451.0	73.8	1.78	5.58
C10012	102	51	1.2	533.0	88.2	2.10	6.15
C10015	102	51	1.5	656.0	110.7	2.62	6.77
C10019	102	51	1.9	840.5	137.4	3.29	8.37
C15010	152	64	1.0	604.8	225.5	2.32	7.03
C15012	152	64	1.2	717.5	264.5	2.89	7.99
C15015	152	64	1.5	902.0	330.1	3.59	8.46
C15019	152	64	1.9	1148.0	414.1	4.51	10.52
C15024	152	64	2.4	1455.5	520.7	5.70	12.88
C20012	203	76	1.2	922.5	574.0	3.50	8.99
C20015	203	76	1.5	1148.0	723.7	4.49	10.04
C20019	203	76	1.9	1455.5	924.6	5.74	12.56
C20024	203	76	2.4	1845.0	1166.5	7.24	15.29
C25015	254	76	1.5	1312.0	1250.5	5.03	13.66
C25019	254	76	1.9	1660.5	1562.1	6.50	14.43
C25024	254	76	2.4	2091.0	1972.1	8.16	17.82
C30019	300	96	1.9	2070.5	2788.0	7.92	22.76
C30024	300	96	2.4	2583.0	3485.0	10.09	29.52
C30030	300	96	3.0	3280.0	4366.5	12.76	36.25
C35030	350	125	3.0	3915.5	7339.0	15.23	44.74

from the regional wind speed V_R multiplied by factors M_d (wind direction multiplier), M_t (topographic multiplier), M_s (shielding multiplier), and $M_{z,\text{cat}}$ (terrain/height multiplier). $M_{z,\text{cat}}$ depends on both the terrain category and the average height of the building.

For example, Frame A, with dimensions as shown in Fig. 9.7, built in the wind region W in Australia with V_R of 49.4 m/s, the multiplier factors of M_d, M_t, and M_s are taken as 1.0 and $M_{z,\text{cat}}$ of 0.87. The design wind speed is calculated as follows:

$$V_{\text{des}} = V_{\text{site}} = V_R \cdot M_d \cdot \left(M_t \cdot M_s \cdot M_{z,\text{cat}}\right) = 42.98\ {}^{\text{m}}\!/_{\text{s}} \tag{9.7}$$

where V_{site} is the site wind speed and V_{des} is the design wind speed, and the basic wind pressure

$$q_u = 0.6V_{\text{des}}^2/1000 = 1.1\ {}^{\text{kN}}\!/_{\text{m}^2} \tag{9.8}$$

where q_u is the ultimate design wind pressure.

The design wind pressure acting on each of the four faces of the frame (AB, BC, CD and DE) is obtained by multiplying q_u by a coefficient of pressure and other related factors. The coefficient of pressure acting on each face is obtained from a combination of the external pressure coefficient C_{pe} and the internal pressure coefficient C_{pi}. The external pressure coefficients C_{pe} should be

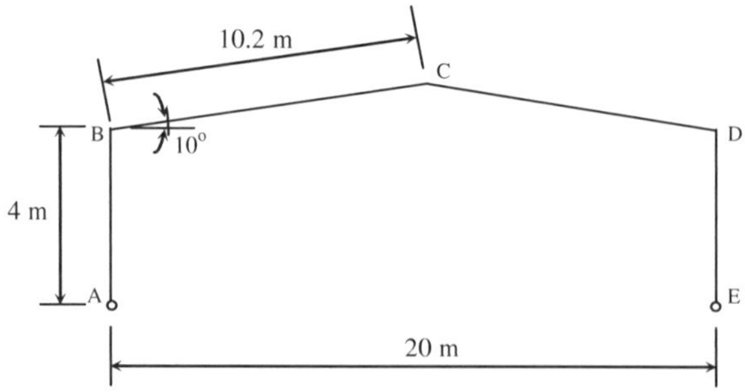

Fig. 9.7 Dimensions of frame from Phan et al. (2013), reprinted by permission of the publisher (Taylor & Francis Ltd, http://www.tandf.co.uk/journals)

Table 9.4 Coefficients of external pressure C_{pe} acting on each face

Wind acting on	AB	BC	CD	DE
Side of frame (WT1)	0.7	–0.3	–0.3	–0.3
Side of frame (WT2)	0.7	–0.7	–0.3	–0.3
End of frame (WL1)	–0.65	–0.9	–0.9	–0.65
End of frame (WL2)	–0.2	0.2	0.2	–0.2

calculated for wind acting on the side and on the end. These values are shown in Table 9.4, calculated based on AS/NZS 1170-2 (2002). For buildings of normal permeability without dominant openings, C_{pi} has a minimum value of –0.3 for suction, and a maximum value of 0.2 for pressure.

The eight wind load combinations for Frame A (WLC1–WLC8), and their corresponding coefficients for both side wind and end wind, are shown in Table 9.5. The coefficients of pressures C_{pe} given by WLC1 are illustrated in Fig. 9.8. The frame will be checked for all eight wind load combinations in the design procedure to be described in Sect. 9.6.3.

In accordance with AS/NZS 1170-0 (2002), the frame will be checked at the ultimate limit state for the following three ultimate load combinations:

$$\mathrm{ULC1} = 1.2G + 1.5Q \tag{9.9a}$$

$$\mathrm{ULC2} = 1.2G + \mathrm{WLC} \tag{9.9b}$$

$$\mathrm{ULC3} = 0.9G + \mathrm{WLC} \tag{9.9c}$$

ULC3 is used for the uplift wind load combination.

A first-order elastic frame analysis program for cold-formed steel sections is used to analyse and design the portal frame. To take into account second-order effects, an amplification factor is applied as described in the Australian code of practice. For each load combination, the bending moment, shear force and axial force are determined. The program is called to analyse each candidate solution in each generation as shown in Fig. 9.1.

Table 9.5 Coefficients of pressure ($C_{pe} + C_{pi}$) acting on each face corresponding to eight wind load cases

Combination	Wind acting on	AB	BC	CD	DE
WLC1	Side, internal pressure	0.7 + 0.2	–0.3 + 0.2	–0.3 + 0.2	–0.3 + 0.2
WLC2	Side, internal suction	0.7–0.3	–0.3–0.3	–0.3–0.3	–0.3–0.3
WLC3	Side, internal pressure	0.7 + 0.2	–0.7 + 0.2	–0.3 + 0.2	–0.3 + 0.2
WLC4	Side, internal suction	0.7–0.3	–0.7–0.3	–0.3–0.3	–0.3–0.3
WLC5	End, internal pressure	–0.65 + 0.2	–0.9 + 0.2	–0.9 + 0.2	–0.65 + 0.2
WLC6	End, internal suction	–0.65–0.3	–0.9–0.3	–0.9–0.3	–0.65–0.3
WLC7	End, internal pressure	–0.2 + 0.2	0.2 + 0.2	0.2 + 0.2	–0.2 + 0.2
WLC8	End, internal suction	–0.2–0.3	0.2–0.3	0.2–0.3	–0.2–0.3

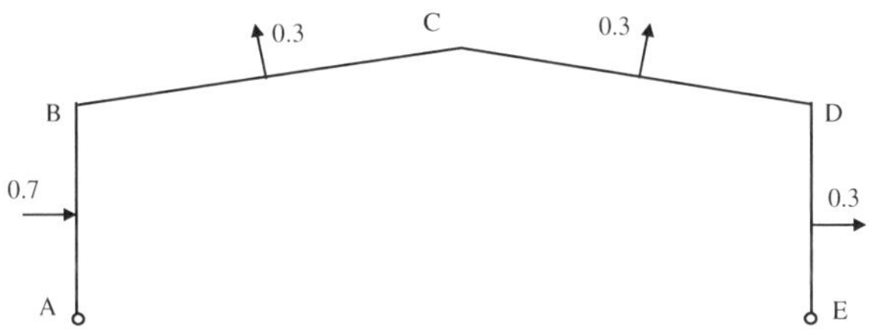

Fig. 9.8 Coefficients of wind pressure for wind load combination 1 (WLC1) for standard frame from Phan et al. (2013), reprinted by permission of the publisher (Taylor & Francis Ltd, http://www.tandf.co.uk/journals)

9.6.3 Member Checks

9.6.3.1 Columns and Rafters

In accordance with AS/NZS 4600 (2005), the columns and rafters are checked for combined axial compression and bending, distortional buckling and combined bending and shear. The combined axial compression and bending check follows Eq. (9.6).

The distortional buckling check is:

$$M^*_{xk} \leq \phi_b M^k_{bx} \tag{9.10}$$

with $M^k_{bx} = Z_c f_c$ and $f_c = M_c/Z_f$, where M^*_{xk} is the design bending moment in member k about x-axis of the effective cross-section, ϕ_b is the capacity reduction factor for bending, M_{bx}^k is the nominal member moment capacity of member k, Z_c is the effective modulus at a stress f_c in the extreme compression fibre, M_c is the critical moment, and Z_f is the full unreduced section modulus for the extreme compression fibre.

The combined bending and shear check is:

$$\frac{M_{xk}^*}{\phi_b M_s^k} + \frac{V_k^*}{\phi_v V_{vk}} \leq 1 \tag{9.11}$$

where M_{xk}^* is the design bending moment in member k about x-axis of the effective cross-section, M_s^k is the nominal section moment capacity of member k about the x-axis, V_k^* is the design shear force in member k, V_{vk} is the nominal shear capacity of the web of member k, ϕ_b is the capacity reduction factor for bending, and ϕ_v is the capacity reduction factor for shear.

9.6.3.2 Eaves Knee Braces

The knee brace is a pin-ended member and is checked for both compression and tension. The compression check is:

$$N_k^* \leq \phi_c N_{ck} \tag{9.12}$$

where N_k^* is the design compressive axial force of member k, N_{ck} is the nominal member capacity of the member k in compression, ϕ_c is the capacity reduction factor for compression.

The tension check is:

$$N_k^* \leq \phi_t N_{tk} \tag{9.13}$$

where N_k^* is the design tensile force of member k, N_{tk} is the nominal section capacity of the member k in tension, and ϕ_c is the capacity reduction factor for tension.

9.7 Optimisation Formulation

The objective of the design optimisation is to determine the portal frame building having the minimum cost, whilst satisfying the design requirements. The cost of the main frame depends on frame spacing, pitch and section sizes. The objective function can be expressed in terms of the cost per unit length of the building as follows:

$$\text{Minimise} \quad W = \frac{1}{b_f} \sum_{i=1}^{n} w_i l_i \tag{9.14}$$

where W is the cost of main frame per unit length of building, b_f is the frame spacing, w_i are the cost per unit length of cold-formed steel sections (see Table 9.3), l_i are the lengths of cold-formed steel structural members, and n is the number of members.

The normalised forms of the design constraints or unity factors given in Eqs. (9.6) and (9.10–9.13) are expressed as follows:

$$g_1 = \frac{N_k^*}{\phi_c N_c^k} + \frac{M_{xk}^*}{\phi_b M_{bx}^k} - 1 \leq 0 \tag{9.15a}$$

$$g_2 = \frac{M_{xk}^*}{\phi_b M_{bx}^k} - 1 \leq 0 \tag{9.15b}$$

$$g_3 = \frac{M_{xk}^*}{\phi_b M_s^k} + \frac{V_k^*}{\phi_v V_{vk}} - 1 \leq 0 \tag{9.15c}$$

$$g_4 = \frac{N_k^*}{\phi_c N_{ck}} - 1 \leq 0 \tag{9.15d}$$

$$g_5 = \frac{N_k^*}{\phi_t N_{tk}} - 1 \leq 0 \tag{9.15e}$$

The design specification is the ultimate limit state (ULS) that constitutes the constraints for the optimisation problem. The relationship known as the fitness function often has the form:

$$F = W\,[1 + \mathrm{C}] \tag{9.16}$$

where W is the objective function being the cost of frame per unit length of building and C is the constraint violation penalty.

In this section, the penalty value is assigned through the maximum level of violation of the unity-factor constraints in Eq. (9.15) as follows:

$$g = \max\,[g_1, g_2, g_3, g_4, g_5] \tag{9.17}$$

In this section, penalty values are imposed empirically, in proportion to the severity of constraint violation. Two levels of violated constraints with the magnitudes as shown in Eq. (9.18) are suitable to eliminate the violated solutions through the evolutionary process, as follows:

$$C = \begin{cases} 0 & \text{if} \quad g \leq 0 \\ g & \text{if} \quad 0 < g \leq 0.5 \\ 10g & \text{if} \quad g > 0.5 \end{cases} \tag{9.18}$$

The proposed optimisation procedure aims to minimise the value of the fitness function F (Eq. 9.16). This is achieved by minimising the cost W and reducing the penalty C to zero. The procedure involves a real-coded GA, frame analysis and cold-formed steel design. As can be seen from Fig. 9.9, the evaluation process computes the fitness function values using the objective function (Eq. 9.14) along with the corresponding penalty values defined in Eq. (9.18). Better (i.e. cheaper) solutions will yield smaller fitness values, and consequently are selected preferentially by the tournament selection operator. The criterion for terminating the program is a predefined total number of function evaluations. This criterion is suitable to investigate the convergence history of the algorithm.

9.8 Design Examples

9.8.1 Frame A with Fixed Topology

Frame A, without knee braces, in which the pitch is 10° and frame spacing of 4 m, is considered. Such a typical pitch and frame spacing are commonly used for cold-formed steel portal frames. This design problem has two discrete decision variables. The optimum sections for the columns and rafters can be determined from the sections shown in Table 9.3 by exhaustive enumeration using the first-order elastic frame analysis program for cold-formed steel sections as mentioned in Sect. 9.6.2. As can be seen from the list of cold-formed steel channel-sections, there are 40 options of cold-formed sections used for members, including both single sections (C) and back-to-back channel-sections (BBC).

Four design options satisfy the member checks (Table 9.6). As can be seen, the smallest unit cost obtained is A$ 513/m with back-to-back channel-section BBC30030 for both columns and rafters. This result will be used to validate the real-coded GA for searching the optimum solution. It is observed that all design constraints, i.e. the member checks, are satisfied. The critical constraint relating to combined axial compression and bending moment for rafter governs the design in the case of ultimate load combination ULC3, with the unity factor being 0.9 against the upper limit of 1.0.

The real-coded GA is used to determine the optimum sections for the members of Frame A as discrete variables, from the 40 alternative of sections for members as mentioned above. The design process (see Fig. 9.1) is repeated with three different population sizes to investigate the possibility of reaching the optimum

Table 9.6 Optimum sections for Frame A with fixed topology

Column section	Rafter section	g_1	g_2	g_3	g_4	g_5	W (A$/m)
BBC30030	BBC30030	−0.10	−0.28	−0.32	−1.0	−1.0	**513**
BBC35030	BBC30030	−0.32	−0.30	−0.35	−1.0	−1.0	547
BBC30030	BBC35030	−0.35	−0.45	−0.59	−1.0	−1.0	599
BBC35030	BBC35030	−0.44	−0.41	−0.58	−1.0	−1.0	633

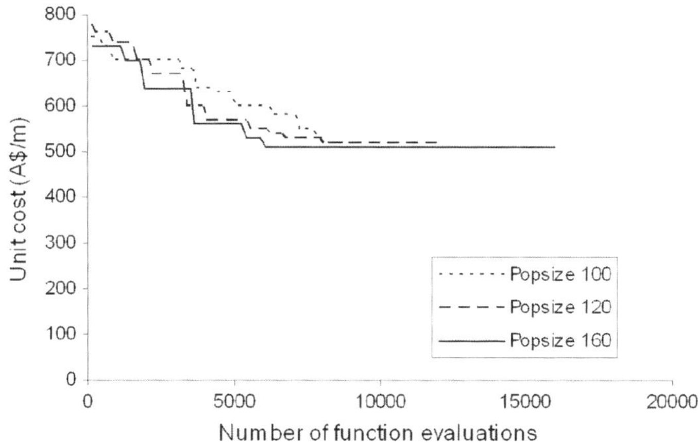

Fig. 9.9 Progress of the GA for frame A with fixed pitch and frame spacing from Phan et al. (2013), reprinted by permission of the publisher (Taylor & Francis Ltd, http://www.tandf.co.uk/journals)

solution. The convergence histories from the GA are shown in Fig. 9.9. As can be seen, the fitness function converges. The three runs in the optimisation process lead to the same minimum-cost solution as the exhaustive enumeration process. This has proved the reliability of the GA with real-coded parameter.

9.8.2 Frame A with Variable Pitch

9.8.2.1 Exhaustive Enumeration, and GA with Fixed Frame Spacing and Variable Pitch

To validate the capacity of the GA in solving the more complicated problem as the effect of the pitch is taken into account, the process of exhaustive enumeration is carried out manually. In this case, the pitch is varied from 5 to 30° in increments of 5° to investigate its effect. The frame spacing is fixed at 4 m. For each pitch, the optimum sections for members of Frame A are determined through exhaustive enumeration by using the first-order elastic frame analysis program manually. The influence of pitch on the cost per unit length of building is shown in Fig. 9.10. The pitch has a clear influence on the cost of portal frame building. The minimum unit cost is A$ 432/m, with a pitch of 20° and sections of BBC30024. However, the constraint of combined axial compression and bending for the rafter check was violated by $g_1 = 0.006$. The logical inference is that the cheapest feasible design can be expected to have a pitch close to 20°.

The optimum pitch and cross-sections for the members of Frame A are determined again using the real-coded GA with a fixed frame spacing of 4 m. In this

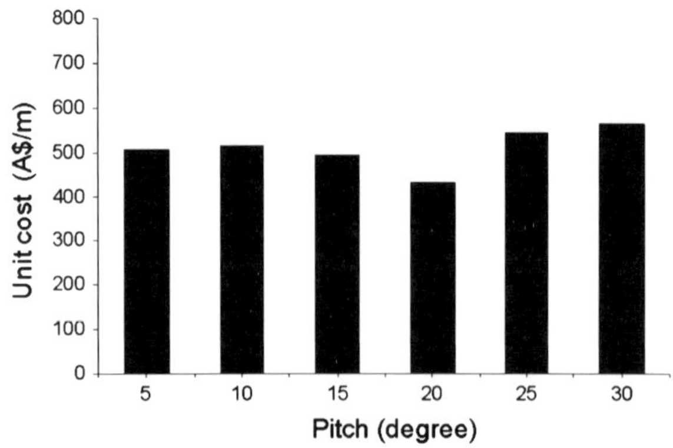

Fig. 9.10 Effect of pitch on unit cost of the Frame A with fixed frame spacing from Phan et al. (2013), reprinted by permission of the publisher (Taylor & Francis Ltd, http://www.tandf.co.uk/journals)

case, the pitch is processed as a continuous variable varying in the range of [5, 90°), whilst sections are discrete, selected from the list (Table 9.3). The design process is also repeated with three different population sizes. The progress of the GA is shown in Fig. 9.11. As can be seen, the minimum unit cost achieved from GA is found to be A$ 433/m with a pitch of 20.5° and section of BBC30024 for both columns and rafters. As expected, the unit cost obtained from the GA is slightly larger than the unfeasible solution with a pitch of 20° from exhaustive enumeration because the GA solution is feasible. The design constraint for the rafter becomes active in the case of the combined actions of axial compression and

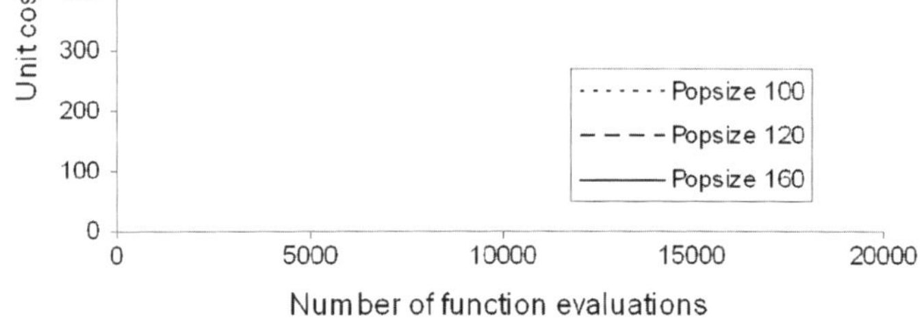

Fig. 9.11 GA progress for frame A with fixed frame spacing from Phan et al. (2013), reprinted by permission of the publisher (Taylor & Francis Ltd, http://www.tandf.co.uk/journals)

moment. Again, this demonstrates that real-coded GA is an effective and reliable method to find the optimum pitch and sections of members in portal frame design.

In addition to the exhaustive enumeration in the beginning of this section, the GA is verified further here by generating the maximum cost feasible design based on a fixed frame spacing of 4 m and variable pitch. To maximise the cost, it is sufficient to multiply the objective function (Eq. 9.14 by -1.0. A population of 100 is used and five random runs in total of the GA are carried out with up to a maximum of 15000 function evaluations per run.

The maximum cost obtained from RC-GA is A\$ 728.5/m with a pitch of 35.5°. The size of both the columns and rafters is BBC35030 which is the largest and most expensive. Only the distortional bucking constraint on the rafter is binding ($g_2 = 0$) under load combination ULC2. The other constraints are slack. This solution was achieved with three GA runs out of five.

The second most expensive solution also has the largest section of BBC35030 for both the columns and rafters. The cost and pitch are A\$ 725/m and 35°, respectively. The critical constraint is also distortional buckling for the rafter with g_2 equal to 0.01 that is a bit higher than the best solution. Taken together with the cost and pitch, this suggests that this solution is a near-optimum. This solution was achieved in one GA run out of five.

One GA run out of the five attempts fails to generate a good maximum cost solution. The cost, pitch and section size for members obtained are A\$ 564/m, 30°, and section of BBC30030, respectively. Also, the distortional buckling constraint ($g_2 = 0.02$) is active for rafter under load combination ULC2. This solution is self evidently not a maximum, but it is still 30 % more expensive than the least cost solution of A\$ 433/m.

Therefore, four attempts out of five in total are successful in the sense that the two solutions in these four attempts are both feasible and satisfactory in terms of the section sizes (i.e. the largest) with at least one binding constraint in each case. The maximum-cost solution is obtained within 7,200 function evaluations. The maximum-cost solution is 68 % more expensive than the corresponding minimum-cost solution.

9.8.2.2 Real-Coded GA with Variable Frame Spacing and Variable Pitch

In this example, design optimisation for Frame A that accounts for effects of both pitch and frame spacing is conducted, using real-coded GA. The pitch and frame spacing are processed as continuous variables. The progress of the optimisation process is shown in Fig. 9.12. As can be seen, the three runs of the optimisation process converge to the same unit cost. The optimum pitch and frame spacing for Frame A obtained from the algorithm are 21° and 3 m, respectively. The optimum section for both the columns and rafters is BBC25024 and the unit cost is A\$ 355/m. The design constraint for the combined actions of axial compression and bending on rafter is active, i.e. $g_1 = 0$.

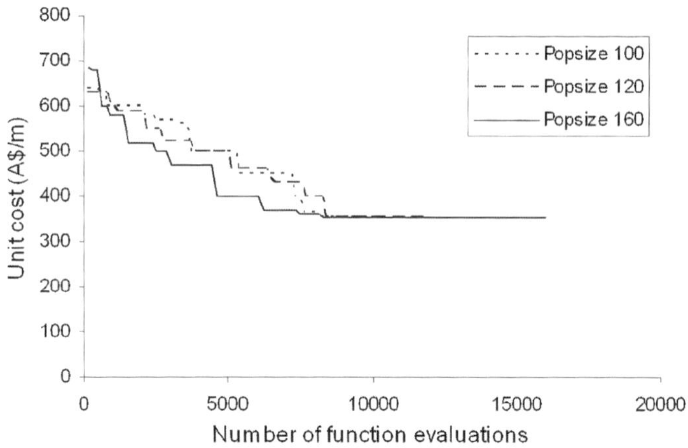

Fig. 9.12 GA progress for frame A with variable pitch and frame spacing from Phan et al. (2013), reprinted by permission of the publisher (Taylor & Francis Ltd, http://www.tandf.co.uk/journals)

It can be seen that when the portal frame design includes the pitch and frame spacing as decision variables, the optimum frame is cheaper than an optimum frame with fixed frame spacing (18 %) or fixed pitch (31 %). This demonstrates the benefit of topology optimisation covering both pitch and frame spacing. It should be emphasised that the above-mentioned cost savings relate to the typical frame geometry, and not previously optimised results described in the literature; the latter are not available.

9.8.3 Frame B with Fixed Pitch and Fixed Frame Spacing

In this example, the design optimisation of Frame B having knee braces with a fixed pitch is considered using the real-coded GA. The pitch is fixed as 10° and the frame spacing is prespecified as 4 m. In this problem, the optimum sections for the columns, rafters and knee braces are again to be selected from the aforementioned list of 40 cold-formed sections that include both single and back-to-back channel-sections (Table 9.3).

The population size of 160 individuals is used for determining the optimum sections for members, since it has been proven in the previous sections to be the appropriate size to search for the optimum solution. The maximum permitted number of function evaluations is set at 8,000. The optimum sections are given in Table 9.7, and corresponding unit cost is A$ 428/m. The design constraint of knee brace buckling becomes active from load combination ULC2. Thus, for the specified pitch of 10° and frame spacing of 4 m, Frame A that does not have knee braces at the eaves is 19.8 % more expensive than Frame B that has knee braces.

Table 9.7 Optimum sections for Frame B with fixed topology

Member type	Cold-formed steel sections	g_1	g_2	g_3	g_4	g_5
Column	BBC 30024	−0.18	−0.31	−0.12	−1.00	−1.00
Rafter	BBC 30024	−0.12	−0.13	−0.10	−1.00	−1.00
Knee brace	C 20015	−1.00	−1.00	−1.00	0	−0.34

9.8.4 *Frame B with Variable Pitch and Variable Frame Spacing*

The optimum design of Frame B taking into account the effect of pitch and frame spacing simultaneously is considered in this example. There are five design decision variables in this problem: pitch and frame spacing considered as continuous variables; column, rafter and knee brace sections as discrete variables. This optimisation problem is more complicated than previous examples as there are more design variables and the solution space is larger. The population size is 160 and the number of function evaluations for terminating the program is 12,800, equivalent to 80 generations.

Accordingly, the real-coded GA is run ten times. Six out of ten runs generate the same optimum solution. The combined actions of bending moment and axial compression on the rafter is active ($g_1 = 0$) in the case of load combination ULC2. The optimum unit cost obtained is A$ 270/m with corresponding parameters as shown in Table 9.8. The optimum pitch (θ_f) is 17.5°, and the optimum frame spacing (b_f) is 4.0 m. This result is 37 % lower than Frame B having fixed topology and 24 % lower than Frame A with both optimum pitch and frame spacing. The effect of knee braces makes the optimum pitch reduce by 3° compared with the optimum pitch of frame without knee braces.

It is interesting to note that the unit cost obtained from four out of ten runs is A$ 283/m which is 4.8 % more expensive than the optimum design for Frame B, with larger frame spacing and sections. In this case, the critical design constraint for combined bending and axial compression for the column becomes active. Such a solution may be suitable for building that requires a larger frame spacing (Table 9.9). The pitch (θ_f) is 17°, and the frame spacing (b_f) is 7.6 m.

Summarising Sects. 9.6–9.8, the real-coded GA can minimise the cost per unit length of a cold-formed steel portal frame building with different frame spacings and topologies, based on the Australian code of practice for cold-formed steel. The

Table 9.8 Optimum solution for Frame B with variable topology

Member type	Cold-formed steel sections	g_1	g_2	g_3	g_4	g_5
Column	BBC 25024	−0.01	−0.03	−0.09	−1.00	−1.00
Rafter	BBC 25024	0	−0.01	−0.07	−1.00	−1.00
Knee brace	C 20015	−1.00	−1.00	−1.00	−0.02	−0.38

Table 9.9 Near-optimum solution for Frame B with variable topology

Member type	Cold-formed steel sections	g_1	g_2	g_3	g_4	g_5
Column	BBC 30030	0	−0.06	−0.04	−1.00	−1.00
Rafter	BBC 30030	−0.02	−0.08	−0.07	−1.00	−1.00
Knee brace	BBC 10015	−1.00	−1.00	−1.00	−0.05	−0.22

influence of a knee brace at the eaves on the optimum topology and unit cost is considered. The real-coded GA program can determine the optimum topology and the most suitable sections for members simultaneously. The frame obtained from the algorithm can be considered as the most economical design in each case, since the critical design constraint in all examples becomes active. The reliability and robustness of the algorithm is clear. In addition, the high consistency of the optimum results is achieved through a number of trials for minimising the objective function.

In the case of optimisation problems having many design variables, both optimum and near-optimum solutions are obtained. Through five design examples, it is shown that the real-coded GA is very efficient in handling the optimisation problems having both continuous and discrete variables. Although the cost is calculated based on material used for the main frames, the reduction is remarkable when the optimum topology is reached. Frames having knee braces result in the most optimum design with the least cost of material. In future research, the position of the purlins and side rails and their cost should be considered, taking into account buckling of the column and rafter members between points of lateral restraint.

9.9 An Efficient Genetic Algorithm for the Design Optimisation

Although GAs have been applied to many engineering problems, the main disadvantage of the conventional GAs is that they often suffer premature convergence and weak exploitation capabilities. Premature convergence, which often leads to a non-optimal solution or a local optimum solution, can occur because of loss of diversity in the population of candidate solutions. This loss of population diversity is due to the tendency of the selection operator in GAs to favour the better solutions when choosing solution to take part in crossover to create the next generation of solutions. In later generations, the best solutions will therefore dominate the population in evolutionary processes.

In an effort to enhance the searching performance and accelerate the convergence speed of the GA, a modification of the distributed GA (DGA) was suggested to use with a number of variable mutation schemes to increase the diversity of the population in the earlier stages (Issa and Mohammad 2010). In addition, the niching techniques were suggested that has been successfully applied to GA in determining the optimum solution of many complicated mathematical functions with multiple constraints (Deb 2001).

In this section, a niching strategy as proposed in Deb (2001) is incorporated into the RC-GA to improve the exploration of the solution space to help determine the optimum solutions, which are building topology as the continuous variables and section sizes as the discrete variables. The proposed optimisation method, to be referred to as real-coded niching GA (RC-NGA), maintains the diversity of the population, thereby increasing the probability of achieving the global optimum solution, by the preferential retention of candidate solutions from regions that are under-represented while simultaneously eliminating some of the candidate solutions from regions that are overcrowded based on the presumption that candidate solutions in the same neighbourhood would tend to be similar.

The results of RC-NGA, in terms of cost of the primary members per square metre of floor plan, are shown to be identical to the benchmark examples presented earlier in this chapter. It is shown that the effectiveness in determining the optimum solution increases significantly in terms of the reliability of the solutions, robustness and computational efficiency. The Australian code of practice is used for demonstration purposes, although any design codes can also be applied. Because there is less snow in many regions in Australia, the large span can be achieved.

9.9.1 Optimisation Formulation and the Niching Strategy in the Selecting Operator

The objective of the overall design optimisation, including the building topology and section sizes of members, is to determine the portal frame building having the minimum cost. The cost of the main frame can be expressed in terms of the cost of the primary members per square metre of the floor area as follows:

$$\text{Minimise} \quad W = \frac{1}{L_f b_f} \sum_{i=1}^{m} w_i l_i \tag{9.19}$$

where W is the cost of main frame per square metre of floor area, w_i are the cost per unit length of cold-formed steel sections (Table 9.3), l_i are the lengths of cold-formed steel structural members, and m is the number of members.

In the real-coded niching genetic algorithm (RC-NGA), tournament selection with a niching technique is applied. The process is conducted by selecting at random two individuals from the current population, namely $\boldsymbol{x}^{(i)}$ and $\boldsymbol{x}^{(j)}$. The normalised Euclidean distance (Deb 2000) between two solutions is:

$$d_{ij} = \sqrt{\frac{1}{n} \sum_{k=1}^{n} \left(\frac{x_k^{(i)} - x_k^{(j)}}{x_k^u - x_k^l} \right)^2} \quad 1 \leq i, j \leq Pop - size \tag{9.20}$$

where d_{ij} is a normalised Euclidean distance between $\boldsymbol{x}^{(i)}$ and $\boldsymbol{x}^{(j)}$, n is the number of decision variables, *Pop-size* is the population size in the RC-NGA, $x_k{}^{(i)}$ and $x_k{}^{(j)}$ are the corresponding kth decision variable in two vectors $\boldsymbol{x}^{(i)}$ and $\boldsymbol{x}^{(j)}$, and $x_k{}^u$ and $x_k{}^l$ are the upper and lower bounds respectively of the kth decision variable.

If this Euclidean distance is smaller than an empirical user-defined critical distance, these solutions are compared using their fitness function values. Otherwise, they are not compared and another solution $\boldsymbol{x}^{(j)}$ is selected at random from the population for comparison. If after a certain number of checks, no solution $\boldsymbol{x}^{(j)}$ is found to satisfy the critical distance, $\boldsymbol{x}^{(i)}$ is selected for the crossover operation. In this way, only solutions in same region (or *niche*) compete against each other for selection and crossover.

9.9.2 Benchmark Examples

9.9.2.1 Portal Frame Without Knee Braces (Type 1)

The design optimisation for Frame Type 1 is considered using RC-NGA. The frame has a span of 20 m and column height of 4 m. This benchmark example was described earlier in this chapter and solved using RC-GA. The design optimisation that accounts for the effect of both the pitch and frame spacing is conducted with the GA parameters and operators described earlier. This problem has four decision variables, viz. pitch and frame spacing processed as continuous variables, and section sizes of the columns and rafters as the discrete ones. The suitable population size is 40. The number of function evaluations is 6000 to terminate the RC-NGA program.

The progress of the optimisation is shown in Fig. 9.13. As can be seen, the RC-NGA converged to the best optimum solution within the predefined number of function evaluations. The most appropriate section size for both the columns and rafters is BBC25024; the optimum pitch is 21°, the optimum bay spacing is 3 m, and the unit cost is A$ 17.75/m^2. These same results are obtained from the RC-GA (earlier in this chapter). The design constraint for the combined actions of axial compression and bending on rafter is critical, with ULC3 load combination. The RC-NGA, with a population size of 40 produces the optimum solution in eight out of ten runs within 6,000 function evaluations.

It should be noted that with the same population size of 40, this problem is solved with the RC-GA routine (earlier in this chapter) with a termination criterion of 8000 function evaluations. RC-GA is trapped at a local optimum solution with a unit cost of A$ 21.55/m^2. The appropriate population size used for RC-GA to obtain the optimum solution is 100 (Fig. 9.13). For RC-GA, only three out of ten runs, with population size of 100, generate the same optimum solution. This means that on average RC-GA requires around 26,667 function evaluations to reach the best solution, compared to RC-NGA which requires 7,500 function evaluations. Based on these results, RC-NGA is approximately 3.5 times more efficient than RC-GA.

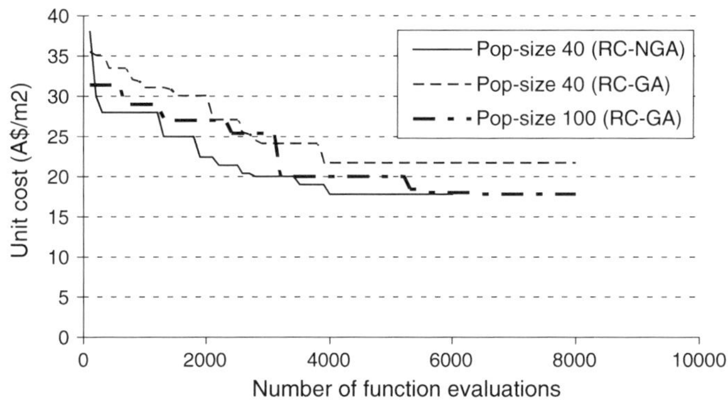

Fig. 9.13 Convergence progress for frame type 1

9.9.2.2 Portal Frame with Knee Braces (Type 2)

The optimum design of portal frame with knee braces at eaves (Type 2) having the same span and column height as in Frame Type 1 is considered. This benchmark example is also solved earlier in this chapter. The same optimisation procedure described earlier is implemented in this section. There are five design decision variables in this problem: pitch and frame spacing as continuous variables, and the column, rafter and knee brace sections as discrete variables.

This optimisation problem is more complicated than the previous example as there are more design variables and the solution space is larger. For RC-NGA, the population size is 50 and the number of function evaluations for terminating the RC-NGA program is empirically 7,500.

With a population size of 50, RC-NGA generates the best-known optimum solution in seven out of ten runs (10,714 function evaluations on average). The RC-GA routine used earlier in this chapter converges prematurely at a local optimum solution in all ten runs with the same population size of 50 (see Fig. 9.14). To improve the performance of the RC-GA, the necessary population size is 120 within 10,000 function evaluations. This enables two runs out of ten to converge to the lowest cost solution found here. This is equivalent to 50,000 function evaluations on average to achieve the best solution. Based on these results, RC-NGA is over 4.5 times more efficient than RC-GA.

9.9.3 Summary

The RC-NGA is developed to minimise the cost of the primary members per square metre of floor plan for cold-formed steel portal frame buildings. The consistency of the optimum solution obtained is improved after a number of runs, conducted with the small population size within more reasonable

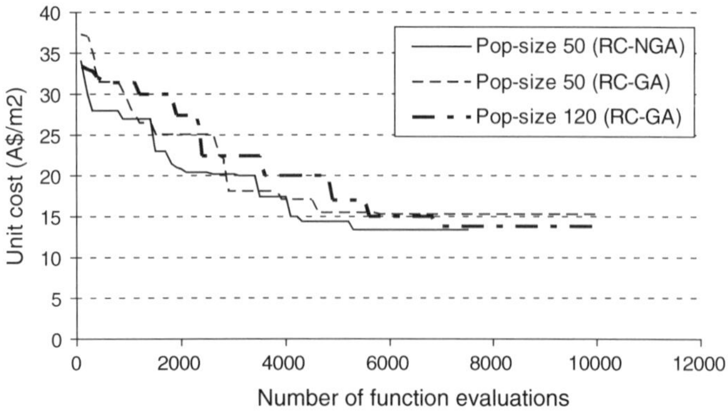

Fig. 9.14 Convergence progress for frame type 2

computing time. This means that the diversity in the population has been maintained, so that the probability of achieving the global optimum result increases effectively.

For each building, the optimisation program aims to determine the optimum topology and the most suitable sections for members. The frame design obtained from the program can be considered as the most economical design in each case, since the critical design constraint in all examples becomes active. The computational efficiency and robustness of the algorithm are also demonstrated. The length of computation time for solving the optimisation problem is therefore reduced significantly.

9.10 Full-Scale Fire Tests on a Cold-Formed Steel Portal Frame Building

9.10.1 Aims of Research

Fires in supermarkets, warehouses and other steel portal frame structures account for damage and revenue losses every year. This research will allow for the evaluation of current design guidance for steel portal frame structures, which is in conflict with recent computational results that suggest this guidance is inadequate and potentially unsafe in certain fire scenarios. This work can be used to improve the existing design guidance, allow for a performance-based approach to design these structures and ensure these structures safe and fit for purpose.

The performance differences between hot rolled I sections and cold formed C sections in portal framed buildings in fire conditions need be established. Current design practices need be evaluated against these materials to determine if they are applicable to both types of steel. Fire tests can be used to validate and improve numerical models of portal frame structures, which are currently mainly theoretical.

Practitioners have long argued that the design of steel portal frames in certain fire-boundary conditions is potentially unsafe. Recent numerical analyses have demonstrated that existing practices for the design of steel portal frames in fire boundary conditions can be unsafe in some conditions and over-prescribed in others. Large-scale fire tests conducted on a cold-formed steel portal frame can evaluate the current design practices and validate the results of the numerical models. The results can be used to develop performance-based design recommendations for future structures that cover both conventional hot-rolled steel and cold-formed steel.

9.10.2 Benefits to Structural Engineering and Impact on Industry

Structural engineers must put building safety and reliability as their top priorities. They have to be satisfied that a building will act as intended under a whole range of possible situations. For structural engineers, the idea of sustainability is probably best captured within performance-based design. More economic buildings meet the aspirations of the engineers in designing buildings that serve the community with reduced waste, transportation costs and lower emissions.

Benefits of this approach to structural engineering include:

- Safer design of portal frames in potentially hazardous situations (e.g. industrial buildings, chemical factories) where it is imperative that the building does not collapse
- More economical design of portal frames in less hazardous situations (e.g. agricultural buildings) with large potential savings in carbon in terms of the size of the concrete foundation and column base design
- Improved safety for the fire service, with a better understanding of the collapse of buildings
- Provide the ground work for a computational design approach for all single-storey steel buildings giving the designer more realistic expectations about building performance allowing for fire protection to be focused and economical
- Improve opportunity for the steel industry to expand, to where the dominance of steel for single storey steel buildings is not repeated, with fire performance often cited as the reason
- Improve understanding of the behaviour of cold-formed steel joints in fire

References

AS/NZS1170-0 (2002) Structural design actions—part 0: general principles. Australian/New Zealand Standard. Australian Institute of Steel Construction, Sydney

AS/NZS1170-1 (2002) Structural design actions—part 1: permanent, imposed and other actions. Australian/New Zealand Standard. Australian Institute of Steel Construction, Sydney

AS/NZS1170-2 (2002) Structural design actions—part 2: wind actions. Australian/New Zealand Standard. Australian Institute of Steel Construction, Sydney

AS/NZS 4600 (2005) Cold-formed steel structures. Australian/New Zealand Standard. Australian Institute of Steel Construction, Sydney

Deb K (2000) An efficient constraint handling method for genetic algorithms. Comput Method Appl Mech 186:311–338. doi:10.1016/S0045-7825(99)00389-8

Deb K (2001) Multi-objective optimization using evolutionary algorithms. Wiley, Chichester

Deb K, Gulati S (2001) Design of truss-structures for minimum weight using genetic algorithms. Finite Elem Anal Des 37:447–465. doi:10.1016/S0168-874X(00)00057-3

Eid MA, Elrehim MA, El-kashef F, Swoboda G (2010) Optimization of ground improvement techniques in tunnelling using genetic algorithms. In: IV European Conference on Computational Mechanics (ECCM 2010), Paris

Issa HK, Mohammad FA (2010) Effect of mutation schemes on convergence to optimum design of steel frames. J Constr Steel Res 66:954–961. doi:10.1016/j.jcsr.2010.02.002

Kameshki ES, Saka MP (2001) Optimum design of nonlinear steel frames with semi-rigid connections using a genetic algorithm. Comput Struct 79:1593–1604. doi:10.1016/S0045-7949(01)00035-9

Pezenshk S, Camp C, Chen D (2000) Design of nonlinear framed structures using genetic optimization. J Struct Eng 126(3):382. doi:10.1061/(ASCE)0733-9445

Phan DT, Lim JBP, Ming CSY, Tanyimboh T, Issa H, Sha W (2011) Optimization of cold-formed steel portal frame topography using real-coded genetic algorithm. Procedia Eng 14:724–733. doi:10.1016/j.proeng.2011.07.092

Phan DT, Lim JBP, Sha W, Siew CYM, Tanyimboh TT, Issa HK, Mohammad FA (2013) Design optimization of cold-formed steel portal frames taking into account the effect of building topology. Eng Optim 45. doi:10.1080/0305215X.2012.678493

Saka MP (2003) Optimum design of pitched roof steel frames with haunched rafters by genetic algorithm. Comput Struct 81:1967–1978. doi:10.1016/S0045-7949(03)00216-5

Tsoulos IG (2008) Modifications of real code genetic algorithm for global optimization. Appl Math Comput 203:598–607. doi:10.1016/j.amc.2008.05.005

Chapter 10
Fire Engineering

Abstract This chapter introduces the basic concepts and methods of fire safety design. The use of computer software in fire engineering is the main theme. Computer software was specially designed for calculating moment capacity. With thermal analysis software, the heat transfer of any new type of steel section in fire can be determined accurately and economically. Shallow floor structures have good inherent fire resistance. However, when the bottom plate of such sections is exposed directly to fire, heat can be transferred along the steel to adjacent compartments. Therefore, although shallow floor construction does not normally need fire protection, partial shielding at compartment corners is still required. Other sections of this chapter are concerned with the fire resistance of composite beams with unfilled voids and temperature modelling with fire flux data and intumescent coating. An objective is to quantitatively predict the amount of additional protection required when the voids between the top flange and the underside of the steel deck are unfilled. The final part of this chapter is to derive a relationship for determining the thickness of intumescent fire protection coating required for shallow floor beams, using existing fire test data. The formula derived can be used for quick, simple design calculation for the required thickness of fire protection.

10.1 Fire Safety Design

10.1.1 Methods of Approach and Design Concept

There are two fire-engineering approaches for steel structures. One is the prescriptive approach, by assessing the performance of structures in fire from tabulated or graphical data derived from standard or fire tests by manufacturers. In this approach, the structure is designed based on ultimate limit state and then the fire protection requirement is calculated from pre-defined data tables and charts in various codes of practice. The other is the calculation approach, by calculating the capacity of members in fire to assess the performance of the structure. In this approach, the limiting temperature and the load ratio, i.e. the ratio between moment capacities of structural members in fire and normal conditions,

W. Sha, *Steels*, DOI: 10.1007/978-1-4471-4872-2_10,

are determined. Both approaches are centred on limiting the steel temperature of structural members, because steel begins to lose its strength at 400 °C and it decreases rapidly with increasing temperatures.

The focus of this section is based on the calculation approach for steel structures. A design concept has been introduced that incorporates fire safety into the structural members themselves instead of designing the structural members and then providing protection to them. There are two fire safety design methods, the limiting temperature/load ratio method and moment capacity method. The latter is used in this section.

If the temperature distribution through a structural member section is known, the reduced strength of all the elements in the cross-section can be calculated. The plastic neutral axis and, hence, the moment capacity in fire conditions can then be determined directly. In the moment capacity method, if the member's moment capacity at the required period of fire is not smaller than the applied moment at the fire limit state, the member may be considered to have adequate fire resistance without protection. Otherwise, the member will require fire protection.

The state-of-the-art in fire engineering is to build fire resistance into the structure themselves. The greatest advantage of this is that it does not require any traditional fire protection (board or spray) or only require minimum protection, that can lead to construction that is considerably more economical. This approach has become more acceptable in Western Europe from the early 1980s, which has resulted in the development of slim floor construction.

10.1.2 Slim Floor Construction and Fire Resistant Steels

Slim floor is a relatively new form of floor construction using steel as the structural member. The development of slim floor construction is an excellent example of achieving built-in fire resistance in building structures. The beam used in slim floors is contained almost totally within the depth of the concrete floor. The floor has a flat appearance similar to a reinforced concrete floor, and has excellent fire resistance, as only the bottom face of the steel section would be exposed to the heat in event of a fire. As the concrete floor protects the steel beam, the structure can withstand fire for longer time without any traditional fire protection material. Originated in Scandinavia, the principles of the slim floor construction were adapted by British Steel (now Tata Steel) and The Steel Construction Institute (SCI), UK, to develop two types of new slim floor beams, the Slimflor (Fig. 10.1) and the ASB Slimdek.

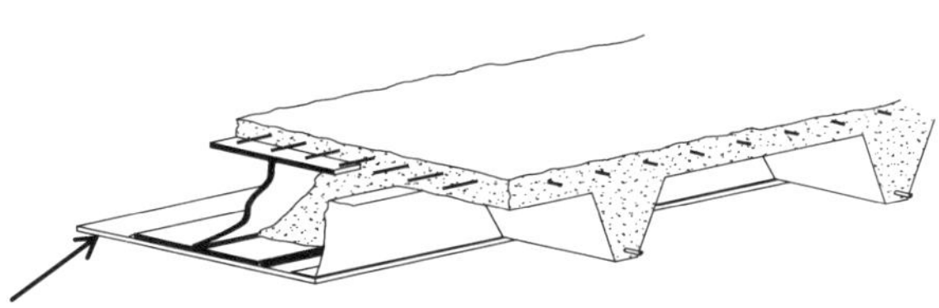

Fig. 10.1 Diagram of typical shallow or slim floor (fabricated Slimflor beam)

A Slimflor beam is a universal column with a steel plate welded to its bottom flange. There are two types of Slimflor beam construction system, the more popular and desirable in situ composite decking system, and the pre-cast unit. In the latter system, the slab rests on the steel plate and the remaining space is filled with in situ concrete. ASB is the transformed steel section of the Slimflor beam, in which the plate and bottom-flange are replaced by a larger bottom flange. The ASB system uses composite construction.

The development of fire resistant steels is another achievement in fire engineering. This type of steel can retain a higher strength in fire condition, and thus increase the fire resistance. This is achieved by developing new compositions and rolling processes. A main effort in developing such steels has been from Nippon Steel of Japan. Several fire resistant steels have also been developed at Queen's University Belfast (Chap. 3).

10.1.3 Software

Three computer programs were developed at SCI to calculate respectively the moment capacity of I-beam, Slimflor and ASB floors in fire conditions. The principal theory is based on plastic analysis. The model is in essence a finite element model. The plastic moment capacity of a beam at elevated temperatures is calculated by dividing the beam into several elements such as bottom flange, lower web, upper web and top flange. In the calculation, the software considers the steels to have reached 2 % strain at failure, and the strength reduction factors at elevated temperatures are applied to the normal strength to obtain the elevated temperature strength. For each element, the area, position of centroid and reduced strength are calculated. The total resistance of all elements is then found. In pure bending, the plastic neutral axis must be found which divides the total resistance into equal tension and compression. After finding the plastic neutral axis, moments are taken about any convenient axis and moment capacity at the temperature is thus obtained.

The elevated temperature in a steel beam section due to the heat transfer in fire determine its moment capacity. The temperatures of each element for each type of beam can be measured in a standard fire test. Although such test give good and representative results of the temperature development in beams and their structural behaviour, it is expensive and therefore there are limited test data on steel beams such as Slimflor and ASB. Thermal analysis software can be used to supplement fire tests. A new steel beam section can be designed more economically by using the software to model the heat transfer in the section, and hence determine its feasibility before manufacturing the section and conducting fire tests. The thermal analysis software used in this section is TFIRE. It is a two-dimensional finite difference heat transfer program developed by SCI and has been verified against fire test data. It was used to model the temperature development in the Slimflor and ASB during the course of developing these sections.

Based on thermal conduction and heat transfer models, TFIRE can be used for any form of structural sections. The program can calculate the heat flow in a steel section and changes of temperature distribution in the section in fire condition. With the temperature data obtained by the software, the moment capacity of the steel section can be determined by using software for calculating that property.

TFIRE uses the following basic heat input expression to model the heat transfer:

$$\frac{dq}{dt} = SV(a\varepsilon_f T_f^4 - \varepsilon_s T_s^4) \tag{10.1}$$

where dq/dt is rate of heat transfer per unit area, S is Stephen-Boltzmann constant, V is view factor of the element, a is surface absorptivity, ε_f is flame emissivity, ε_s is surface emissivity, T_f is furnace or gas temperature, and T_s is element temperature.

The consideration of the interface resistance between nominally touching surfaces is important in the modelling. For normal conduction between elements in perfect contact, the following expression applies:

$$\frac{dq}{dt} = \frac{T_2 - T_1}{\frac{1}{K_1} + \frac{1}{K_2}} \tag{10.2}$$

If an interface resistance exists, this expression becomes:

$$\frac{dq}{dt} = \frac{T_2 - T_1}{\frac{1}{K_1} + \frac{1}{K_2} + \frac{1}{K_i}} \tag{10.3}$$

where K_1 and K_2 are thermal conduction terms, and K_i is interface resistance coefficient. Fig. 10.2 shows an example of approximate cross-section for a slim floor beam used in TFIRE.

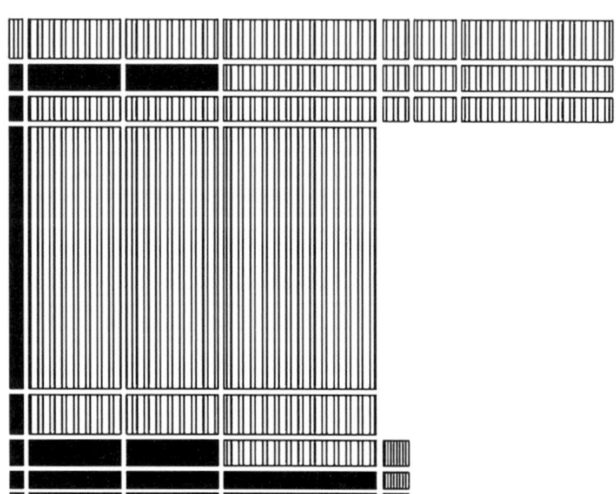

Fig. 10.2 Approximated cross-section based on user input data

10.1.4 Moment Resistance of Unprotected Slim Floors in Fire

First, TFIRE is used to calculate the temperature distribution in a floor section when the compartment underneath is in fire. Second, its moment resistance is calculated. For the fabricated Slimflor beam, it is divided into six rectangular elements, plate, bottom flange, lower half of the web (lower web), upper half of the web (upper web), top flange and concrete compression flange. For asymmetric beams, the division is more detailed, and, for the beams designed for British market (280 ASB 100 and 300 ASB 153), consideration is given for the fillet at the junctions of the web and the bottom flange. The width and height of each element are given in Table 10.1, for three beam sizes.

For fabricated beam, the calculation is based on a 254 × 254 × 89 universal column (UC) section, with flange plate thickness 15 mm and width 455 mm. The concrete depth above the top flange is 30 mm. Decking is made with ComFlor CF210 deck. S355 steel grade is used, and the concrete strength is taken as 20.1 MPa. Temperature distribution at 60 min in fire is calculated using TFIRE, with fire temperature taken from the standard fire curve. The calculation give a flange plate temperature of 822 °C, bottom flange 611 °C and lower web 433 °C. All other parts of the steel section are below 400 °C, therefore full strength being preserved. With this temperature distribution, the fire moment resistance is 352 kNm, compared with 523 kNm for moment resistance at room temperature. These give a load ratio of 0.67.

For asymmetric beams, the bond strength limits between the steel sections and concrete when hot are taken as 0.9 and 0.4 MPa, for British (280 ASB 100 and 300 ASB 153) and European versions, respectively. The beam configurations and calculation results are detailed in Table 10.2. The temperatures of those elements not included in the table are below 400 °C and therefore they are at full strength. The moment resistance data at room temperature are included for comparison. The moment resistance in fire is based on fire temperatures at 60 min.

The values of the load ratio, i.e. the ratio between the load resistance at the fire limit state and that under the normal 'cold' condition, are centred on 0.5, which is the normal design fire load ratio. This confirms that the slim floors have up to

Table 10.1 Width and depth of ASB elements (from bottom to top, all dimensions in mm)

Element	280 ASB 100		300 ASB 153		European	
	Width	Depth	Width	Depth	Width	Depth
1	300	16	300	24	375	20
2	44.7	6.4	55.9	7.2	10	10
3	31.8	6.4	41.4	7.2	10	30
4	19	7.2	27	5.6	10	85
5	19	20	27	20	10	85
6	19	20	27	20	10	20
7	19	184	27	202	200	20
8	190	16	190	24		

Table 10.2 Beam parameters, temperatures in fire at 60 min and moment resistance of ASB

Beam designation		280ASB100	300ASB153	300ASB153	European
Top flange width (mm)		190	190	190	200
Bottom flange width (mm)		300	300	300	375
Beam depth (mm)		276	310	310	270
Flange thickness (mm)		16	24	24	20
Web thickness (mm)		19	27	27	10
Decking		CF210	CF210	CF225	CF210
Concrete depth above beam flange (mm)		30	30	0	40
Steel temperature (°C) of element (Table 10.1)	1	786	747	747	804
	2	684	635	635	612
	3	643	597	597	485
	4	599	564	564	≤400
	5	532	509	509	≤400
	6	422	410	410	≤400
Moment resistance (kNm)	Fire	253	461	429	190
	Room temperature	554	889	739	418
Load ratio		0.46	0.52	0.58	0.45

1-h fire resistance without traditional fire protection. Although the calculations are based on five case parameters, the general conclusion should be applicable to other beam sizes and floor configurations.

In summary, based on modelling using computer software for calculating the steel beam section temperatures and moment resistance in fire, without any applied fire protection, Slimflor and asymmetric beam (ASB) slim floors can support around 50 % of the designed load in fire at 60 min. This is due to the inherently good fire resistance in these structures offered by the protection of concrete surrounding steel beams.

10.2 Heat Transfer in Steel Structures and Their Fire Resistance

10.2.1 Heat Transfer in Fire Across a Wall in Shallow Floor Structure

In shallow floor structures, the supporting steel floor beam is contained within the depth of the floor deck, and the structure therefore has inherently good fire resistance (Fig. 10.1). For most applications, it requires no fire protection for fire resistance ratings up to 1 h.

This offers great advantage. However, a steel beam can span through the top of a wall. When there is a fire in a compartment on one side of the wall, heat can transfer along the bottom flange. Problems can arise from the exposed beam (Fig. 10.3). When a conventional beam is used, the beam in the other side of the wall would be covered by the protection, and so there would be no increase in temperature. In shallow floor structures, the temperature rise of the beam in the compartment adjacent to the one on fire may be excessive. The problem is not structural integrity due to the loss of strength, but with the possibility of ignition,

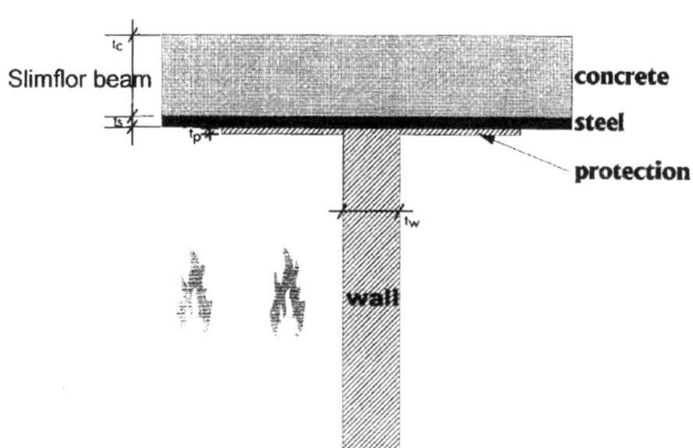

Fig. 10.3 Schematic configuration of Slimflor beam across wall (2D section at extended part of bottom plate)—t_c = thickness of concrete; t_s = thickness of steel plate; t_p = thickness of protection; t_w = thickness of wall

and therefore spread of fire, and personal safety. In this section, the heat transfer along a fabricated shallow floor (Slimflor) beam across a wall is modelled using the TFIRE software developed at The Steel Construction Institute, UK.

The calculation is based on the configuration shown in Fig. 10.3. The TFIRE program is essentially a 2D model. In the calculations, the sectioning required by the 2D model is made just inside the edge of the bottom plate welded to the I-beam in the Slimflor beam, but outside the bottom flange of the I-beam. The temperature during a fire is the highest at this section point (the point of the arrow in Fig. 10.1). Two plate thicknesses (t_s in Fig. 10.3), 15 and 22 mm, are examined here. A concrete thickness t_c of 260 mm is used; the value of this thickness is not critical as the temperature within the concrete block decreases rapidly away from the steel, due to the good insulation property of concrete. Normal weight concrete is used.

The effect of using fire resistant board protection at the bottom of the steel plate is examined here. Three conditions are tested: no protection, 150 × 10 mm protection on each side, and 300 × 10 mm protection on each side (thickness $t_p = 10$ mm). It is assumed that the protection material and the wall have the same material properties: density of 0.165 g/cm^3, no moisture, and thermal conductivity 0.1 W/ °C m. The assumed low wall density should not have a large effect on the results.

The temperature profile in the fire compartment follows the standard fire curve. The results are given in Figs. 10.4, 10.5. The wall thickness t_w is taken as 100 mm for cases where protection is used (Fig. 10.5).

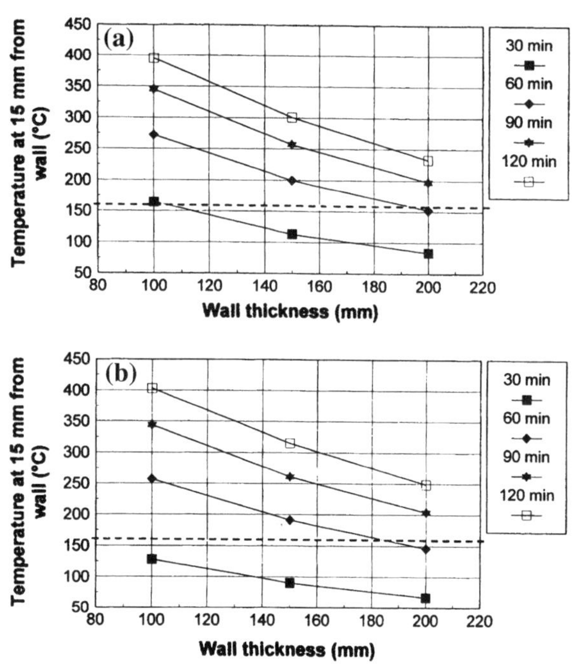

Fig. 10.4 Temperatures of steel plate (no protection) 15 mm from wall in compartment adjacent to fire: **a** plate thickness 15 mm; **b** plate thickness 22 mm

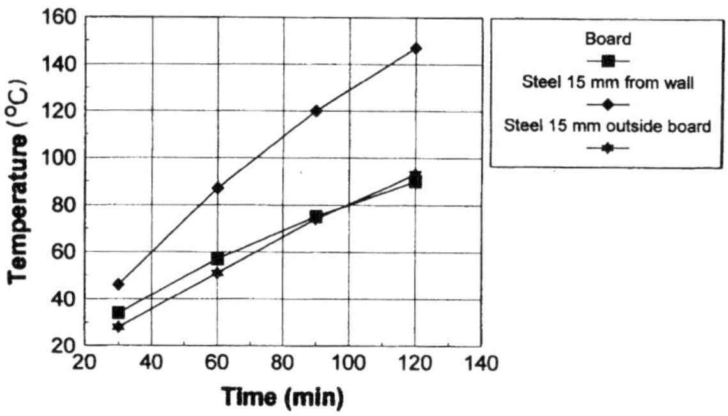

Fig. 10.5 Temperatures of protection board and steel plate in compartment adjacent to fire; 150 × 10 mm protection on each side of wall; plate thickness 22 mm

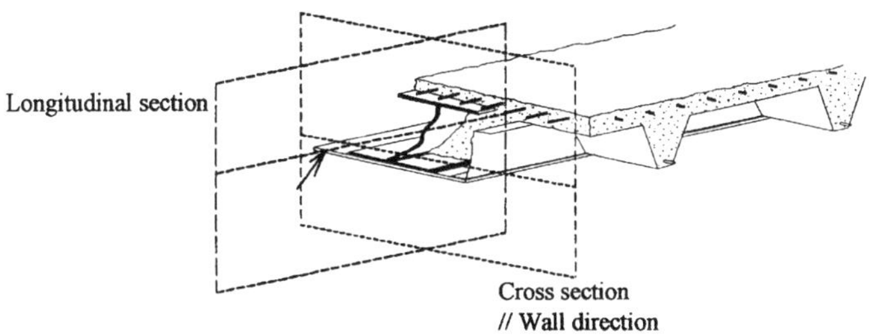

Fig. 10.6 Different sections

The heat transfer from the plate edge to the web does not significantly affect the temperature distribution within the steel beam. The TFIRE model is 2D. For the above calculations, the sectioning is made on the hanging edge of the steel plate welded to the bottom flange (i.e. a longitudinal section) (Fig. 10.6). Calculation is carried out to examine temperatures of the steel directly exposed to the fire away from the wall. The results from the longitudinal section are always close to those predicted from the cross-sectional 2D configuration. Calculation based on longitudinal sectioning gives a temperature of 832 °C, using the method described in this section.

As normal practice, the temperature rise in a compartment adjacent to a fire is restricted to 140 °C above its normal temperature (i.e., 160 °C if the normal room temperature is 20 °C). The above calculations show that, without protection, the temperature of the bare steel beam in a compartment adjacent to the one on fire would be above this limit in most cases (Fig. 10.4). However, with 150 × 10 mm protection on each side of the wall, the temperature rise in an adjacent compartment is confined to acceptable values. In Fig. 10.5, steel temperatures from two positions are taken. The first position is 15 mm from the wall. The second position is 15 mm outside the protection board, which itself extends for 150 mm from the wall.

Because the modelling is 2D, the result is applicable to the newer asymmetric beam shallow floor for bottom flanges of the same thickness values. The longitudinal sections of the fabricated shallow floor and the asymmetric beam are the same.

In summary, without protection, the temperature in the bare steel section in compartments adjacent to the fire would rise significantly to an unacceptable level. With some minimal fire protection for sections near the wall, any temperature rise is much reduced.

10.2.2 Fire Resistance of Composite Beams with Unfilled Voids

This section investigates composite floor structures, as a continuation of the theme of the programme of research on fire engineering modelling of steel building structures of this chapter. Modern multistorey steel framed buildings widely use composite construction for floor slabs and beams. A concrete topping is designed to act in a composite way with a steel deck to form the floor slab. The floor slab is then designed to act compositely with the steel beams to form composite beams.

Two types of deck profile are used, the reentrant dovetail deck and the open trapezoidal deck (Fig. 10.7). For this type of structures, steel beams require fire protection. For dovetail decks, the voids formed by the deck profile are small, and it is normally not necessary to fill them to maintain the required fire resistance. However, for trapezoidal decks, the voids can result in a significantly increased rate of heating of the steel section effectively unprotected above its top flange. To fill these voids has severe economic disadvantages because it is very labour intensive. This serves against the use of trapezoidal decks with board-type fire protection. The purpose of this section is to quantitatively evaluate the effect of not filling the voids formed with open trapezoidal decks on the fire resistance of the structure.

The essential parameter is the load ratio, the ratio between the load capacities at the fire limit state and under normal 'cold' condition. For most applications, a load ratio of 0.6 is adequate. The approach taken is as follows:

(1) Obtain the bottom flange temperature, T_1, that will give a load ratio of 0.6. Temperature values of various parts of the beam are required for this calculation, which are obtained from real temperature distributions from fire test data.

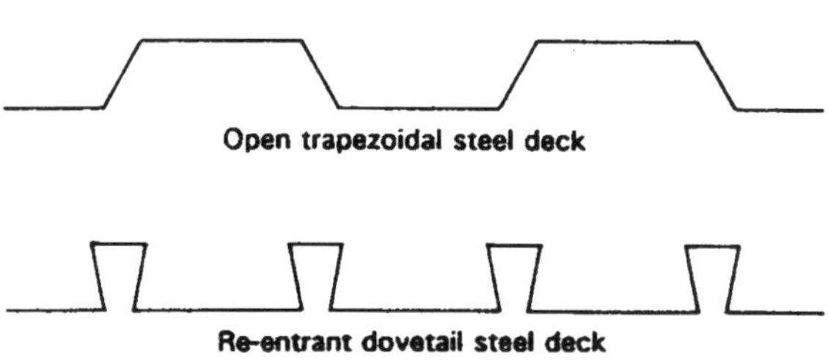

Fig. 10.7 Types of profiled steel decks

(2) The thickness required of fire protection board is proportional to $(T_{\text{bottom flange}}-140)^{-1.3}$. Therefore, the thickness to achieve a load ratio of 0.6 (d_1) can be calculated using:

$$\frac{d_1}{d_0} = \left(\frac{T_0 - 140}{T_1 - 140}\right)^{1.3} \tag{10.4}$$

where T_0 is the bottom flange temperature measured in the tests with protection thickness d_0 and T_1 is the bottom flange temperature to give the load ratio of 0.6.

(3) Repeat (1) and (2) for the situation of voids being unfilled and obtain the protection thickness (d_{1u}) to achieve the load ratio of 0.6. Temperatures corresponding to unfilled voids are indicated by a subscript u.

(4) Obtain the percentage increase in protection thickness (d_{1u}/d_1) for the same load ratio (0.6). The calculations are based on the actual parameters used in the fire tests (Table 10.3).

The bottom flange temperatures and the corresponding calculated load ratios are given in Table 10.4. For 60 min fire resistance, the ratio of fire protection thickness required achieving a same load ratio of 0.6 without and with the voids being filled can be obtained:

$$\frac{d_{1u}}{d_1} = \frac{\frac{d_{1u}}{d_0}}{\frac{d_1}{d_0}} = \frac{\left(\frac{T_{0u}-140}{T_{1u}-140}\right)^{1.3}}{\left(\frac{T_0-140}{T_1-140}\right)^{1.3}} = \frac{\left(\frac{575-140}{555-140}\right)^{1.3}}{\left(\frac{528-140}{584-140}\right)^{1.3}} = \frac{1.063}{0.839} = 1.27 \tag{10.5}$$

Therefore, the thickness of the protection need to be increased by 27 %.

For 90 min fire resistance, similar calculation can show that the thickness of the protection board also need to be increased by 27 %, although the temperature values (T_0 and T_{0u}) are very different from those for 60 min. It is therefore expected that the same level of increased protection will be necessary for other fire

Table 10.3 Beam parameters for composite beams with unfilled voids

Variable	Value
Beam	305 × 102 × 33 UBS275
Slab depth	125 mm
Deck depth	60 mm, open trapezoidal
Concrete	Normal weight, grade 30

Table 10.4 Bottom flange temperature and corresponding load ratio

Time (min)	Voids	$T_{\text{bottom flange}}$ (°C)	Load ratio
60	Filled	528 (T_0)	0.71
	Unfilled	575 (T_{0u})	0.54
90	Filled	706 (T_0)	0.28
	Unfilled	775 (T_{0u})	0.14
–	Filled	584 (T_1)	0.6
	Unfilled	555 (T_{1u})	0.6

resistance periods such as 120 min. Further, although the above calculations are based on a load ratio of 0.6, it is reasonable to assume that similar amount of extra protection would be required for other load ratios.

The calculation gives a quantitative answer regarding the amount of extra fire protection needed when the voids between the top flange and the underside of the steel deck in a composite beam structure are unfilled. The result essentially provides the design engineer with an option of either filling the voids, or increasing the thickness of the board protection (by 27 %) and leaving the voids unfilled. Given the labour and materials consumed in filling the voids, the latter is perhaps more economical and definitely offers a more speedy construction.

In summary, for composite floor using I-beam, when the voids between the top flange and the underside of the steel deck are not filled, an extra 27 % protection is required. This percentage of extra protection is the same for different fire resistance periods.

10.2.3 Temperature Modelling with Fire Flux Data

This section involves computer modelling of temperature development in steel sections during fire using heat flux data measured in fire tests. In TFIRE, a model developed at SCI, the standard consideration is to calculate the rate of heat transfer from fire to steel section using the temperature difference between the two elements. In a standard fire test, the fire temperature, measured by thermocouples in the test furnace, is made to follow a standard fire curve. An alternative is to measure directly the heat flux into the steel during a standard fire test (Fig. 10.8).

Thermal calculations are then made with TFIRE program to obtain the bottom flange temperature of an asymmetric slim floor beam (Fig. 10.9). This opens

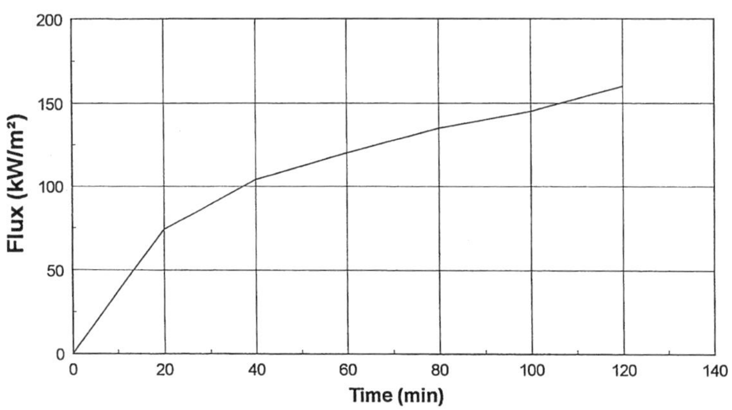

Fig. 10.8 Measured heat flux as a function of time in standard fire test

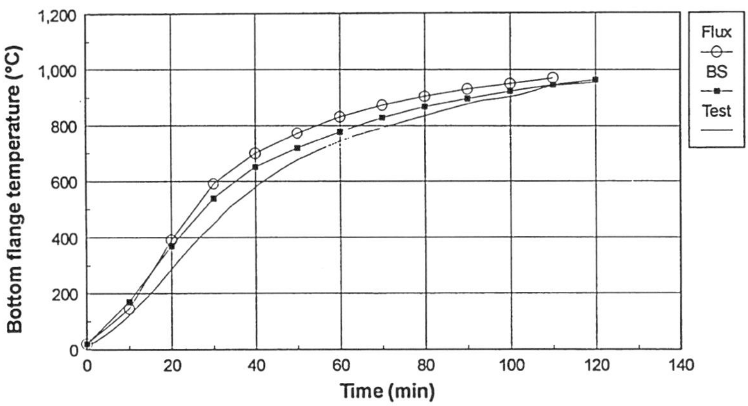

Fig. 10.9 Bottom flange temperatures of asymmetric beam in fire. Temperatures are calculated using measured heat flux data ('Flux') and standard fire temperatures ('BS'). Actual test temperatures ('Test') are included for comparison

the possibility of modelling section temperatures under real fire conditions, with any kind of (uneven) fire temperature distribution, so long as the local heat flux is measured.

In summary, thermal modelling calculation using measured heat flux data gives reasonable section temperature, showing the accuracy of both the flux measurement and the model.

10.2.4 Intumescent Coating

TFIRE is a program that can consider the behaviour of intumescent coating in the thermal analysis. The intumescent swelling thickness, thermal conductivity, emissivity and absorptivity are needed for constructing analysis software for calculating the temperature development. These are the subject of this section.

Intumescent material is extensively used in steel frame buildings to protect the steel surface from exposure to heat by providing a physical barrier. It responds to heat with endothermic chemical reactions generating multiple tiny bubbles causing the material to swell into a thickness of 5–100 times that of the original material. The low thermal conductive bubbles solidify into a thick multicellular char layer.

Nullifire System S has a long record in both laboratory tests and real fire situations. Its fire resistance can last up to 120 min depending on the thickness applied. System S is differentiated for external or internal use. In general, sections with their section factor H_p/A greater than 90 m^{-1} will not achieve fire resistance longer than 30 min. Once applied with 0.3 mm of intumescent material, 30 min can be achieved easily.

S605 is a single-pack aromatic solvent-based intumescent coating for fire protection of both internal and external structural steelwork, particularly for the

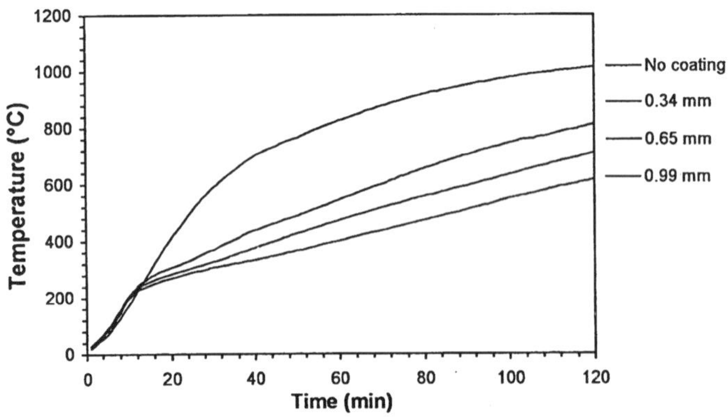

Fig. 10.10 Temperature-time curves from fire tests of bottom steel plate in a Slimflor structure with S605 coating at different thickness applied

Fig. 10.11 Coating thickness as a function of temperature. The furnace temperature follows that in a standard fire curve

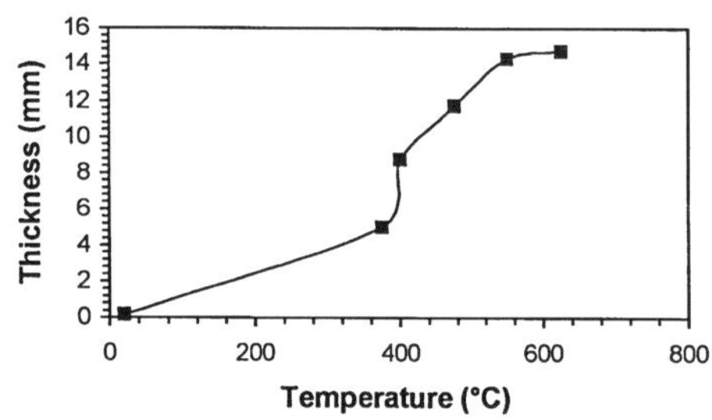

protection of exposed structural steel. S607 is a water borne thin film intumescent coating for internal structural steel work. It is complemented by water borne acrylic top seal.

Fire tests have been carried out to measure the performance of intumescent coatings in fire. Temperature–time curves given in Fig. 10.10 of slim floor flange plate for different thickness of coating show the significant fire protection effect of these coatings.

Laboratory tests were carried out with thin steel plate of 200 × 100 × 3 mm painted with intumescent coating all round with varying thickness. The furnace temperature was made to follow that in a standard fire curve. One series of tests was made with Nullifire S607 system with a dry thickness of 0.17 mm (300 g/m^2). At 300 °C, the coating starts to emit smoke and fume. Significant expansion starts at 375 °C, and continues until 625 °C. Fig. 10.11 shows the measured coating thickness as a function of furnace temperature. The expanded coating remains black until 750 °C when it starts to turn white through combustion and to crack. At 900 °C the coating is totally white and starts to drop off from the steel plate. At 1050 °C, the coating almost completely drops off and the heat separation effect is lost. White powder is left.

Thicker coating can expand more. When the coating thickness is 0.5 mm (900 g/m^2), it expands to approximately 55 mm after being heated up to 840 °C within 30 min time. At 1005 °C it is almost fully combusted.

10.3 Intumescent Fire Protection Coating Thickness for Shallow Floor Beams

Providing adequate fire protection to steel structure (columns and beams) is the traditional way to prevent failure and to ensure that the structural steel member can meet the required fire resistance standard. The most common types of fire protection in building structures are board protection, concrete encasement and intumescent coating. However, these protections can increase the cost of construction by about 50 % of the structural section member and the depth of the section by up to some 6 cm. Sometimes different types of fire protection can be used together to achieve a better fire resistance.

The development of shallow floor construction is an excellent example of the use together of different concepts of fire protection to give a better performance of fire resistance. In this type of construction, a beam is contained almost totally within the depth of a concrete floor. Only the bottom flange or a plate welded to the bottom flange is exposed outside of the concrete floor. Therefore, intumescent coating can be applied on the exposure area at the bottom flange.

Shallow floor construction provides a solid, flat-slab appearance similar to reinforced concrete construction. The fabricated Slimflor beam is based on a universal column section, with a single horizontal plate welded to its bottom flange, shown in Fig. 10.1. The asymmetric Slimflor beam (ASB) does not require welding of an additional plate and achieves optimum properties for design. The Top-Hat beam is formed by four plates welded together to make a shape like a hat. The floor can be constructed of pre-cast concrete units or long span composite slabs with deep profiled steel decks.

Because the Slimflor beams are almost totally contained within the floor slab, they have inherently good performance in fire and in most cases can achieve 60 min fire resistance without applied protection. However, if the fire resistance requirement is more than 60 min then protection must be applied to the bottom plate. Thin film intumescent coatings are ideal in this respect because they add a negligible amount to the depth of the section. Typically, thickness of about 1 mm is used for protecting conventional steelwork. For shallow floor (Slimflor) beams, the required thickness is expected to be smaller due to the intrinsically good fire resistance of the construction.

In building standards, a formula is given to calculate the required thickness of fire protection to regular steel section depending on the fire resistance rating and on the limiting temperature of the member. However, this formula does not apply to the three new shallow floor beams. The aim of this section is to derive a new formula to calculate the required thickness of fire protection for the new shallow floor beams from fire test data.

Intumescent coatings are paint-like materials which are inert at normal temperature but, when subjected to heat, swell to form an insulating char, stable over a limited period of time in fire conditions. Their thermal conductivity value varies with temperature, and is around 0.5 W/m °C. These materials provide thermal protection to the underlying surface through the absorption of heat by endothermic chemical reactions and through the insulating properties of the final char. The two commonly used intumescent coatings manufactured by Nullifire Ltd., England are S605 and S607. S605 has been marketed for a number of years and is a hydrocarbon solvent-based system. It contains 10–25 % xylene, 2.5–10 % C9 alkylbenzenes and 50–75 % proprietary resins, pigments and fillers. It has been fully assessed for conventional beams and columns using the procedures of the Association of Specialist Fire Protection (ASFP). Specifically, it has been fire tested utilising structural sections after extensive accelerated and natural ageing. The assessment of S605 for shallow or slim floor systems can therefore gain confidence from the earlier assessment. S607 is a newer water-based product developed from S605. It contains proprietary resins, water and fillers. It has been tested in Norway on a shallow floor system and performed in a very similar manner.

The three types of shallow floor beams are the Slimflor beam, asymmetric Slimflor beam and Top-Hat beam. By using the previous fire tests result data (Sha and Lau 2001) a formula can be derived to calculate the required thickness of fire protection for these beams. Fire test data include the sizes of the shallow floor beams tested, the coating thickness, the testing time and the temperatures measured at the bottom flange or at the plate welded to the bottom flange. The important data from the fire tests are shown in Table 10.5.

Nullifire Ltd. carried out fire tests on four unloaded Slimflor beams and a Top-Hat beam protected using S605 and S607 intumescent coatings manufactured by it. The test furnace was approximately 1 m cube. Protection was only applied on the plate welded to the bottom flange (Slimflor) or the bottom flange (Top-Hat). This was directly exposed to fire and the rest of the section was buried in sand, Fig. 10.12 (Sha and Lau 2001). Temperatures were measured along the beams at

Table 10.5 Data after fire tests at Nullifire, SINTEF and Warrington

Beam	H_p (m)	A (m^2)	Coating	Thickness (mm)	Temperature (°C) 60 min	90 min	120 min
Slimflor	0.436	0.01372	S607	0.166	687	802	903
	0.436	0.01372	S605	0.343	559	714	818
	0.436	0.01372	S605	0.651	483	605	718
	0.436	0.01372	S605	0.993	388	498	609
ASB	0.316	0.01267	S605	0.285	469	630	717
	0.316	0.01267	S605	0.774	–	462	587
Top-hat	0.470	0.01104	S607	0.179	721	823	915
	0.470	0.01104	S607	0.179	718	811	868
	0.470	0.01104	S607	0.537	532	673	747
	0.420	0.00893	S607	0.418	–	–	800

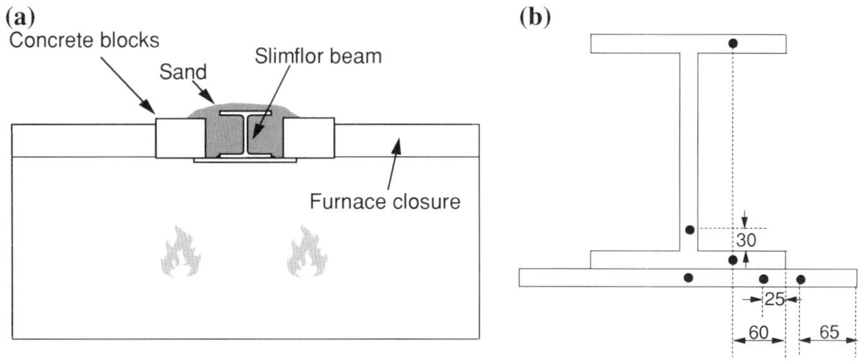

Fig. 10.12 Test at Nullifire Ltd. **a** Beam cross-section in furnace showing sand; **b** beam cross-section with positions of thermocouples

two sections, each approximately one-third of the exposed beam length (approximately 1 m) from the end of the beams.

In Table 10.5, the last three Top-Hat beams were tested at Norwegian Fire Research Laboratory (SINTEF). All tested Top-Hat beams have a lower-flange thickness of 10 mm. The bottom face of the beams was protected using the S607 intumescent coating. The furnace temperature was measured with thermocouples.

An unloaded asymmetric Slimflor beam was tested at Warrington Fire Research Centre, England. The bottom face of the beam was protected using the S605 intumescent coating. In the test, only the bottom flange was directly exposed to fire, and the rest of the section was surrounded by concrete. Half of the beam, the left-hand side, was coated with a relatively thin layer of S605 coating, at a density of 500 g/m^2 (thickness = 0.285 mm). The coating for the other half, the right-hand side, was relatively thick, 1,500 g/m^2 (thickness = 0.774 mm). The furnace at Warrington Fire Research Centre was much larger than the one used in Nullifire Ltd. The furnace was 3.6 m long and 0.95 m wide. Figure 10.13 shows the elevation and section views of the test assembly. The Warrington data are also used in this section (Table 10.5), undistinguished from the Nullifire data. It is difficult to directly compare the results with the tests at Nullifire in a smaller furnace as the beams were different, but other work has demonstrated that tests at the two laboratories give comparable outcome (Chap. 11).

As mentioned above, these beams are almost contained within the floor slab. They have very good performance in fire and in most cases can achieve 60 min fire resistance without applied protection. For more than 60 min fire resistance, protection must be applied to the bottom flange or the plate welded to the bottom flange. Therefore, the fire test data at 60 min are not used for this section.

The required thickness of fire protection to the steel section depends on the rate of heating of the section and on the limiting temperature of the member. For

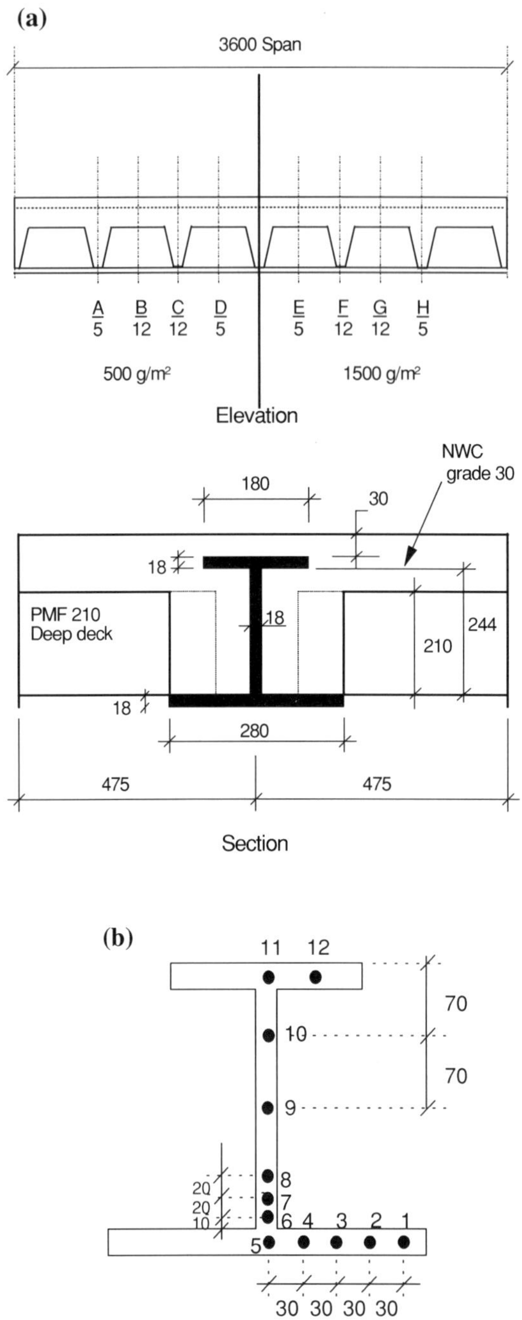

Fig. 10.13 The ASB test beam. **a** Elevation and section views; **b** position of thermocouples

enclosed members with box or profile protection, the heat flow depends on the section factor (H_p/A) of the member and on the properties of the fire protection material. The introduction of the section factor concept enables a broader use of fire test data to be made, leading to economy while maintaining the required performance standard. A steel section with a large perimeter (H_p) will receive more heat than one with a smaller perimeter. In addition, the greater the cross-sectional area (A) of the section, the greater is the heat sink. It follows therefore that the increase in temperature for a small thick section will be slower than for a large thin one. Section factor is thus a measure of the rate at which a section will heat up in a fire. The higher its value the greater will be the protection thickness required.

The thickness of fire protection for the regular beams, d_i, is determined from the formula:

$$d_i = \left(H_p/A\right) \lambda_i \left[t_e/(40(\theta_s - 140))\right]^{1.3} F_w \quad (10.6)$$

where λ_i is thermal conductivity of the fire protection material, t_e is fire resistance (minutes), θ_s is limiting temperature of the section, F_w is modification factor (< 1) due to the density of the protection material, H_p is perimeter of section exposed to fire (m), and A is cross sectional area of the steel member (m^2).

Because the shallow floor beams have a different cross sectional shape compared to the regular steel section the thickness of fire protection required might be different from what is obtained using the above formula. To derive the formula for shallow floor beams the first task is to calculate the section factors of the shallow floor beams used in the previous fire tests. Now rearranging the formula like the following:

$$d_i/\left(H_p/A\right) = \lambda_i \left[t_e/\left(40\left(\theta_s - 140\right)\right)\right]^{1.3} F_w \quad (10.7)$$

As d_i, H_p, A, θ_s and t_e are known the plot of $d_i/(H_p/A)$ against $[t_e/(\theta_s-140)]^{1.3}$ should yield a linear relationship. θ_s is the section temperature measured at time t_e in the fire test, t_e being 90 and 120 min. The values of these and the other parameters are listed in Table 10.5. From this, the combined coefficients of λ_i, F_w and $40^{-1.3}$ can be found.

Table 10.5 includes the values of d_i, H_p and A calculated with the beam condition used in the fire tests. Figure 10.14 shows the graph of $d_i/(H_p/A)$ against $[t_e/(\theta_s - 140)]^{1.3}$. The 19 data points in the figure are from the 19 temperature values in the last two columns of Table 10.5. It may be noted that in the last row of Table 10.5, there is no datum collected in the fire test corresponding to 90 min, resulting in the odd number of the total data points. Both coatings, from the three test facilities, are shown in the figure, and their performance has little difference. The figure also displays a best-fit line. As mentioned before, these shallow floor beams are almost contained within the floor slab. They have very good performance in fire and in most cases can achieve 60 min fire resistance without applied protection. The graph in Fig. 10.14 only includes the values of $d_i/(H_p/A)$ plotted

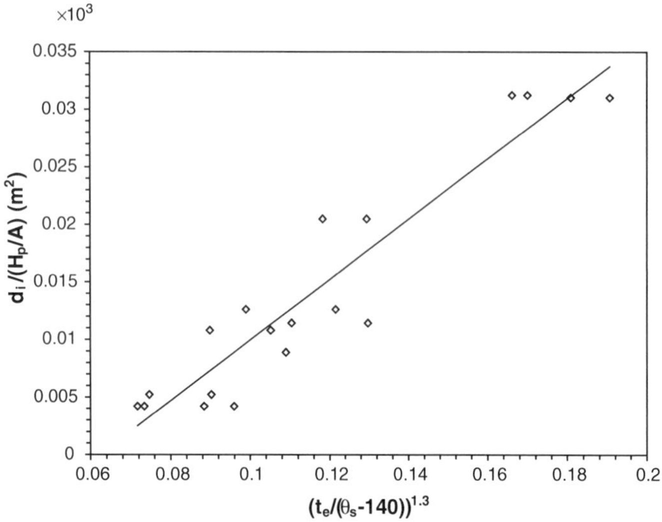

Fig. 10.14 $d_i/(H_p/A)$ against $[t_e/(\theta_s - 140)]^{1.3}$ from fire test data

against the values of $[t_e/(\theta_s-140)]^{1.3}$ at 90 and 120 min fire resistance. The equation for the best-fit line represents the required thickness of fire protection to the shallow floor beams:

$$d_i/\left(H_p/A\right) = 0.263\,[t_e/\,(\theta_s - 140)]^{1.3} - 0.016 \tag{10.8}$$

Obviously the combined coefficients of λ_i, F_w and $40^{-1.3}$ are represented in 0.263. The significant difference between this formula and the original formula for the regular section (Eqs. 10.6 and 10.7) is that this formula has a negative constant (–0.016) added into it. This indicates that the required thickness of fire protection for the shallow floor beams is thinner than for the regular steel beams. This formula is useful for the engineers for quick determination of the required thickness of fire protection.

The steel temperatures at various parts of a shallow floor beam can also be calculated from heat transfer models (Sect. 10.2.1). This was carried out with a Slimflor beam of the same dimension as used in the fire tests (Sha and Lau 2001). Figure 10.15 shows the calculation results, in the form of the plot of $d_i/(H_p/A)$ versus $[t_e/(\theta_s-140)]^{1.3}$. The thermal conductivity of the intumescent coating used in the calculations was 0.42 W/ °C·m. The typical variance in the thermal conductivity (λ) of steel is included in the program, in the form of $\lambda = 54 - 0.033\,(T - 20)$ where T is the steel temperature in Celsius. When T is greater than 829 °C, λ remains at 27.3 W/ °C·m. The relationship between $d_i/(H_p/A)$ and $[t_e/(\theta_s - 140)]^{1.3}$ for plate is:

$$d_i/\left(H_p/A\right) = 0.186\,[t_e/\,(\theta_s - 140)]^{1.3} - 0.021 \tag{10.9}$$

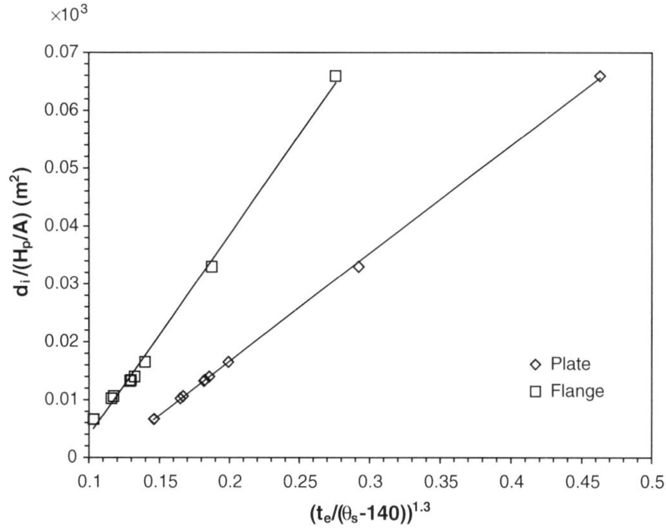

Fig. 10.15 $d_i/(H_p/A)$ against $[t_e/(\theta_s - 140)]^{1.3}$ from heat transfer model calculation

and for flange:

$$d_i / \left(H_p / A\right) = 0.345\left[t_e / \left(\theta_s - 140\right)\right]^{1.3} - 0.031 \tag{10.10}$$

The value 140, embedded in Eqs. 10.9 and 10.10, is an assumption.

In summary, the required thickness of intumescent coating for fire protection of shallow floor beams is thinner than for regular steel beams because of a negative constant. The formula can be used for quick, simple design calculation for the required thickness of fire protection. For design purposes there should be a safety factor used in conjunction with the formula.

Reference

Sha W, Lau NC (2001) Temperature development during fire in slim floor beams protected with intumescent coating. In: Zingoni A (ed) Structural engineering, mechanics and computation, vol 2. Elsevier Science, Oxford, pp 1103–1110

Chapter 11
Fire Resistance of Protected Slim Floors

Abstract The insulating effectiveness of intumescent coatings for fire protection is evaluated, using computer models and software, based on fire test data. The procedure include: (i) examination of testing set-up and data, and correction of temperature data obtained from small furnace tests on unloaded beams; (ii) evaluation of beam temperature distribution as a function of protective coating type and thickness; (iii) calculations of moment resistance and load ratio for any given coating thickness and the determination of minimum coating thickness required for different periods of fire resistance up to 120 min; (iv) extension of calculation results for different beam sizes; (v) examination of the effect of service holes; (vi) assessment for different types of coatings on asymmetric beams by comparison with test data on Slimflor beams. The coating thickness sufficient for all types of slim floor construction is obtained.

In slim floor construction, the supporting floor beam is contained within the depth of the floor deck. This provides a solid flat slab appearance similar to reinforced concrete construction. The fabricated Slimflor beam, developed by British Steel (now Tata Steel) and The Steel Construction Institute (SCI), is based on a universal column section, with a single horizontal plate welded to its bottom flange (Fig. 10.1). The floor can be constructed of pre-cast concrete units or long span composite slabs with deep profiled steel decks. The purpose of this chapter is to evaluate the thickness required for specified fire resistance, for two commonly used intumescent coatings.

The asymmetric beam (ASB), developed by British Steel (now Tata Steel) and The Steel Construction Institute (SCI), does not require welding of an additional plate, as the Slimflor beam does, and achieves optimum properties for design. The floor is constructed of long span composite slabs with deep profiled steel decks (Fig. 11.1). Because the asymmetric beams are almost totally contained within the floor slab, they have inherently good performance in fire and in most cases can achieve 60 min fire resistance without applied protection. However, for more than 60 min fire resistance, protection must be applied to the bottom flange. Such fire protection can effectively be provided in the form of intumescent coating, which will add a negligible amount to the depth of the section. The required thickness, however, would be much smaller than that needed for normal universal

W. Sha, *Steels*, DOI: 10.1007/978-1-4471-4872-2_11,

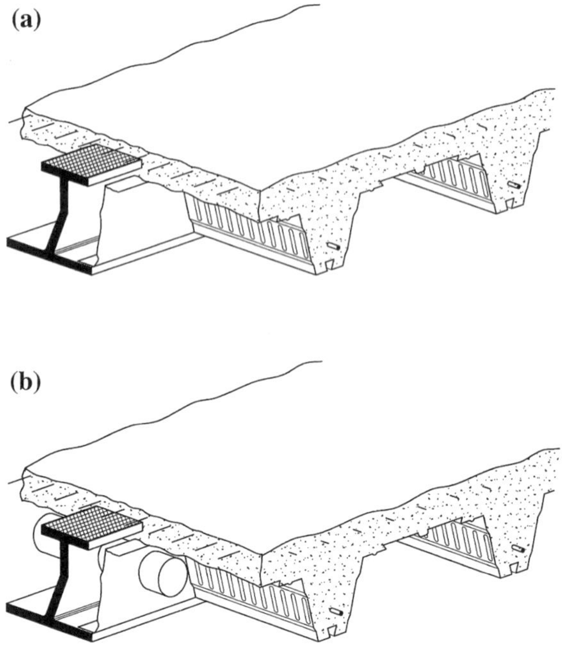

Fig. 11.1 Diagrams of typical slim floors using asymmetric beams. **a** Without service holes; **b** with service holes. From Sha (2001b)

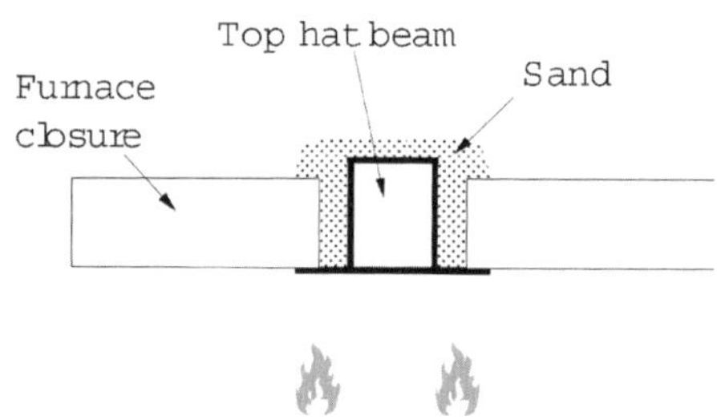

Fig. 11.2 Schematic of SINTEF tests. From Sha (2001a)

beams. The purpose of Sect. 11.7 is to evaluate the coating thickness required for specified fire resistance, for a commonly used intumescent coating.

11.1 Fire Tests

A series of fire test were carried out at Norwegian Fire Research Laboratory (SINTEF) to examine the intumescent paint system S607 for fire protection of the Scandinavia design of Top-Hat or HQ slim floor beam (Fig. 11.2). Two beam sizes were used, with lower flange thickness of 10 and 30 mm, respectively. By comparison, it was found that the temperatures of the former are closer to those of the UK Slimflor system, and results from this size are therefore used here. The relevant data are summarised in Table 11.1. Temperatures in brackets are the corresponding

Table 11.1 Temperature (°C) at bottom flange and lower part of web of Top-Hat slim floor beam at 60, 90 and 120 min from a test conducted in SINTEF programme

Coating thickness (g/m^2)	Time (min)	Bottom flange	Lower web
300	60	718 (721)	488 (529)
	90	811 (823)	595 (617)
	120	868 (915)	663 (688)
900	60	532	331
	90	673	459
	120	747	545

temperatures measured in the Nullifire furnace. Coating thickness in this chapter is always quoted by the wet weight per square metre of surface area (g/m^2). These will be used to evaluate the effect of coating thickness of S607 on beam temperatures.

Six fire tests were carried out at Nullifire Ltd. on unloaded Slimflor beams protected using S607 and S605 intumescent coatings. The furnace temperature was made to follow the standard fire curve. In all tests UC 203 × 203 × 60 beams were used and the bottom plates were 15 mm thick and approximately 200 mm wider than the flange of the UC section. The first test carried out was on an unprotected slim floor beam. This was to enable comparison to be made with Warrington Fire Research Centre (WFRC) test data. In the tests, only the bottom plate, where the intumescent protective coatings were applied, was directly exposed to fire, and the UC section welded to the plate was buried in sand (Fig. 10.12a). This made the experiments considerably easier to conduct as casting and drying of concrete were avoided, although the beam temperature development would be somewhat different compared with situations where the beam is covered by concrete. The correction of temperatures due to this factor will be discussed in detail in Sect. 11.2.1. The positions of thermocouples in each section are shown in Fig. 10.12b.

In addition to the tests on Slimflor beams one test was conducted on section of beam previously tested at SINTEF. The purpose of this test was to show that the Nullifire furnace had similar heating characteristics to the SINTEF furnace. The results of this test are included in Table 11.1, in brackets. The data from this test has been treated as informative only because the assessment in this chapter is for Slimflor beams rather than the HQ beam. However, the temperatures compared well with the Nullifire temperature being slightly higher than the SINTEF values. This is an indication that the Nullifire furnace is more severe than the SINTEF furnace.

Although the detailed temperature–time profiles for all 12 thermocouples on each test run are available, data directly useful for the modelling are average temperatures of plate, bottom flange and the lower part of the web (referred to as lower web throughout this chapter) at 60, 90 and 120 min. These are summarised in Table 11.2. The top flange temperature is always below 400 °C and therefore the steel is at full strength.

Fire tests at Warrington Fire Research Centre (WFRC) were carried out in a large furnace on loaded bare beams. Among a number of tests made, two are particularly relevant to the modelling here, in terms of their value in temperature

Table 11.2 Temperatures (°C) at various locations of Slimflor beams at 60, 90 and 120 min from tests conducted at Nullifire Ltd

Coating type	Thickness (g/m^2)	Time (min)	Plate	Bottom flange	Lower web
Unprotected	0	60	836	701	480
		90	953	858	608
		116	1,004	942	688
		120	1,016	949	698
S607	300	60	687	588	414
		90	802	708	522
		120	903	806	598
S605	600	60	559	418	336
		90	714	557	446
		120	818	655	529
S605	1,200	60	483	406	279
		90	605	532	377
		120	718	641	465
S605	2,000	60	388	282	216
		90	498	377	291
		120	608	477	368

Table 11.3 Temperatures (°C) at various locations of Slimflor beams at 60, 90 and 120 min from tests conducted at WFRC

Test	Beam in	Time (min)	Plate	Bottom flange	Lower web
UC 203 × 203 × 60	Sand	60	838	693	496
		90	934	848	652
		110	990	897	697
		116 (test ended)	1,023	946	761
UC 254 × 254 × 73	Concrete	60	803	603	339
		90	932	806	506
		110 (test ended)	994	878	578

corrections. The first was carried out on a beam identical to those tested at Nullifire Ltd., with the space above the plate covered by sand. The second test was on a slightly larger beam, a UC 254 × 254 × 73 Slimflor, covered by concrete, to simulate a situation closer to real practice. The test results shown in Table 11.3 were derived from tests and therefore form a basis for the corrections of the data obtained from Nullifire tests.

11.2 Data Treatment and Numerical Modelling

The approach adopted is to first show that for protected sections the performance in the Nullifire furnace is close to that at SINTEF. Second, the performance in the Nullifire furnace is compared with that measured at WFRC.

The comparison with SINTEF was only used in a qualitative sense as the tests at SINTEF were on a Top-Hat beam which, although being a slim floor beam, is not the same as the Slimflor beam studied in this chapter.

11.2.1 Temperature Corrections of Nullifire Data

The moment resistance of the Slimflor beams in fire conditions can be calculated knowing the beam temperature distribution data. Here, three different kinds of temperature corrections are made before such calculations. These are:

(a) correction between tests at Nullifire and WFRC;
(b) correction due to testing in sand;
(c) correction from pre-cast condition to decking.

Tests with bare steel Slimflor beams at a same size (203 × 203 × 60) were conducted in both Nullifire and WFRC, their results included in Tables 11.2 and 11.3. The differences in measured temperatures are taken as correction factors for temperature readings obtained in Nullifire tests (Table 11.4). There are at least two reasons for this correction to be necessary. First, the WFRC tests were carried out on longer and loaded beams, and therefore represent a more realistic situation. Second, the effectiveness of the modelling package was validated against experimental results from WFRC.

The derived corrections for up to 90 min are comparatively small indicating that the Nullifire furnace is similar to the WFRC furnace. For greater than 90 min, the only significant difference is in the lower web.

All Nullifire tests were carried out with the space around the I-beam above the bottom plate filled with sand. The difference in thermal behaviour between sand and concrete would influence the beam temperature, and this is taken into account by using a temperature correction factor. This factor is obtained by comparing the temperatures between the two tests carried out at WFRC (Table 11.3). Although the tests were conducted on beams of slightly different size, the thickness of the bottom flange of UC 203 × 203 × 60 and UC 254 × 254 × 73 is identical (14.2 mm). In such a case the beam temperature distribution would be very similar and the beam size effect is negligible. The correction factors are given in Table 11.5.

In practice, a Slimflor may be used with pre-cast concrete or constructed using a composite slab with deep decking. While a pre-cast concrete floor completely shelters the steel beam except the bottom plate from fire, a decking system leaves

Table 11.4 Temperature correction factors due to different testing furnace used (°C, + indicating that the WFRC measured temperatures are higher)

Time (min)	Plate	Bottom flange	Lower web
60	+2	–8	+16
90	–19	–10	+44
116[a]	+19	+4	+73

[a]Time closest to 120 min where data are available and used as correction for temperatures at 120 min

part of the UC covered with only a relatively thin layer of concrete. Therefore, the latter represents a more severe condition regarding structural response to fire. However, all Nullifire tests were conducted in a pattern essentially analogous to a pre-cast system. As the eventual design tables for the required coating thickness will be for the most severe fire situation, a temperature correction is made to derive temperature distributions of beams in a decking system (Table 11.6). This correction is based on work by SCI and is related to the observed differences between Slimflor beams constructed using pre-cast flooring and deep decking.

Total correction factors are obtained by summing up the correction factors for each cause shown in Tables 11.4, 11.5 and 11.6. This is listed in Table 11.7. All further modelling and analyses are made with data corrected using this table.

11.2.2 Temperature Distribution as a Function of Coating Thickness

Temperatures at plate, bottom flange and lower web as functions of coating thickness are required for predicting the effectiveness of any given thickness. For S607 coating, temperature reduction factors due to increased coating thickness are derived from both Nullifire and SINTEF results. For S605 coating, this is achieved

Table 11.5 Temperature correction factors due to testing in sand (°C, – indicating that the measured temperatures with sand were higher)

Time (min)	Plate	Bottom flange	Lower web
60	–35	–90	–137
90	–2	–42	–146
110[a]	+4	–19	–119

[a]Time closest to 120 min where data are available and used as correction for temperatures at 120 min

Table 11.6 Temperature correction factors due to decking (°C, + indicating that temperatures with decking are higher than temperatures with pre-cast units)

Position	Plate	Bottom flange	Lower web
Pre-cast	788	578	447
Decking	829	628	467
Correction	+41	+50	+20

Table 11.7 Total temperature correction factors (°C, + indicating that the measured temperatures in the Nullifire furnace should be increased)

Time (min)	Plate	Bottom flange	Lower web
60	+8	–48	–101
90	+20	–2	–82
120	+64	+35	–26

Table 11.8 Temperature decrease (°C) per 100 g/m^2 increase of S607 coating thickness

Location	Plate		Bottom flange		Lower web	
Thickness range (g/m^2)	0-300	>300	0-300	>300	0-300	>300
60	49.7	31.0	37.7	31.0	22.0	26.2
90	50.3	23.0	50.0	23.0	28.7	22.7
120	37.7	20.2	47.7	20.2	33.3	19.7

Table 11.9 Temperature decrease (°C) per 100 g/m^2 increase of S605 coating thickness

Location	Plate			Bottom flange			Lower web		
Thickness range (g/m^2)	0–600	600–1,200	>1,200	0–600	600–1,200	>1,200	0–600	600–1,200	>1,200
60	46.2	12.7	–	47.2	2.0	–	24.0	9.5	–
90	39.8	18.2	–	50.2	4.2	–	27.0	11.5	–
120	33.0	16.7	13.8	49.0	2.3	20.5	28.2	7.1	12.1

with the Nullifire test data for the bare beam, and those coated with different thickness. The method, to calculate the temperature decrease per 100 g/m² increase of coating, involves basically the interpolation of the test data, of the temperature with thinner coating and the temperature with thicker coating.

For S607, coating thickness dependence of section temperature is estimated using the SINTEF and Nullifire data in Tables 11.1 and 11.2, respectively, for different thickness ranges. This is obtained in the form of temperature decrease for every increase of 100 g/m^2 of coating thickness, as summarised in Table 11.8. As a Top-Hat beam does not have a plate welded to its bottom flange, the temperature characteristics for its bottom flange are used for both the plate and bottom flange in a Slimflor beam.

For S605, coating thickness dependence of section temperature is estimated using the Nullifire data in Table 11.2, for different thickness ranges. This is obtained in the form of temperature decrease for every increase of 100 g/m^2 of coating thickness, as summarised in Table 11.9.

11.3 Design Tables for 60, 90 and 120 min Fire Resistance

The reduced moment resistance in fire conditions is calculated using plastic theory, as permitted. The basic methodology adopted is to obtain load ratios for the given temperature distribution (i.e. temperatures in plate, bottom flange and lower web) which is in turn determined by the type of coating (S607 or S605) and its thickness. The load ratio is defined as the ratio of load applied at the fire limit state and load capacity under normal 'cold' conditions.

The calculations are conducted assuming that the beam is designed at room temperature as a composite beam using a deep steel deck with 85 mm concrete cover

Table 11.10 Load ratio as a function of coating thickness with the minimum safe thickness indicated in bold

Fire resistance (min)	S607		S605	
	Thickness (g/m^2)	Load ratio	Thickness (g/m^2)	Load ratio
60	0	0.49	0	0.49
	100	0.55	100	0.55
	200	0.64	**200**	0.66
90	500	0.53	500	0.58
	600	0.57	**600**	0.68
	700	0.62	700	0.71
120	1,300	0.56	1,300	0.53
	1,400	0.60	1,400	0.57
	1,500	0.64	**1,500**	0.60

above the top of the beam. The shear connection percentage is taken as 40 % in ‘cold’, increased to up to 100 % in fire. All calculations are carried out for the beam size used in Nullifire fire tests, i.e. 203 × 203 × 60 with 15 mm thick bottom plate, with no service holes. The fire performance of Slimflor beams designed as composite beams is not as good as beams designed as non-composite so, these conditions represent the most severe fire situation that could be encountered in normal construction, and therefore the results would in fact give additional safety factors for many practical cases. Except where tests were made for the given coating condition, fitted temperature distribution functions given in Tables 11.8 and 11.9 are used.

The detailed results are given in Table 11.10. The minimum thickness corresponds to a load ratio of 0.6, which is adequate for almost all applications.

11.4 Larger Beams

As stated earlier, the design table (Table 11.10) is based on calculations for a relatively small beam size, 203 × 203 × 60. The question remains as to the behaviour of larger beams in a fire condition. To clarify this, further modelling work using TFIRE, a computer program developed at The Steel Construction Institute, is conducted to calculate the temperature of beams at different sizes. The details of physical models used in this program and its good accuracy are described elsewhere (Chap. 10).

In TFIRE, the protection effect of normal fire protection board can be modelled with good confidence. The approach used here starts from finding a suitable board thickness that will give a similar temperature distribution after 120 min in a standard fire, for a 203 × 203 × 60 Slimflor beam. Then, temperature distributions for larger beams protected with a same board thickness are calculated (Table 11.11), together with calculated safe load ratio for each temperature distribution.

It is clear from the table that the larger beams would have lower temperatures at a given time. This is reasonable as the larger volume of the section will certainly

Table 11.11 Temperature distribution of different beams with the same thickness of fire protection board at 120 min and the corresponding safe load ratio

Beam	Average temperature (°C)		Safe load ratio
	Plate	Bottom flange	
203 × 203 × 60	769	620	0.57
254 × 254 × 73	767	627	0.60
254 × 254 × 107	739	588	0.68
305 × 305 × 283	663	492	0.87

need more heat to heat up, and the heat within the section will dissipate more quickly, too. Therefore, the small beam of 203 × 203 × 60 represents a more severe fire proposition, and the calculation given in Table 11.10 should be safe when applied to larger beams.

11.5 Effect of Service Holes and Stickability

For a Slimflor beam with service holes in its web, its temperature during a fire will be higher. The extent of this temperature increase was measured experimentally in a fire test carried out on a bare Slimflor beam at Warrington Fire Research Centre. The test was carried out on a 254 × 254 × 73 UC beam with a 460 × 15 plate. The temperature increase as a result of increased exposure to fire due to service holes is 72 and 139 °C for plate and bottom flange, respectively, at 60 min, and 64 and 142 °C for plate and bottom flange, respectively, at 90 min. The configuration of the test is equivalent to a practical situation where the holes are left empty, and therefore usually most vulnerable to the heat, in a fire.

The basic approach in evaluating the effect of service holes is as follows:

(1) Obtain temperature values at plate and bottom flange for a given coating thickness using test data given in earlier sections and Tables 11.8 and 11.9;
(2) Increase the temperatures to a level that will be experienced for beams with service holes using data earlier in this section and as discussed below; this approach is regarded as conservative;
(3) Calculate the moment resistance from such a temperature distribution taking into account of the reduction in web area.

In stage 2, the temperature at the remaining lower part of the web is taken as that at the bottom flange minus 35 °C, as proved to be reasonable from earlier calculations. Also, as there are no data as regard to the temperature increase at 120 min due to holes in the web, data for 90 min are used. The part of the UC section above the hole is assumed to be always below 400 °C. Although the part of the web immediately above the hole is likely to reach a higher temperature, it would contribute very little to the overall strength in any case.

In stage 3, the UC beam size used in calculation is 203 × 203 × 60, as for the above calculations for the beam without service holes. The relatively small

Table 11.12 Load ratio as a function of coating thickness for Slimflor with service holes with the minimum safe thickness indicated in bold

Fire resistance (min)	S607		S605	
	Thickness (g/m^2)	Load ratio	Thickness (g/m^2)	Load ratio
60	400	0.59	300	0.52
	500	0.67	**400**	0.65
90	1,000	0.55	800	0.59
	1,100	0.61	**900**	0.62
120	1,800	0.56	2,000	0.58
	1,900	0.61	**2,100**	0.63

size difference between the test beam and the beam adopted for calculation would only influence the final load ratio results to a minimal extent. A hole diameter of 160 mm is used, with the remaining web height below the hole being 15.8 mm.

With this method, the load ratio resistance, i.e. the ratio between moment resistance at the fire limit state and that under normal 'cold' conditions (both with web holes), is given in Table 11.12. The minimum thickness corresponds to a load ratio of 0.6.

Comparing this table with Table 11.10, it can be seen that increases of between 200 and 600 g/m^2 are required to fire protect Slimflor for same periods of times, with the actual amount generally in proportion to the thickness for a normal beam. The calculations here, like previous calculations for normal beams, represent a conservative estimate. In design guide for Slimflor construction, the hole diameter is limited to 0.6 of the depth of the UC section. Due to the parameter setups in the program, the service hole diameter of 160 mm is above 0.6 of the depth of the 203 × 203 × 60 beam. Therefore, the calculation results should cover all sizes of service holes used in practice.

For S605, the value of 2,100 g/m^2 corresponding to 120 min fire resistance represents an extrapolation as the maximum thickness tested was 2,000 g/m^2. The thickness of 2,000 g/m^2 would allow a load ratio of 0.58 to be achieved on a 203 section. This thickness would be adequate for virtually all circumstances, as service holes are normally only installed in beams more than 250 mm deep.

The assessment carried out above is based on the insulating properties of the coatings. An assessment of the stickability of the products by Fire Safety Engineering Consultants Ltd., UK has concluded that the 'stickability' of the coatings should be adequate for the assessed fire resistances and protection thickness.

11.6 Summary for Slim Floors Protected Using Intumescent Coatings

The effectiveness of two Nullifire intumescent coatings, S607 and S605, for the fire protection of Slimflor beams is evaluated using a combination of experimental fire tests and numerical data treatments, modelling and calculation.

A number of fire tests, carried out at three locations, i.e. Nullifire, Warrington Fire Research Centre, and Norwegian Fire Research Laboratory (SINTEF), are examined and their data compared. Based on corrected temperature data, and the interpolated as well as extrapolated coating thickness dependence of beam temperature distribution, load ratio that can be afforded by protection of any given thickness of coating can be obtained.

We give the required coating thickness adequate for all types of construction using Slimflor. For a fire resistance of 60 min, a nominal thickness of 200 g/m^2 of either S607 or S605 will be enough. For fire resistance of 90 min, the required thickness is 700 and 600 g/m^2 for S607 and S605, respectively. For 120 min fire resistance, a thickness of 1,400 g/m^2 of S607 or 1,500 g/m^2 of S605 is needed. An increase in coating thickness of between 200 and 600 g/m^2, depending on coating type and fire resistance period, is required for beams with service holes in the web.

An assessment of the stickability of the products by Fire Safety Engineering Consultants Ltd., UK has concluded that the 'stickability' of the coatings should be adequate for the assessed fire resistance and protection thickness.

11.7 Asymmetric Slim Floor Beams

11.7.1 Tests

Several fire tests were carried out at Warrington Fire Research Centre. An unloaded asymmetric beam whose bottom face was protected using the S605 intumescent coating (Fig. 10.13a) was included in the tests. In the test, only the bottom flange was directly exposed to fire, and the rest of the section was surrounded by concrete. Half of the beam, the left-hand side, was coated with a relatively thin layer of S605 coating, at a density of 500 g/m^2. The coating for the right-hand half was relatively thick, 1,500 g/m^2. The number below the section symbol indicates the number of thermocouples at each section (A–H).

The positions of thermocouples in each section are shown in Fig. 10.13b. In sections with only five thermocouples, they are at positions 1, 3, 5, 8 and 11.

In the following, concerning the protected beam, temperatures at sections B and G were used, as these sections:

(a) are away from the mid-span position where the coating thickness changes;
(b) have full sets of 12 thermocouples;
(c) should have higher temperatures due to heat passing through the end diaphragm.

Although the detailed temperature–time profiles for all thermocouples are available, data directly useful are average bottom flange temperatures and temperatures at various positions in the lower part of the web at 90 and 120 min (Table 11.13). Note that each bottom flange temperature is the average of five

Table 11.13 Measured temperatures (°C) at various locations of the asymmetric beam at 90 and 120 min in the fire test

Coating thickness (g/m^2)	Time (min)	Bottom flange	TC5	TC6	TC7	TC8
500	90	630	558	528	464	384
	120	717	643	612	547	465
1,500	120	587	527	506	445	378

thermocouple readings (positions 1–5). The upper part of the web (positions 9 and 10) and the top flange temperatures (positions 11 and 12) are always below 400 °C and therefore the steel is at full strength.

11.7.2 Numerical Modelling and Load Ratio Calculation

The reduced moment resistance in fire conditions is calculated using plastic theory, as permitted. The steel section is split into eight elements as shown in Fig. 11.3. The width and height of each of the eight elements are given in Table 11.14. The reduction of strength of each element is obtained from factors (based on 2 % strain), depending on the temperature of the element, which was obtained from measurements during the test. The concrete is split into four elements. The basic methodology adopted is to obtain load ratios for the given temperature distribution which in turn is determined by the coating thickness.

The temperatures of elements 2 and 4 are interpolated from the temperature readings from adjacent thermocouples. Taking into account of the distances between the centroid positions and adjacent thermocouples does this, and the formulae are given in Table 11.14. Small differences exist for the flange widths but these should have only marginal effect on load ratio calculations.

Based on the method described above, the load ratio, which is the ratio between the load applied at the fire limiting state and the load resistance under

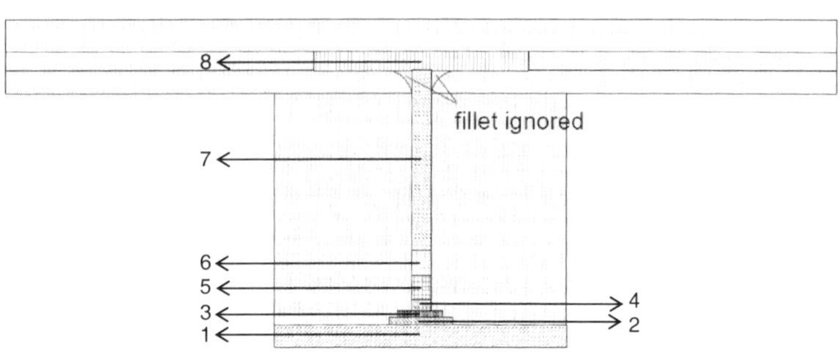

Fig. 11.3 Cross-section split into elements for moment resistance calculation. From Sha (2001b)

Table 11.14 Steel element width and depth (mm) used in the model (from bottom to top in Fig. 11.3)

Element	Width	Depth	Temperature corresponding to test
1	300	18	(TC1 + TC2 + TC3 + TC4 + TC5)/5
2	43.7	6.4	0.64·TC6 + 0.36·TC5
3	30.8	6.4	TC6
4	18	7.2	0.68·TC6 + 0.32·TC7
5	18	20	TC7
6	18	20	TC8
7	18	184	$\leq$400
8	190	18	$\leq$400

normal 'cold' conditions, is obtained. Calculated load ratios for the asymmetric beam protected with 500 g/m² S605 coating are 0.71 and 0.53, at 90 and 120 min, respectively. In the calculation, bond stress limits (bond strength) of 0.9 and 0.6 N/mm² are used for hot and 'cold' conditions, respectively, in accordance with experimental testing result.

The temperatures in the half of the beam protected with 1,500 g/m² coating are considerably lower than those with 500 g/m² coating (*compare* data in Table 11.13: 1,500 g/m² at 120 min and 500 g/m² at 90 min). Therefore, the load ratio that will be afforded in the former condition should be well above 0.71. The load ratio for an unprotected asymmetric beam is 0.50.

11.7.3 Coating Thickness Required for 120 min Fire Resistance

From Sect. 11.7.2, a coating thickness of 500 g/m² can only protect for load ratios up to 0.53 for 120 min fire resistance, while a thickness of 1,500 g/m² overprotects. In the following, the appropriate coating thickness for achieving a load ratio of 0.6, which, for ASB, is adequate for all applications, is given.

Coating thickness dependence of section temperatures at 120 min is estimated using test data in Table 11.13, for the thickness range between 500 and 1,500 g/m². This is in the form of temperature decrease for every increase of 100 g/m^2 of coating thickness. The temperature decrease per 100 g/m^2 increase of S605 coating thickness at 120 min is 13.0, 10.9, 10.6, 10.5, 10.2, and 8.7, for elements 1-6 from bottom of beam, respectively.

Using temperature distributions interpolated based on these, the load ratios achievable are 0.57, 0.60 and 0.63, for coating thickness of 700, 800 and 900 g/m², respectively, for 120 min fire resistance. The minimum safe thickness corresponds to a load ratio of 0.6.

11.7.4 Effect of Service Holes

In an asymmetric beam with service holes in its web (Fig. 11.1b), the temperature will be higher during fire. The extent of this temperature increase was measured experimentally on an bare asymmetric beam at Warrington Fire Research Centre, in a fire test carried out at the same time as the one described in Sect. 11.7.1. The temperature differences between hole and hole-free regions are summarised in Table 11.15. The configuration of the test is equivalent to a practical situation where the holes are left empty, and therefore usually most vulnerable to the heat, in a fire.

The basic approach in evaluating the effect of service holes is as follows:

(1) Obtain temperature values at various elements (Fig. 11.4) in bottom flange and lower web for a given coating thickness using test data given in earlier sections and Sect. 11.7.3;
(2) Increase the temperatures to a level that will be experienced for beams with service holes using data in Table 11.15 and as discussed below; this approach is regarded as conservative;
(3) Calculate the moment resistance from such a temperature distribution taking into account of the reduction in web area (Fig. 11.4).

In stage 2, temperature interpolations using same formulae as used before are used for temperature increase factors. Also, as there are no data as regard to the temperature increase at 120 min due to holes in the web, data for 90 min are used. The part of the web above the hole and the top flange are assumed to be always below 400 °C. Although the part of the web immediately above the hole is likely

Table 11.15 Temperature increase (°C) at various locations as a result of increased exposure to fire due to service holes

Time (min)	Bottom flange	TC5	TC6	TC7
60	40	59	81	137
90	38	74	100	169

Fig. 11.4 Steel cross section with hole. From Sha (2001b)

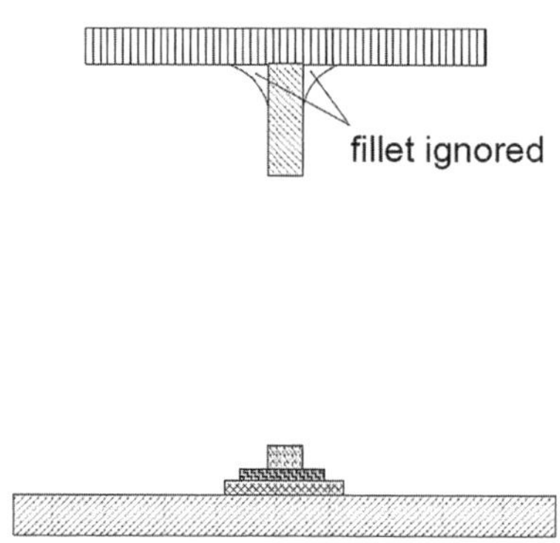

Table 11.16 Load ratio as a function of coating thickness for asymmetric beam with service holes with the minimum safe thickness indicated in bold

Fire resistance (min)	Thickness (g/m^2)	Load ratio
60	**500**	0.82
90	1,000	0.59
	1,100	0.64
120	1,500	0.48
	1,800	0.58
	1,900	0.62

to reach a higher temperature, it would contribute very little to the overall strength in any case.

In stage 3, the beam size used in calculation is the same as for the above calculations for the beam without service holes. A hole-diameter of 160 mm is used, with the remaining web height below the hole being 20 mm.

With this method, the load ratio resistance, i.e. the ratio between moment resistance at the fire limit state and that under normal ‘cold’ conditions (both with web holes), is given in Table 11.16. The minimum thickness corresponds to a load ratio of 0.6.

Comparing this table with the calculations for the beam without holes (Sect. 11.7.3), it can be seen that increases of 600 or 1,100 g/m² are required to fire protect for 90 and 120 min, respectively. For 60 min fire resistance, the assumed minimum thickness practically applicable, 500 g/m², is enough. The calculations here, like previous calculations for normal beams, represent a conservative estimate as a hole diameter of 160 mm is the maximum permissible in the design guide, while the beam used in the calculations is small in the range of asymmetric beams.

Although there are no fire test data of ASB slim floor protected with S607, comparative data do exist for Slimflor beams protected with S605 and S607. These have been extensively evaluated in previous sections in this chapter, revealing only marginal difference in the behaviour of the two types of coatings. Here, the required thickness of S607 to fire protect ASB is obtained by using the results for S605 discussed in previous sections and taking into account of the differences between the two coatings as calculated from Slimflor test data. S607 coating thickness required for all types of ASB construction is 500 and 700 g/m² for up to 90 min and 120 min fire resistance, respectively, in the case of no service hole. With service hole, this will be 500, 1,300, and 1,700 g/m², for 60, 90, and 120 min fire resistance, respectively.

11.7.5 Summary

The effectiveness of the Nullifire intumescent coating S605 for the fire protection of asymmetric slim floor beams is evaluated using a combination of experimental fire testing and numerical modelling and calculation. A fire test carried out at Warrington Fire Research Centre is examined. Based on the temperature data, and the extracted information on coating thickness dependence of beam temperature distribution, load ratio that can be afforded by protection of any given thickness of coating can be obtained.

For fire resistance of up to 90 min, a nominal thickness of 500 g/m^2 will be enough. For 120 min fire resistance, a thickness of 800 g/m^2 is needed. For beams with service holes in the web, coating thickness of 500, 1,100 and 1,900 is required for 60, 90 and 120 min, respectively.

It is important to note that the assessment here is based on the apparent insulating properties of the coatings. It has not considered the 'stickability' of the coatings, or has made any judgement as to whether the assessed thickness is practical.

References

Sha W (2001a) Fire resistance of slim floors protected using intumescent coatings. In: Topping BHV (ed) Proceedings of the eighth international conference on civil and structural engineering computing. Civil-Comp Press, Stirlingshire, Paper 65. doi:10.4203/ccp.73.65

Sha W (2001b) Fire resistance of protected asymmetric slim floor beams. In: Topping BHV (ed) Proceedings of the eighth international conference on civil and structural engineering computing. Civil-Comp Press, Stirlingshire, Paper 67. doi:10.4203/ccp.73.67

Index

W. Sha, *Steels*, DOI: 10.1007/978-1-4471-4872-2,

Printed by Printforce, the Netherlands